SKYLAB

Pioneering Space – the Next Frontier

By TD Barnes

Copyright TD Barnes 2024

Contents

Glossary

NASA Centers and Agencies

NASA: National Aeronautics and Space Administration

NASA Hq: NASA Headquarters

MSFC: George C. Marshall Space Flight Center

KSC: John F. Kennedy Space Center

MSC: Manned Spacecraft Center (later became JSC)

JSC: Lyndon B. Johnson Space Center

LaRC: Langley Research Center

Goddard Space Flight Center: A NASA center for space science

Ames Research Center: A NASA research center in California

Lewis Research Center: A NASA center focused on aeronautical research

Jet Propulsion Laboratory (JPL): A NASA center specializing in robotic space missions

Aerospace Contractors

Martin Marietta: Martin Marietta Corporation

Grumman: Grumman Aerospace Corporation

McDonnell Douglas: McDonnell Douglas Corporation

McDonnell Douglas, Huntington Beach, California: Eastern facility

McDonnell Douglas, St. Louis: Western facility

North American: North American Aviation, Incorporated (later North American Rockwell and Rockwell International)

The Boeing Company: Aerospace contractor

Lockheed Aircraft Corporation: Aerospace contractor

Bendix Corporation: Aerospace contractor

GLOSSARY OF ABBREVIATIONS AND ACRONYMS

AAP: Apollo Applications Program

AAPO: Apollo Applications Program Office

AES: Apollo Extension System

ALEM: Apollo Lunar Exploration Mission

GSFC: Goddard Space Flight Center

JPL: Jet Propulsion Laboratory

JSC: Johnson Space Center

KSC: Kennedy Space Center

LaRC: Langley Research Center

MSC: Manned Spacecraft Center (later Johnson Space Center)

MSFC: Marshall Space Flight Center

OMSF: Office of Manned Space Flight

OSSA: Office of Space Sciences and Applications

OWS: Orbital Workshop

SL: Skylab

Aerospace and Space Terminology

AM: Airlock Module

ATM: Apollo Telescope Mount

BeV: Billion Electron Volts

BTU: British Thermal Unit

CC: Cubic Centimeters

CM: Command Module

CSM: Command and Service Modules

cu m: Cubic Meters

DOD: Department of Defense

EREP: Earth Resources Experiments Package

ERTS: Earth Resources Technology Satellite

EVA: Extravehicular Activity

g: Gram; Gravity

gal: Gallon

HF: High Frequency

HSCC: Historical Services and Consultants Company

IU: Instrument Unit

LC: Launch Complex

LEM: Lunar Excursion Module

LM: Lunar Module

MDA: Multiple Docking Adapter

MeV: Million Electron Volts

MHz: Megahertz (Million Cycles Per Second)

mi: Miles

MOL: Manned Orbiting Laboratory

MORL: Manned Orbital Research Laboratory

mph: Miles Per Hour

MW: Megawatts

nm: Nautical Miles

PSF: Pounds Per Square Foot

psi: Pounds Per Square Inch

psia: Pounds Per Square Inch Absolute

RF: Radio Frequency

sec: Seconds

SEB: Source Evaluation Board

SLA: Spacecraft Lunar Module Adapter

SM: Service Module

SSESM: Spent-Stage Experimental Support Module

V: Volts
VHF: Very High Frequency
W: Watts
Units of Measurement
°C: Degrees Celsius (Centigrade)
°F: Degrees Fahrenheit
cm: Centimeter
km: Kilometer
kg: Kilogram
m: Meter
mm: Millimeter
mo, mos: Month, Months
sec: Seconds
wk: Week
yd: Yard
Yr: Year

Foreword

The evolution of space stations was a testament to humanity's relentless pursuit of knowledge and the continuous refinement of our technological capabilities. This journey, marked by groundbreaking achievements and unwavering determination, began in earnest in the early 1970s, a period of significant strides in space exploration.

The story of modern space stations truly began on April 19, 1971, with the launch of Salyut 1 by the Soviet Union. As the world's first space station, Salyut 1 was more than a technological triumph; it embodied human ingenuity and ambition. Its mission was to demonstrate the feasibility of prolonged space habitation and to conduct pioneering scientific experiments in the vacuum of space. This historic milestone heralded the dawn of sustained human presence in orbit, laying the groundwork for the future of space exploration.

Building on this success, the Soviet Union launched Salyut 2 on April 3, 1973. Unlike its predecessor, Salyut 2 was designed for military purposes, focusing on reconnaissance and defense-related tasks. However, technical difficulties prematurely ended its mission, limiting its contributions to space exploration.

The military focus continued with Salyut 3, launched on June 24, 1974, which was notably equipped with a space-based cannon, underscoring its intended role in military research and observation. The subsequent station, Salyut 4, launched on December 15, 1974, marked a shift back to scientific endeavors. It played a crucial role in advancing our understanding of biology, materials science, and astronomy, contributing invaluable data to these fields.

As Soviet expertise in space station operations grew, Salyut 5 was launched on June 22, 1976. Although it retained a military orientation similar to Salyut 3, it also facilitated various experiments, bridging the gap between defense and scientific research. This trend of blending scientific inquiry with military applications continued with Salyut 6, launched on September 29, 1977. Salyut 6 focused on a wide range of scientific research, including Earth observation and biological studies, significantly advancing our knowledge in these areas.

The Salyut program culminated with the launch of Salyut 7 on April 19, 1982. This final station continued the scientific traditions of its predecessors, concentrating on Earth observations and biological research, further cementing the Soviet Union's legacy in space station development.

The 1980s marked a new era in space stations with the launch of Mir on February 20, 1986. Mir represented a significant leap forward in space habitation technology with its modular design, allowing continuous expansion and adaptation. It became a cornerstone of long-duration human spaceflight and international scientific collaboration, serving as a bustling hub of microgravity research where scientists worldwide conducted experiments and explored the possibilities of sustained human presence in space.

While the Soviet Union advanced its space station program, the United States embarked on its journey into space station technology. The seeds of this endeavor were sown in 1963 amidst the fervor of the Apollo program. NASA began exploring options for post-Apollo crewed space missions, with the Office of Manned Space Flight cautiously navigating budget constraints and competition from the Air Force. By 1965, the Apollo Applications Program was established, overseeing the transition from lunar exploration to Earth-orbital missions.

Despite initial ambitions, the program evolved into a more focused mission plan by 1969, culminating in the launch of Skylab on May 14, 1973. As America's first space station, Skylab was designed to conduct scientific research in Earth orbit. It hosted experiments in solar astronomy, Earth observations, and life sciences, providing invaluable insights into the effects of long-term spaceflight on the human body. Although its operational life was relatively short, Skylab set the stage for future space habitation efforts and laid the groundwork for subsequent space station programs.

In the following decades, the International Space Station (ISS) emerged as a symbol of international cooperation and technological advancement. The ISS's journey began with the launch of its first module, Zarya, on November 20, 1998. This collaborative effort among NASA, Roscosmos, ESA, JAXA, and CSA created a state-of-the-art laboratory for scientific research across multiple disciplines, including astrobiology, astronomy, meteorology, and physics. The ISS continues to advance our knowledge and capabilities in space, fostering global collaboration and pushing the boundaries of what was possible in human spaceflight.

China also made significant strides in space station development, beginning with the launch of Tiangong 1 on September 29, 2011. This initial module was China's first space station endeavor, designed for scientific experiments, docking tests, and space technology demonstrations. It was succeeded by Tiangong 2, launched on September 15, 2016, which built upon its predecessor's achievements by providing a platform for further scientific research, technology testing, and crewed missions. The most recent chapter in space station history was China's Tiangong Space Station (CSS), which began with the launch of its first module, Tianhe, on April 29, 2021. The CSS represented China's long-term commitment to space exploration, designed as a modular station to support extensive scientific research and technology demonstrations.

The development of the Skylab Space Station can be chronologically organized into three distinct phases: early space station activities, the Apollo Applications era, and the development and operations of Skylab.

The first phase traces the origins of space station concepts, beginning with Hermann Oberth's seminal 1923 study on crewed space stations, which he presented to the scientific community. This early vision laid the groundwork for future exploration and spurred interest in understanding the effects of extended space habitation on humans. Over the decades, the scientific community sought to explore techniques for conducting experiments from space and successfully constructing and launching a space station. A significant milestone in this journey was the crewed space station symposium in Los Angeles in 1960, where leading aeronautical and aerospace scientists presented forty papers addressing these pivotal issues.

The second phase of the chronology spans from July 1965 to February 1970, encompassing the Apollo X Apollo Extension System and the Apollo Applications Program. During this time, space station concepts evolved significantly, informed by experiences from the Mercury, Gemini, and Apollo programs. This era saw the refinement of concepts, issuance of contracts, and the development of the Orbital Workshop, which eventually reached its final "dry" configuration. In February 1970, the program was rebranded from the Apollo Applications Program to the Skylab Program, marking a new chapter in its development.

The third phase covers the period from the rebranding to the conclusion of the Skylab program, including its post-operational phase. This era focused on the final stages of construction and operational activities, including completing equipment and experiment preparations, the launch and flight of Skylab, and subsequent recovery and evaluation. This phase was critical in bringing the space station from concept to reality, marking the culmination of years of planning and development.

From the early days of Salyut to Skylab, the evolution of space stations mirrors humanity's relentless quest for knowledge and exploration. Each station's unique objectives and achievements have contributed to our understanding of space and paved the way for future endeavors beyond our home planet.

In 1966, the dawn of the United States' first space station program began with a vision to unfold into an extraordinary journey of discovery and innovation. Initially, the concept was rooted in repurposing hardware from previous programs, modest in scale but rich in ambition. As the Skylab project evolved, it became clear that this endeavor would transcend its humble beginnings to become a trailblazing mission of remarkable scope and significance.

The Skylab program demanded creativity and resourcefulness that matched its bold ambitions. The mission required unprecedented innovation from its inception through the rigorous design, development, and testing phases. As the specifications for flight hardware solidified and operational plans crystallized, Skylab emerged not just as a space station but as a pioneering beacon of human ingenuity and exploration.

Skylab's achievements were profound and far-reaching. The space station provided an unparalleled vantage point for scientific inquiry, delivering groundbreaking solar and stellar astronomy advancements. Through Skylab's observations, scientists gained invaluable insights into the Sun's corona and solar winds, expanding our understanding of solar physics and laying the foundation for future solar astronomy research. The station's unique position in orbit also allowed comprehensive studies of Earth, revealing details about our planet's surface and atmosphere that were previously inaccessible.

One of Skylab's most revolutionary contributions was its exploration of materials processing in a microgravity environment. Experiments aboard the station demonstrated the potential for growing crystals with exceptional structural perfection and uniformity—qualities unattainable on Earth. Similarly, studies of metal alloys and composites in weightlessness opened new avenues for materials science and technological advancement.

However, the most remarkable aspect of Skylab's legacy was its demonstration of human adaptability and problem-solving in space. Shortly after Skylab's launch, the mission faced severe challenges, including losing the micrometeorite shield. This led to a critical loss of thermal balance and a significant reduction in solar power generation. These setbacks threatened to derail the mission, but the Skylab team—consisting of astronauts, engineers, technicians, and support staff—rose to the occasion. Their ability to turn these crises into opportunities not only salvaged the mission but also showcased humanity's extraordinary resilience and resourcefulness in space.

Skylab's impact extended far beyond its role as a sophisticated orbiting laboratory. It provided a wealth of data on human adaptation to long-duration spaceflight, deepening our understanding of how living organisms can thrive in microgravity. The station's Earth-focused research contributed valuable insights into environmental phenomena, ranging from geological and meteorological observations to studies of ecological systems and human impacts on the planet.

In addition to its contributions to Earth and space science, Skylab functioned as a space-based factory. It enabled astronauts to manufacture metal alloys, grow uniform crystals, and conduct experiments.

Cosmonaut Valeriy V. Polyakov, who boarded Russia's Mir space station on January 8, 1994, looks out Mir's window during rendezvous operations with the Space Shuttle Discovery.

Left: Soviet cosmonauts Georgi M. Grechko, left, and Yuri V. Romanenko during their record-breaking 96-day mission aboard Salyut 6. Right: NASA astronaut Norman E. Thagard during his American record-breaking 115-day flight aboard Mir.

Chapter 1 - Conception

The concept of an earth-orbiting space station designed to study celestial phenomena and our planet has intrigued visionaries since the early days of space travel speculation. As the United States space program began to take shape in the 1950s, space stations emerged as one of many ambitious projects under consideration. However, following President John F. Kennedy's decisive commitment in 1961 to land a man on the Moon by the decade's end, the focus shifted predominantly to Project Apollo, and space stations were momentarily set aside.

Project Apollo, a hallmark of the 1960s space race, demanded an unwavering national commitment. With its monumental goal of reaching the Moon, Apollo eclipsed other space endeavors. As the decade progressed, the future of space exploration beyond Apollo remained uncertain. Social and political upheavals in the latter half of the 1960s challenged the nation's priorities, placing the space program under significant scrutiny. Despite these pressures, those charged with America's space ambitions recognized the necessity of planning beyond Apollo to maintain the nation's leadership in space exploration. Yet, the prevailing national mood, which had previously fueled the space program, was shifting.

In the summer of 1965, NASA took a pivotal step by establishing the office to eventually spearhead the Skylab program. This new office initiated Skylab, starting as a conceptual design study. Between 1965 and 1969, meticulous efforts were dedicated to defining the spacecraft's design and the program's objectives. Skylab remained in the shadow of Apollo's lunar achievements throughout this period. It was not until the realization dawned that America's space program could not sustain the same level of urgency and focus that Apollo had commanded that Skylab began to emerge as a vital component of crewed spaceflight.

Skylab, therefore, became an essential platform for continuing human space exploration. As the space program shifted away from Apollo's intense focus, Skylab was conceived to bridge the gap, providing a new avenue for research and maintaining a presence in space while preparing for the next generation of missions and spacecraft.

In early 1952, long before NASA had been conceived under President Dwight Eisenhower's administration, the visionary Dr. Wernher von Braun was already articulating his bold dreams about space exploration. In an influential series published in Collier's Magazine, von Braun, who would later become the first director of NASA's Marshall Space Flight Center in 1960, laid out his ambitious vision for the future of human spaceflight. He famously predicted, "Development of the space station was as inevitable as the sun's rising; man has already poked his nose into space, and he was not likely to pull it back."

Von Braun's vision was nothing short of extraordinary. He envisioned a massive space station, a colossal wheel 250 feet in diameter, orbiting Earth at 1,075 miles. This rotating structure would create artificial gravity, allowing its inhabitants to live and work in a simulated environment akin to Earth's gravity. This concept was not just a flight of fancy, but a well-thought-out plan supported by the striking illustrations of artist Chesley Bonestell. These artistic renderings helped bring von Braun's futuristic ideas to life and later inspired the space station depicted in Stanley Kubrick's film 2001: A Space Odyssey.

Von Braun's grand scheme also included constructing this station before human missions to the Moon. In a subsequent Collier's article, he outlined how this space station would serve as a "watchful guardian of the peace" and act as a crucial springboard for what he termed one of the most outstanding scientific achievements of the era—the lunar journey. His vision for the space station was a hybrid of a fortress and a research laboratory designed to accommodate up to eighty people.

He and other experts in the field projected that the station could be built by 1967 at approximately $4 billion in 1952 dollars. This ambitious plan, dubbed "the next long step in space," was envisioned as a precursor to lunar missions, which von Braun and his colleagues anticipated would occur by 1977. Von Braun's forward-thinking concepts captured the public's imagination. They set the stage for the eventual advancement of human space exploration, reflecting his profound impact on the field and his unyielding belief in humanity's future among the stars.

His design included a rotating structure to simulate artificial gravity, intended to support a crew of 80 members. This space station was envisioned as a multifunctional hub, housing astronomers to operate telescopes, meteorologists to forecast space weather, and soldiers for surveillance. Von Braun projected that the construction of such a station would require a fleet of 7,000 short tons (6,400 metric tons) of space shuttles. He saw this space station as a research center and a launch pad for future missions to the Moon and Mars.

As technology advanced through the 1950s and early 1960s, innovations such as the transistor, solar cells, and enhanced telemetry began to reshape the landscape of space exploration. These developments enabled the creation of uncrewed satellites capable of photographing weather patterns and monitoring potential threats, such as enemy nuclear weapons. The emergence of these technologies diminished the immediate necessity for a large space station, as these tasks could now be efficiently managed with smaller satellites. Consequently, NASA's Apollo program, which focused on lunar exploration, chose a mission approach that did not involve the assembly of a large space station in orbit.

Despite this shift in priorities, the concept of a space station continued to be of significant scientific interest. The idea evolved toward a more practical design—a smaller, single-rocket-launched station dedicated to research and scientific purposes. This evolving vision set the stage for future space station developments, leading to the creation of Skylab, America's first space station, which became a reality in the early 1970s.

In 1959, Wernher von Braun, then head of the Development Operations Division at the Army Ballistic Missile Agency, submitted his final plans for Project Horizon to the U.S. Army. Project Horizon's primary aim was to place humans on the Moon, a goal that NASA would later pursue. However, von Braun's vision extended beyond lunar exploration. He proposed the idea of an orbiting laboratory, utilizing an upper stage from the Saturn V rocket. This concept would later influence the design of Skylab, reflecting von Braun's lasting impact on space station design.

Throughout the early 1960s, NASA centers undertook various studies to explore different space station designs. These studies considered platforms that could be launched using the Saturn V rocket, with potential crewed missions employing the Saturn IB in conjunction with an Apollo command and service module or a Gemini capsule launched by a Titan II-C. The latter option proved more cost-effective, especially when cargo was not required. Proposals ranged from a modest Apollo-based station designed to accommodate two to three astronauts to a larger, rotating station capable of housing 24 crew members for extended periods. One notable proposal from 1962, put forward by the Douglas Aircraft Company, involved adapting a Saturn S-IVB stage into a crewed space laboratory, demonstrating the varied and innovative approaches considered in the quest to establish a functional space station.

NASA vs. DOD

The collaboration between the Department of Defense (DoD) and NASA was pivotal in shaping the early space exploration efforts. In September 1963, both agencies agreed to join forces in the ambitious project of constructing a space station. However, as the collaboration progressed, the DoD initiated its parallel effort, leading to the announcement of the Manned Orbital Laboratory (MOL) in December 1963.

The MOL was designed to focus on photo reconnaissance and was equipped with large telescopes intended to be operated by a two-person crew. The design dimensions of the MOL were comparable to those of a Titan II upper stage, and it was planned to be launched with the crew aboard a modified Gemini capsule. This capsule featured a hatch cut into its heat shield, establishing an adaptation to accommodate the MOL's specific requirements.

Over the next five years, the MOL competed with NASA's space station plans for financial support. The rivalry between the two programs led to frequent political and administrative suggestions that NASA should collaborate with or adapt the MOL design. In response, NASA adjusted its plans to ensure its space station project diverged from the MOL's approach, establishing distinct paths for the two initiatives.

The successful Apollo 11 Moon landing in 1969 presented NASA with a new challenge: the potential loss of the 400,000 workers who had played critical roles in the Apollo program. Wernher von Braun, then head of NASA's Marshall Space Flight Center, championed developing a smaller space station to address this concern and provide continued work for his team. His vision was to maintain momentum beyond the Saturn rockets, which were anticipated to be completed early in Project Apollo.

In response, NASA established the Apollo Logistic Support System Office to explore ways to utilize Apollo hardware for scientific endeavors. Initially, this office considered a range of projects, including extended lunar missions to require two Saturn V rockets, a "lunar truck" based on the Lunar Module (LM), a large crewed solar telescope using the LM as quarters, and small space stations utilizing LM or Command and Service Module (CSM) components. Although the office's focus was broad at first, it increasingly centered on the concept of space stations.

In 1963, as the Office of Manned Space Flight (OMSF) began studies to chart the future of crewed space exploration, the foundational ideas that eventually led to Skylab started to take shape. Despite the Apollo lunar landing program being in its early stages, it became evident that the complex and lengthy development processes for space projects necessitated early planning to ensure the continuity of human spaceflight.

Post Apollo Plans

The summer of 1965 was a pivotal period for the American space program, marked by intense activity and significant advancements. In its seventh year of operation, the National Aeronautics and Space Administration (NASA) was deeply engrossed in the Gemini program, its second series of crewed missions in Earth orbit. This followed the conclusion of the Mercury program on May 16, 1963. For the subsequent 22 months, while preparations were underway for the two-person Gemini spacecraft, no American astronauts ventured into space. However, the first crewed Gemini mission was successfully launched on March 23, 1965, following two uncrewed test flights.

The Mercury program had provided NASA with essential insights into the basics of crewed spaceflight. Even before the first Mercury astronaut orbited Earth, President John F. Kennedy had challenged NASA with an ambitious goal: to land a man on the Moon and return him safely by the end of the decade. This monumental task required extensive groundwork, including developing and testing rockets, ground support facilities, and launch complexes. The Gemini program was crucial in this preparatory phase, focusing on key objectives such as spacecraft rendezvous—bringing two spacecraft together in orbit—and assessing human capability to function and survive in the weightlessness of space.

This summer, the American public became increasingly familiar with the Staffed Spacecraft Center (MSC) in Houston, Texas, central to the Gemini program's operations. Established on the flat Texas coastal plain, approximately 30 kilometers southeast of dt Houston, MSC was NASA's newest field center. It was inaugurated in 1964 and became the hub for managing the Gemini missions. Before this, the Space Task Group, based at Langley Research Center in Hampton, Virginia, had overseen the Mercury program. The transition to MSC marked a new chapter in the American space endeavor, with the center tasked with spacecraft design, testing, crew training, and mission control.

By mid-1965, MSC was bustling with activity. Nearly 5,000 civil servants, alongside approximately 10,000 aerospace-contractor employees, were engaged in various facets of the space program. The center was under the leadership of Robert R. Gilruth, who had been a pivotal figure since the early days of NASA. Gilruth, who joined Langley Research Center in 1937, had a distinguished career in aeronautics research and guided missile development. In 1958, he and his team transitioned to the newly formed Space Task Group, tasked with designing the spacecraft necessary for crewed spaceflight. The same principles of research and design that characterized the Space Task Group continued at MSC, with engineers overseeing the construction of the Mercury and Gemini spacecraft, which McDonnell Aircraft Company in St. Louis built.

As summer progressed, MSC was fully engaged in its mission. By mid-June, two crewed Gemini missions had been completed, with a third in preparation. The center was also busy with astronaut training, which included the first six scientist-astronauts. In September, buoyed by the program's success, NASA announced plans to recruit additional flight crews.

In parallel, MSC engineers made significant strides in preparing for the Apollo program. By 1962, the decision had been made to adopt a lunar orbit rendezvous mission mode, necessitating developing two separate spacecraft. North American Aviation, Inc., in Dey, California, was tasked with building the command and service module to carry astronauts to lunar orbit and return them to Earth. Meanwhile, across the country in Bethpage, New York, the other components of the Apollo spacecraft were being constructed.

At the heart of America's ambitious lunar exploration program in the 1960s, Grumman Aircraft Engineering Corporation was busy developing the lunar module—a delicate, spidery spacecraft designed to land two astronauts on the Moon's surface and return them safely to the command module orbiting the Moon. This module was a vital component of the Apollo missions, and its development was a collaborative effort between Houston's engineers, who established the fundamental design, and the contractors who built and tested the spacecraft.

From the beginning of the Gemini 4 mission, Houston's Mission Control Center (MCC) had taken responsibility for all spaceflights once the booster had cleared the launch pad. The center's role was central to the mission's execution, focusing on guidance and navigation, propulsion, attitude control, life support, and environmental control systems. This was where the drama of spaceflight was most visible, capturing the public's imagination with the inherent risks faced by the astronauts. However, this focus often overshadowed the equally critical aspects of the missions—the launch vehicles and the complex operations at the launch site.

The Saturn launch vehicles, essential for the Apollo missions, were developed under the auspices of NASA's largest field center, the George C. Marshall Space Flight Center, located ten kilometers southwest of Huntsville, Alabama. Marshall was established around the team led by Wernher von Braun, a pioneering rocket scientist whose vision for spaceflight dates back to his school days. By 1965, under von Braun's direction, the Marshall Space Flight Center was advancing the Saturn V rocket—a colossal three-stage vehicle designed to carry astronauts to the Moon.

Marshall's approach to rocket development was shaped by its unique history. Rocket research was initiated in Germany in the 1930s and was supported by the German army, which transitioned to the U.S. army's supervision after World War II. By 1950, the team had relocated to Redstone Arsenal near Huntsville, where they designed and built rockets much like an army arsenal. This hands-on, end-to-end approach characterized their work, producing numerous Redstone and Jupiter missiles. Von Braun proudly asserted in 1962 that his organization could manage the entire lifecycle of a space vehicle, from concept to testing. This capability was evident in their production of eight out of ten Saturn I first stages.

The Apollo program required a division of labor between Marshall and Houston, often described as "above and below the instrument unit," referring to the electronic control center between the booster's uppermost stage and the spacecraft. Despite their separate focuses, the centers collaborated closely, as illustrated by von

Braun's remark: "They built a damned good rocket, and we built a damned good spacecraft." Von Braun envisioned a broader role for Marshall in the future of space exploration, including crewed operations and space station construction. However, such ambitions were still distant as long as Saturn development was the primary focus.

By mid-1965, the Saturn V program was progressing smoothly. The final test flights of the Saturn I were completed, and preparations were underway for the Saturn IB's test flights. In August, each of the Saturn V's three stages successfully underwent static tests, demonstrating their capability to operate at full thrust and duration. The third stage's successful test of its restart capability was a crucial milestone, simulating the injection of the Apollo spacecraft into its lunar trajectory. Though flight testing remained, these achievements marked significant progress.

Amid this success, there was an undercurrent of concern at Marshall. The center's future was uncertain beyond the Saturn V program, with no new projects and many positions slated for transfer to Houston. Von Braun's memo from August 1965 reflected this uncertainty, acknowledging the need to address Marshall's future role in space exploration. Despite its groundbreaking work, Marshall was left contemplating its next challenge as it awaited a new mission to engage its expertise.

Meanwhile, some 960 kilometers southeast, the John F. Kennedy Space Center, situated adjacent to Cape Canaveral on Florida's Atlantic coast, was undergoing rapid expansion. Originally the Launch Operations Directorate of Marshall, it had evolved into a bustling center with a workforce of 20,000 people by 1965. Construction teams were busy finishing the enormous Vehicle Assembly Building, a vital component of the center's growing infrastructure.

In the early 1960s, the Kennedy Space Center (KSC) embarked on a monumental undertaking to cement its place in the annals of space exploration history. Central to this endeavor was the Assembly Building, a colossal structure designed to accommodate the assembly of the Saturn V rockets. Standing at an impressive 110 meters in height, this building was where the massive rockets were pieced together indoors, shielded from the elements and ensuring a controlled environment for assembly.

Two months following the completion of the Assembly Building, attention shifted to the road tests for the crawler-transporter, a behemoth vehicle engineered to transport the fully assembled Saturn V rockets from the Assembly Building to one of the two launch pads located twelve kilometers eastward on Cape Canaveral. This gargantuan machine, capable of moving the towering rockets upright and complete, played a crucial role in the launch sequence, symbolizing the scale and complexity of the operations at KSC.

During this period, NASA's launch teams were intensely focused on wrapping up the Saturn I flight program while simultaneously managing the Gemini missions in collaboration with the Air Force. Under the guidance of Kurt Debus, a prominent figure who had arrived from Germany alongside Wernher von Braun in 1945, the KSC was tasked with a spectrum of responsibilities that extended well beyond launching rockets.

KSC's role encompassed the comprehensive integration of booster stages and spacecraft. Although each component was meticulously tested by its respective manufacturer, it was the responsibility of KSC's engineers to ensure these components functioned seamlessly together. This involved rigorous checks and a detailed verification process to confirm that NASA's intricate "interface control" system was fully operational. If discrepancies arose between components produced by different contractors across various states, KSC's engineers were charged with diagnosing and rectifying these issues.

The scale of this task required close coordination not only with other NASA centers but also with significant contractors across the nation. The Office of Manned Space Flight (OMSF) at NASA Headquarters in Washington orchestrated this vast network of collaboration and oversight. OMSF, one of NASA's three primary program offices, was responsible for overseeing and integrating the efforts of the field centers and their associated contractors. Reporting directly to

Robert C. Seamans, Jr., the Associate Administrator, and NASA's third-ranking official, OMSF's significance was underscored by its substantial share of the agency's budget and public attention. Since the Apollo commitment in 1961, OMSF dominated the public eye and NASA's financial resources, reflecting its pivotal role in the space agency's ambitious goals and projects.

In 1965, the Office of Manned Space Flight (OMSF) at NASA was under the direction of George E. Mueller, pronounced "Miller." A distinguished electrical engineer with a doctorate in physics, Mueller brought a wealth of experience from academic and industrial research to his role. Before becoming the associate administrator for crewed spaceflight in 1963, Mueller had served as vice president at Space Technology Laboratories, Inc., in Los Angeles. There, he played a significant role in the Air Force's Minuteman missile program, which honed his skills in managing complex aerospace projects.

Upon assuming leadership of OMSF, Mueller faced the formidable task of reorganizing the office and steering it toward his vision for achieving the Apollo program's ambitious objectives. His first year in Washington was dedicated to reshaping OMSF and familiarizing the field centers with his management approach. Mueller believed that centralized control was crucial for the success of Apollo, and he established an administrative structure that assigned primary responsibility for policymaking to NASA Headquarters while delegating substantial authority to the field centers. This approach was designed to streamline decision-making and enhance coordination, though it required careful navigation given the centers' historical preference for autonomy.

Early in his tenure, Mueller boldly decided to prove pivotal for the Apollo program. Convinced that the original testing protocols would not meet the ambitious timeline, he implemented "all-up" testing for the Saturn V rocket. This new procedure called for testing the complete vehicle with all stages functioning simultaneously, a significant departure from the traditional stage-by-stage testing method used by NASA and its predecessor, the National Advisory Committee for Aeronautics (NACA). Inspired by his experiences with the Minuteman missile, where all-up testing had demonstrated its efficacy, Mueller believed this approach would be more time-efficient and cost-effective. However, it posed substantial risks to reliability and quality control.

Implementing all-up testing was challenging, as it required overcoming resistance from the field centers, which had established procedures and were skeptical about the new approach. Nevertheless, when all-up testing proved successful, it not only put Apollo back on schedule but also created the possibility that some of the Saturn V rockets could be repurposed for other uses.

Mueller's decision to pursue the moon landing as a primary goal marked a significant shift from conventional space exploration strategies. Historically, European pioneers such as Konstantin Eduardovich Tsiolkowski and Hermann Oberth had envisioned a permanent orbiting space station as the first step in exploring the cosmos. Such a station was seen as an essential vantage point for Earth observation and a staging area for deeper space expeditions. Wernher von Braun, who influenced these European ideas, championed the concept of an Earth-orbiting space station in a popular magazine article in the early 1950s.

The rationale behind establishing an orbiting space station was grounded in practical considerations. Given the limitations of rocket technology, it was deemed more feasible to assemble a space station in orbit using multiple smaller rockets rather than developing the enormous rockets required for a direct journey to distant space destinations. Additionally, an orbiting station would provide a critical platform for studying many unknowns related to human spaceflight, including the effects of weightlessness on the human body. This forward-thinking approach underscored the importance of incremental steps in advancing space exploration, setting the stage for future missions and discoveries.

In the early days of space exploration, the concept of a space station emerged as a crucial step in humanity's quest to understand and utilize the cosmos. The urgency and ambition of this pursuit were evident in the immediate aftermath of the Soviet Union's launch of Sputnik, which ignited the so-called "space race." This metaphor symbolized the intense competition and drive within the United States to achieve dominance in space exploration despite the unclear nature of the ultimate goal.

By late 1958, the House Select Committee on Space began extensively reviewing the nation's space capabilities and future objectives. The committee interviewed leading scientists, engineers, corporate executives, and government officials to explore possibilities beyond the Mercury program. Their findings were encapsulated in the report titled The Next Ten Years in Space, which identified a space station as the next logical advancement in human spaceflight.

In parallel, Wernher von Braun and his team at the Army Ballistic Missile Agency advocated for similar goals during their briefings to NASA. Their recommendations and NASA Deputy Administrator Hugh Dryden's report to Congress in February 1959 included a space station and a crewed lunar landing as critical objectives.

Later, in 1959, NASA established a Research Steering Committee on Manned Space Flight to investigate the potential for post-Mercury missions. This committee, often credited as the precursor to the Apollo program, initially prioritized developing a space station over a lunar landing. During its early meetings, the committee debated the scientific and practical merits of a space station versus a moon landing, with a space station being considered a fundamental step in a long-term space exploration strategy.

Despite these debates, NASA Administrator T. Keith Glennan remained cautious, refraining from committing to either goal. However, by early 1960, Glennan did acknowledge that, after the Mercury program, the moon should become the primary objective of crewed spaceflight.

The justification for a crewed orbital station persisted, particularly considering the technical challenges and uncertainties surrounding lunar missions. Robert Gilruth, a key figure in NASA, addressed a symposium on crewed space stations in the spring of 1960, describing the agency's flight missions as a compromise between ideal ambitions and practical capabilities. Gilruth concluded that while multi-man Earth satellites appeared achievable, more ambitious programs like crewed lunar landings should not be pursued immediately.

Heinz H. Koelle, head of the Future Projects Office at the Marshall Space Flight Center, echoed this sentiment. He argued that developing a small laboratory in Earth orbit was the next logical step, with plans for a larger, more complex station to follow as rocket payload capacities improved. This perspective was commonly held by the Marshall team until 1962.

Despite the ongoing advocacy for a space station, the momentum for crewed lunar missions continued to grow throughout 1960, eventually shaping NASA's focus and priorities in the years to come.

In the fiscal 1961 budget hearings, discussions about space stations were notably sparse. Unlike the previous year's budget proposal, which had sought funds for preliminary studies, the 1961 budget made no such request. This shift indicated a broader focus change within NASA's long-range planning. The ambitious goal of establishing a permanent space station by 1969, which had previously guided NASA's vision, was dropped. Instead, the Space Task Group contemplated a significantly smaller laboratory that could fit within the adapter section designed to support the Apollo spacecraft on its launch vehicle.

The trajectory of NASA's ambitions took a decisive turn in May 1961 when President John F. Kennedy announced the moon landing as America's primary goal in space exploration. Kennedy's proclamation was a challenge of unprecedented clarity and ambition. It was a concise, definite, and measurable goal, offering a clear benchmark for success or failure. This singular focus on the lunar landing meant that NASA and the aerospace industry had to narrow their efforts toward this monumental objective, effectively sidelining other projects like space stations.

Had the United States been committed to a long-term, 20-year program of methodical space development, Wernher von Braun's 1952 concept for a space station might have been pursued more vigorously. However, with only eight years to achieve the lunar landing, the idea of a space station, particularly one of substantial size or complexity, became increasingly improbable due to time and budget constraints.

The decision to prioritize the moon landing did not wholly rule out the possibility of a space station, but it did render such an undertaking less feasible. The focus was now on making the lunar mission a reality. During the subsequent year, von Braun's Marshall Space Flight Center team advocated for a moon mission via earth-orbit rendezvous. This approach entailed sending "tankers" into orbit around Earth to refuel a lunar-bound vehicle. This method was considered safer and more practical than other mission modes, such as direct flight or lunar-orbit rendezvous. Despite its merits, NASA ultimately chose lunar-orbit rendezvous in June 1962 for the Apollo missions, which effectively closed off extensive earth-orbital operations as a precursor to the lunar landing.

From mid-1962 onward, space stations were relegated to the realm of advanced studies. These studies were crucial for identifying the needs of the space program and pinpointing areas requiring research and development. Much of this future-oriented work was undertaken by aerospace contractors, as NASA's resources were heavily invested in the Apollo program. The early 1960s marked a period of significant imaginative and exploratory thinking fueled by Congress's generous funding. Between 1962 and 1965, Congress allocated $70 million for future studies, leading to over 140 contracts for studying earth-orbital, lunar, and planetary missions and the spacecraft needed for these endeavors.

In the early years of space exploration, Langley Research Center established itself as a pioneering force in space station studies. By the summer of 1959, Langley had developed a concept for a space station that anticipated many of the functions and purposes that Skylab would later fulfill. This initial design explored the innovative idea of utilizing a spent rocket stage, demonstrating a forward-thinking approach to influence future space station development.

Langley's early efforts laid crucial groundwork to develop Skylab, setting the stage for what would become the United States' first space station.

During this period, the concept of creating much larger space stations began to take shape. A key area of interest was artificial gravity, which would be achieved by rotating the station. Langley researchers determined that an optimal rotation rate would be four revolutions per minute with a radius of 25 meters. They experimented with various configurations and ultimately favored a hexagonal wheel design with spokes extending from a central control module. This design envisioned a station capable of enclosing nearly 1,400 cubic meters of workspace and accommodating between 24 and 36 crew members. At launch, this station would weigh approximately 77 metric tons.

Designing a station of this magnitude posed significant challenges, particularly in launching and assembly. The initial approach, advocated by Wernher von Braun a decade earlier, involved launching the station in sections and assembling it in orbit. However, this method raised concerns about potential complications. Langley engineers explored alternative solutions, such as inflatable structures, although these were ultimately deemed too vulnerable to meteoroid impacts. Instead, they proposed a collapsible structure that could be deployed in orbit in an umbrella-like fashion and awarded North American Aviation a contract to develop this concept further.

In July 1962, Langley presented their preliminary findings in a symposium. The findings addressed many issues associated with large rotating stations, including life support, environmental control, and waste management. The engineers at Langley felt they had made significant progress but were concerned that their proposed solutions might be too ambitious for NASA's immediate needs.

Concurrent with Langley's efforts, similar studies were underway at the Manned Spacecraft Center (MSC) in Houston. By early 1962, MSC had begun planning a sizeable rotating space station to be launched aboard the Saturn V rocket. Like Langley, MSC's objectives included evaluating the challenges of living in space and conducting scientific and technological research. The plan included resupply missions and relief crews transported by the smaller Saturn IB and an Apollo spacecraft modified to accommodate six astronauts, double the usual capacity. MSC aimed to have the station in orbit within four years.

By the fall of 1962, as the Apollo program's immediate demands began to ease, NASA Headquarters turned its attention to future programs. In late September, officials encouraged the centers to continue their technical studies despite uncertainty about when a space station might be realized. Rising costs associated with the Apollo program threatened to limit funding for future projects, prompting a call for simplicity and fiscal restraint. Langley and MSC responded to this challenge, although they diverged in their vision for the space station's mission. Langley envisioned the station as a laboratory for advanced technology, with significant input from NASA's offices of space science and advanced technology. In contrast, MSC saw the station primarily as a base for crewed missions to Mars.

In the summer of 1963, Joseph Shea, then Deputy Director for Systems in the Office of Manned Space Flight, sought to crystallize the future objectives for crewed spaceflight. Recognizing the potential of an orbiting laboratory, Shea reached out to NASA's field centers and Headquarters program offices to solicit recommendations on various related issues. In his correspondence, Shea highlighted several considerations under examination by the Office of Manned Space Flight (OMSF), emphasizing the need for substantial justification to secure approval for such a venture.

Shea sought detailed input on the proposed space station's potential purposes, configurations, and specific scientific and engineering requirements. Two principal points framed this request: first, the significance of a space station program to scientific advancement, technological progress, or national objectives; and second, the unique attributes of this station that distinguished it from the capabilities of existing programs like Mercury, Gemini, Apollo, or uncrewed spacecraft.

During the following six months, public statements and internal communications underscored NASA's commitment to developing a space station to fulfill pressing national needs. By mid-1963, NASA had articulated a clear rationale for an Earth-orbiting laboratory. The primary mission of early flights was to determine whether humans could live and work effectively in space for extended periods. The phenomenon of weightlessness in space presented a unique challenge that could not be accurately simulated on Earth except for brief periods using aircraft.

The effects of prolonged exposure to weightlessness and the implications of a sudden return to normal gravity were unknown. Addressing these biomedical concerns was critical, but it also served a broader objective: preparing for future interplanetary travel. The first-generation space station would provide a platform for developing and testing various systems, structures, and operational techniques necessary for an orbital launch facility or a larger space station. Moreover, such a crewed laboratory would have significant potential for conducting scientific research in astronomy, physics, and biology.

As NASA progressed in defining the mission objectives and configuration of the space station, it became evident that the design did not need to be dictated solely by experimental needs. The agency explored different approaches, including the possibility of using Gemini hardware for various increasingly extended missions, a cost-effective strategy that appealed to Washington. Alternatively, a large rotating station was considered to simulate varying levels of artificial gravity for further experimentation.

In August 1963, Shea reflected on these options during a conference. He deliberated whether a minimal Apollo-type Manned Orbiting Laboratory (MOL) would suffice for significant biomedical experiments or if the advantages of a multi-purpose MOL justified more extensive development. Modifying the spacecraft's mission would involve considerable adjustments but no fundamental structural changes. Proposed upgrades included replacing the conventional hydrogen-oxygen fuel cells with solar cells to reduce weight and adopting a nitrogen-oxygen atmosphere instead of pure oxygen to mitigate respiratory issues. Furthermore, a more compact and regenerable molecular sieve system was suggested instead of the bulky lithium hydroxide canisters for carbon dioxide absorption.

Drawing upon earlier studies, the study group meticulously prepared a list of essential medical experiments, detailing their approximate weights, volumes, and the power, time, and workspace required for their execution. This thorough analysis revealed that the command module alone could not support more than a minimal subset of these experiments. Even with the addition of a laboratory module and a third crew member, time constraints would still prevent the completion of all desired tests.

North American's analysis concluded that all three proposed concepts—varying in their configurations—were technically feasible and capable of fulfilling the required mission objectives. Among these, the command module alone was the least costly option; however, the limitation of a two-man crew introduced operational challenges. The addition of a laboratory module, while beneficial, increased costs by 15-30% and introduced significant weight issues. The payload with a dependent module was very close to the Saturn IB's weight-lifting capacity, and the independent module would exceed this limit.

Given that NASA anticipated increasing the Saturn's thrust by 1967, this issue, though not a reason to dismiss the concept, was expected to remain a challenge until 1969. To address these concerns, North American recommended that any further studies focus on the Apollo configuration with a dependent module, as this option demonstrated the most significant applicability across all proposed missions.

The study's findings were positively received at NASA Headquarters, where the funding situation for post-Apollo programs remained uncertain. North American was asked to continue its investigation into 1964, specifically addressing the technical problems associated with extending the life of Apollo subsystems.

Several proposed schemes called for a larger crewed orbiting laboratory capable of supporting four to six crew members for up to a year, with ample space for conducting experiments. Among these designs was the medium-sized laboratory, which often featured a zero-gravity environment that could be adapted to provide artificial gravity. A notable example of this type of laboratory was Langley's Manned Orbiting Research Laboratory,

a study initiated in late 1962. This design envisioned a four-man module, 4 meters in diameter and 7 meters in length, equipped with its life-support systems.

The laboratory's design incorporated advanced technologies, including molecular sieves. These sieves contained highly absorbent minerals, such as zeolite, whose structure featured a three-dimensional lattice with regularly spaced channels of molecular dimensions. Molecules small enough to enter these channels, such as carbon dioxide, were absorbed and could later be released through heating, allowing the sieves to be regenerated for continued use. This innovation was a critical component in managing the life-support systems of the proposed laboratories.

As the ambitious vision for space exploration expanded during the early 1960s, the concept of an orbiting space station began to take shape with considerable technical and logistical considerations. The Skylab project, an early manifestation of this vision, was rooted in the desire to leverage existing technologies and reduce complexity while achieving groundbreaking scientific objectives in space.

The decision to utilize proven hardware for the station's deployment was at the heart of this effort. The Saturn IB rocket and the Air Force's Titan III were identified as a reliable launch vehicle capable of delivering the laboratory into orbit. The Gemini spacecraft was chosen to transport crews to and from the space station. This selection was driven by Gemini's established operational track record, meaning there would be no need to solve new operational challenges associated with a new spacecraft design. The Skylab module, designed to be launched in its final, fully assembled configuration, avoided the complications of in-orbit assembly and deployment, establishing a significant advantage in streamlining the mission.

However, despite these advantages, the initial costs associated with the Skylab program were deemed unfavorable. The complexities of crew rotation and the logistical challenges of maintaining a space station were considered significant drawbacks. During this period, there was considerable interest in larger station concepts, such as the Project Olympus by the Manned Spacecraft Center (MSC). These concepts envisioned a station requiring a Saturn V booster for launch, with additional spacecraft for ferrying crews and logistics. Such designs anticipated a crew size ranging from 12 to 24 astronauts, with the station having a planned operational lifespan of up to five years. These large laboratories were proposed to span between 46 and 61 meters in diameter, offering around 1400 cubic meters of space. They typically featured continuous rotation to simulate artificial gravity alongside non-rotating central hubs for docking and zero-gravity experiments. Despite the promising potential of these designs, they also implied a significant increase in cost and development time.

Despite the appeal of the Apollo program as a temporary laboratory, Houston's Manned Spacecraft Center was increasingly inclined toward developing a larger, more permanent space station. In June 1963, MSC initiated two crucial studies to explore viable designs. One study, conducted by Douglas Aircraft Company, focused on a zero-gravity station, while the other, led by Lockheed, explored a rotating station concept. The study specifications included using a Saturn V booster, a hangar for accommodating a 12-man ferry craft, and provisions for a 24-man crew.

Douglas's design, submitted in February 1964, proposed a 31-meter-long cylindrical space station equipped with pressurized compartments for living, recreation, and research. It included a command center, a laboratory with a one-man centrifuge for simulating gravity, and a hangar capable of servicing four Apollo spacecraft. Although this design was considered feasible within future capabilities, it was ultimately deemed unjustifiable due to its size and associated costs.

In contrast, Lockheed's concept appeared more promising for eventual adoption. This design featured a Y-shaped module with a central hub for zero-gravity activities and a hangar for ferry and logistics spacecraft. The radial arms of the Y-shaped structure could accommodate up to 48 astronauts, with varying levels of artificial gravity provided along the arms. This approach was favored by MSC engineers not only for its physiological benefits but also for its efficiency in long-duration missions. As one engineer remarked, living in an environment where gravity was simulated could simplify the daily experience for astronauts, making tasks like finding objects and maintaining orientation more intuitive.

As these studies progressed, NASA focused on operationalizing a more immediate solution. In 1964, plans were set to repurpose the Extended Apollo spacecraft as the first space laboratory. George Mueller's decision in November 1963 to streamline testing for the Apollo program had already suggested the possibility of surplus hardware. While officials were reluctant to predict the exact number of excess Saturn rockets, the 1964 plans anticipated that ten or more Saturn V rockets could potentially be available for repurposing into space laboratories. This strategic pivot underscored NASA's commitment to leveraging existing resources to advance its goals in space research.

In the early 1960s, budgetary constraints began to overshadow the once-ambitious goals of NASA, influencing the trajectory of its space programs. In 1963, after two years of substantial financial support, Congress decided to trim NASA's budget for fiscal year 1964 from $5.7 billion to $5.1 billion. This reduction marked a pivotal shift in NASA's planning and aspirations. Wernher von Braun, a leading figure in space exploration, expressed his concerns to Heinz Koelle in August 1963. The harsh financial realities tempered Von Braun's usually optimistic outlook; he remarked, "I'm convinced that in view of NASA's overall funding situation, this space station thing will not get into high gear in the next few years. Minimum C-IB approach [Saturn IB and Extended Apollo] was the only thing we can afford at this time."

This uncertainty continued to influence NASA's strategy in the subsequent year. By April 1964, Koelle informed von Braun that NASA Administrator James Webb had instructed planners to develop various alternative objectives and missions along with their associated costs and implications rather than focusing on a detailed long-term program. Von Braun's wry response captured the agency's predicament: "Yes, that's the new line at Hq., so they can switch the tack as the Congressional winds change." During the FY 1965 budget hearings in February 1964, discussions regarding advanced crewed missions suggested a gradual evolution from Apollo-Saturn hardware to more advanced spacecraft. NASA had yet to commit to a concrete post-Apollo space station plan at this stage.

However, in June 1964, Michael Yarymovych, Director of Earth-Orbital Mission Studies, presented NASA's evolving plans at the First Space Congress in Cocoa Beach, Florida. Yarymovych outlined that Extended Apollo would initially serve as a laboratory and later as a logistics system within an expanding Earth-orbital program. He indicated that, in the future, NASA would select a more sophisticated space station from various medium and large-scale concepts under consideration. This strategic shift was underscored by NASA's commitment to enhance the capabilities of the Apollo system, with Yarymovych being assigned to this special task.

During this period, a new project emerged to become Skylab's principal rival for the next five years: an Air Force orbiting laboratory. Since the launch of Sputnik, the U.S. Air Force and NASA had been engaged in a competitive race for dominance in space. Initially, NASA held the upper hand due to the Space Act of 1958, emphasizing that space activities should be dedicated to peaceful purposes. Consequently, NASA's Mercury program was selected over the Air Force's "Man in Space Soonest" as the United States' first crewed space endeavor.

Despite this, the Space Act also allocated the Department of Defense (DoD) responsibilities for military operations and weapon system development. As a result, the Air Force pursued studies over the next three years to explore space bombers, crewed spy satellites, interceptors, and command and control centers. In congressional briefings following the 1960 elections, Air Force representatives highlighted the emerging notion that "military space, defined as space out to 10 Earth diameters, was the battleground of the future." This assertion underscored the growing interest in and competition for military applications of space technology, setting the stage for the development and rivalry to shape the future of space exploration.

Despite the Air Force's persistent advocacy, convincing civilian authorities of the necessity for space to become the next strategic battleground proved elusive. In a pivotal moment, Congress allocated $86 million to the Air Force to develop the Dyna-Soar, a crewed space glider. However, Secretary of Defense Robert S. McNamara decisively blocked the expenditure. The Department of Defense's Director of Defense Research and Development testified before a congressional committee, asserting that at that juncture, there was "no definable need" or "military requirement" for a crewed military space program. While acknowledging the potential future military applications of space technology, he suggested that NASA could develop much of this technology independently, if not most of it. The budget allocations of 1962 highlighted this shift: NASA's authorization of $3.7 billion for space activities dwarfed the Air Force's allocation, which had previously been more comparable.

The Cold War era's dynamic was such that Soviet advancements often catalyzed American actions. In August 1962, a Soviet space achievement heightened the debate. The USSR successfully launched two spacecraft, Vostok 3 and Vostok 4, into orbits that came as close as 6 kilometers. This feat prompted American reports suggesting a potential rendezvous and docking. The Air Force saw significant military implications in this Soviet accomplishment, which led McNamara to reassess the Air Force's plans for space.

The critics, however, questioned the effectiveness of communication between NASA and the Air Force regarding technical and managerial issues. In response to these concerns, James Webb, then NASA Administrator, created a new position within NASA—the Deputy Associate Administrator for Defense Affairs. He appointed Admiral Walter F. Boone, USN (ret.), to this role in November 1962. Meanwhile, the Pentagon's focus shifted during the Cuban Missile Crisis in the fall of 1962, and the urgency for a space program waned somewhat. However, when the matter resurfaced, McNamara displayed a notable change in stance.

In early 1962, Air Force officials began discussing the "Blue Gemini" program—a proposal to utilize NASA's Gemini hardware for preliminary training missions in rendezvous techniques and to support a military space station. Some NASA officials viewed this as an opportunity to expand the Gemini program and secure additional Department of Defense funds. Nonetheless, when Webb and NASA's Associate Administrator for Space Flight, Robert C. Seamans Jr., sought to increase Air Force participation in December 1962, McNamara countered by proposing that his department take control of all American crewed spaceflight programs. NASA officials successfully resisted this bid for dominance but reluctantly agreed to McNamara's condition: neither agency would initiate a new crewed program in near-Earth orbit without mutual consent.

The debate over space program responsibilities persisted for several months. During this period, the Air Force attempted to assert control over various aspects of the program. The landscape of space exploration continued to evolve, marked by significant events such as the launch of Mariner 2 toward Venus on August 27, 1962, followed by two Explorer missions and Walter M. Schirra's Mercury flight in October. By November, NASA had conducted its third successful Saturn I test flight, further fueling the discussions and decisions to shape the future of American space endeavors.

In September, NASA and the Department of Defense (DoD) reached a crucial agreement regarding advanced space station studies. This arrangement was designed to safeguard NASA's ability to pursue its space station ambitions while enhancing collaboration through the Aeronautics and Astronautics Coordinating Board, the principal channel for formal communication between the two agencies. The agreement's preamble emphasized the desirability of merging the requirements of both agencies into a unified space-station project wherever feasible.

The impetus for this agreement was partly driven by the Air Force's dissatisfaction with the Gemini program. The notion of a "Blue Gemini" had lost traction by 1963, and the Dyna-Soar space glider, once a promising project, began to lose its appeal. With the Gemini program scheduled to precede Dyna-Soar by two years, Secretary of Defense Robert McNamara decided to terminate the Dyna-Soar project on December 10, 1963. This decision redirected part of the Dyna-Soar funding toward a new initiative—the Manned Orbiting Laboratory (MOL). The MOL program aimed to establish a military role for crewed space missions but was constrained by budgetary limitations, necessitating a cost-effective approach.

The Manned Orbiting Laboratory (MOL) was a pivotal yet ultimately unrealized component of the United States Air Force's (USAF) human spaceflight program during the 1960s. Emerging from the USAF's early concepts of crewed space stations designed for reconnaissance, the MOL was intended as a successor to the canceled Boeing X-20 Dyna-Soar military reconnaissance spaceplane. The project's vision evolved into a single-use laboratory, where crews would be launched on 30-day missions and return to Earth aboard a modified Gemini B spacecraft. This spacecraft, derived from NASA's Gemini program, was paired with the laboratory for a comprehensive mission profile.

Publicly announced on December 10, 1963, the MOL was presented as an inhabited platform to demonstrate the military utility of placing humans in space. However, its primary mission as a reconnaissance satellite was a highly classified black project. Seventeen astronauts were selected for this groundbreaking program, including Major Robert H. Lawrence Jr., the first African American astronaut. The spacecraft was contracted to the McDonnell Aircraft Corporation, while the Douglas Aircraft Company was tasked with constructing the laboratory. Although the Gemini B bore a strong resemblance to NASA's Gemini spacecraft, it featured significant modifications, including a circular hatch through the heat shield to facilitate crew passage between the spacecraft and the laboratory. Launches into polar orbit were planned from the newly developed Vandenberg Space Launch Complex 6 (SLC-6).

As the decade progressed, the escalating Vietnam War increasingly diverted funding away from the MOL program, resulting in repeated delays to its first operational flight. Concurrently, advancements in automated systems began to diminish the perceived advantages of a crewed space platform over an unmanned one. The MOL program achieved a significant milestone with the uncrewed test flight of the Gemini B spacecraft on

November 3, 1966. Despite this progress, the program was canceled in June 1969 without any crewed missions being flown.

After the MOL's cancellation, seven of its astronauts transferred to NASA in August 1969, joining NASA Astronaut Group 7. These astronauts would go on to contribute significantly to space exploration, all eventually flying on Space Shuttle missions between 1981 and 1985. The Titan IIIM rocket, developed specifically for the MOL, never saw flight; however, its UA1207 solid rocket boosters found a second life as part of the Titan IV, and the materials, processes, and designs pioneered for these boosters informed developing the Space Shuttle's Solid Rocket Boosters.

Several elements of the MOL program had enduring impacts on NASA and other space initiatives. For instance, the spacesuits initially developed for MOL influenced subsequent NASA designs. Moreover, MOL's waste management system was later utilized aboard Skylab, and other MOL equipment was found to be applicable to NASA Earth Science missions. Although SLC-6 was refurbished to support military Space Shuttle launches, these plans were ultimately abandoned following the Space Shuttle Challenger disaster in January 1986.

The Air Force planned to employ existing hardware to achieve this goal, utilizing the Titan IIIC launch vehicle—initially developed for Dyna-Soar—and a modified Gemini spacecraft. The only new elements in the system would be the laboratory itself and its associated test equipment. The Titan IIIC could carry up to 5,700 kilograms, with approximately two-thirds allocated to the laboratory and the remainder to test equipment. The MOL was designed to offer 30 cubic meters of space, roughly equivalent to a medium-sized trailer. The laboratory and spacecraft would be launched together, with two astronauts transferring from the Gemini spacecraft to the laboratory upon reaching orbit, where they would spend approximately one month. The Air Force estimated the cost for four MOL flights to be $1.5 billion, with the first mission projected for 1968.

The MOL decision immediately raised concerns regarding its potential impact on the cooperative development of an orbital space station, as outlined in the NASA-DoD agreement.

Some observers interpreted the MOL initiative as a potential setback to the Webb-McNamara agreement. However, both NASA and the DoD characterized MOL as a distinct military endeavor rather than a broader space program. They agreed not to consider it part of the National Space Station program—a separate project under joint review. In March 1964, NASA and the DoD established the National Space Station Planning Subpanel, tasked with recommending a space station project to follow MOL.

Despite Air Force press releases suggesting that the military might assume primary responsibility for space stations, NASA officials maintained that MOL would complement their post-Apollo plans. The similarity between the two programs prompted engineers at Langley Research Center and the Manned Spacecraft Center (MSC) to revise their designs to differentiate them from MOL.

However, McNamara's announcement did not constitute formal program approval. For the next 20 months, MOL faced challenges in gaining recognition and securing adequate funding. Although planning and some contracts progressed in 1964, political realities influenced the program's development. In September 1964, Congressman Olin Teague, chairman of the House Subcommittee on Manned Space Flight and NASA Oversight, suggested that the DoD repurpose Apollo for its needs. Shortly after the 1964 elections, Senate Space Committee chairman Clinton Anderson opposed MOL, arguing that its cancellation could save over a billion dollars, which could be redirected toward an Extended Apollo station. Despite such opposition and speculation about MOL's potential cancellation, the FY 1966 budget proposal tentatively allocated $150 million for the program.

Reluctant to approve potentially overlapping programs, the Bureau of the Budget allocated funds to MOL in December 1964, contingent upon further studies and a review scheduled for May 1965. During this period, the DoD would define the military experiments to be conducted while NASA explored Apollo configurations that could meet military requirements. A joint study would also assess MOL's suitability for non-military missions. A NASA-DoD news release on January

25, 1965, emphasized the need to avoid program overlaps and indicated that both agencies would use existing or actively developed hardware and facilities for their crewed spaceflight programs as much as possible.

In February 1965, a NASA committee undertook a three-month study to evaluate Apollo's potential as an Earth-orbiting laboratory and to define key scientific experiments for a post-Apollo orbital program. Although the committee worked closely with an Air Force team, its recommendations had little impact on the MOL program, which remained largely unchanged. Nevertheless, the study helped NASA clarify its post-Apollo objectives.

Support for MOL grew in late 1964, with advocates urging the DoD to accelerate development. The House Military Operations Subcommittee recommended initiating full-scale development without delay, and a House Committee on Science and Astronautics member advocated for a rapid launch of MOL within 18 months. Advances in Russian and American space missions, including multi-crewed flights and spacewalks, made a military role for space more plausible. On August 25, 1965, President Lyndon B. Johnson finally approved the MOL program.

A Pentagon spokesperson explained that evaluating man's role in space required more comprehensive studies than could be conducted on the ground, as the Gemini spacecraft was deemed too cramped for this purpose. The Air Force's abandonment of Apollo was justified by the argument that Apollo's lunar capabilities exceeded what MOL required.

Presidential Review of NASA's Plans

In early 1964, NASA embarked on a comprehensive reassessment of its space exploration plans, prompted by a request from the White House. President Lyndon B. Johnson, having been a key supporter of the U.S. space program since his tenure as Senate majority leader, recognized that the post-Apollo era promised to be both costly and complex. Seeking clarity on future space objectives and the supporting research and development programs, he tasked NASA with providing a detailed statement.

NASA Administrator James E. Webb formed an ad hoc Future Programs Task Group to address this challenge. After five months of deliberation, the group's report presented no groundbreaking proposals. Instead, it acknowledged the significant demands the Gemini and Apollo programs placed on NASA's financial and human resources. The report recommended that NASA focus its efforts on these ongoing programs and delay large-scale new missions until further study could be conducted. It suggested that the United States leverage the Saturn and Apollo programs' capabilities to maintain its space leadership well into the 1970s. The report also advocated for the early planning of intermediate missions using existing hardware and proposed a modest investment in Extended Apollo missions, which could potentially begin by 1968. Additionally, it advised continuing long-term planning for space stations and crewed missions to Mars in the 1970s.

This report, which Webb utilized extensively in subsequent congressional hearings, was generally well-received within NASA. Robert Seamans, in particular, was eager to extend the capabilities of the Apollo program beyond lunar landings. However, the report faced criticism from various quarters. The Senate Space Committee deemed it "somewhat obsolete," arguing that it provided less information than anticipated and lacked essential details. Consequently, they recommended a 50% reduction in funding for Extended Apollo, asserting that enough studies had already been conducted. Some NASA officials and members of Congress had hoped for a more ambitious proposal, possibly including a recommendation for a Mars landing as the next crewed endeavor. At the Marshall Space Flight Center in Huntsville, some officials felt the plan did not present a significant challenge, particularly once Apollo was completed.

At that point, NASA officials were acutely aware of the limited experience gained in crewed spaceflight. The longest Mercury mission had lasted less than 35 hours. Webb and Seamans argued before congressional committees that the outcomes of the Gemini missions, which would provide longer-duration flights, might significantly influence future planning. They advocated delaying decisions on significant new programs until after the lunar landings. The issue of funding was also a significant concern. By fiscal year 1964, NASA's budget had plateaued at $5.2 billion, barely sufficient for Gemini and Apollo. Without an increase in funding, any new crewed programs would have to wait until after 1966, when Apollo's spending would decrease. The Johnson administration and Congress were reluctant to increase NASA's budget, with Great Society programs and the Vietnam War demanding significant resources. Additionally, there was competing interest in the Air Force's space program, with some advocating for the Manned Orbital Laboratory (MOL) as the nation's first space laboratory.

Despite these challenges, there were strong arguments for initiating Extended Apollo. A new program, even one based on Saturn and Apollo hardware, would require three to four years of lead time. If a new program did not commence in 1965 or early 1966, the gap between the end of Apollo and the start of its successor would negatively impact the Apollo workforce. With skilled engineers nearing the completion of their tasks, this potential hiatus was a pressing concern, particularly for Marshall, given that Saturn IB-Apollo flights were expected to end by early 1968.

In the fall of 1964, a Future Projects Group, appointed by Wernher von Braun, began biweekly meetings to consider Marshall's future. Meanwhile, George Mueller explored ways to keep the Apollo team intact in Washington. By 1968 or 1969, the U.S. aerospace industry would be capable of producing and flying eight Apollos and twelve Saturns annually. However, Mueller faced a difficult situation: the buildup of the Apollo industrial base left him with insufficient funds to employ it effectively after the lunar landing.

Until mid-1965, Extended Apollo (Apollo X) remained classified as an advanced study project. That summer, Mueller advanced it to the project definition phase, establishing the Saturn-Apollo Applications Program Office alongside the Gemini and Apollo offices at NASA Headquarters. Maj. Gen. David Jones, an Air Force officer on temporary assignment to NASA, led the new office, with John H. Disher serving as deputy director, a position he would hold for the next eight years. The office's establishment on August 6, 1965, was marked by minimal fanfare, overshadowed by the ongoing Apollo and Gemini programs. Nonetheless, it represented a significant milestone, following six years of space station studies and three years of post-Apollo planning.

The new program faced several substantial challenges, including securing fiscal support from the Johnson administration and Congress, defining new relationships between NASA centers, and coordinating Apollo Applications with the ongoing Apollo missions. Despite these obstacles, Mueller's commitment to the Extended Apollo studies remained strong, driven by the needs of the Apollo team and the belief in the value of continued space exploration. The Apollo Applications Program (AAP) would require all the confidence and motivation it could muster in the demanding years ahead.

In the early days of the Apollo Applications Program (AAP), the landscape of American space exploration was shaped by a mixture of ambition and skepticism. Within a month of establishing the Apollo Applications Office, its director, Dr. George E. Mueller, presented its preliminary plans to congressional committees. Despite recognizing the need for continued crewed spaceflight beyond the lunar landings, these committees displayed a notable lack of enthusiasm. Extending Apollo's capabilities through various routine, long-duration flights seemed to them more like an exercise in "boring holes in the sky" rather than a meaningful advancement of American space leadership. To these skeptics, the program was a mere continuation of existing activities rather than a groundbreaking leap forward.

Mueller's challenges were not confined to Congress. NASA's top officials, including Administrator James E. Webb, also approached the AAP with caution. Webb was hesitant to initiate a new, costly program without assurance that the Apollo missions would achieve their objectives. Mueller, however, faced a pressing need to address two concerns: he required alternatives to safeguard against potential delays in the Saturn and Apollo programs, and he needed a compelling program to maintain the coherence and morale of the crewed spaceflight division.

Mueller was acutely aware of the possibility of unforeseen delays with the Saturn rockets and Apollo missions. Consequently, he sought a parallel program to sustain the momentum of crewed space exploration. Conversely, should the Apollo program proceed smoothly, he wanted to maximize the benefits of the substantial investment being poured into Apollo hardware. This was particularly urgent for the Marshall Space Flight Center, where the end of the Saturn program meant that no major new launch vehicles were on the horizon.

Wernher von Braun, a prominent figure at Marshall, recognized the center's need to diversify beyond just launch vehicles. Between 1962 and 1965, Huntsville's Future Projects Office explored several concepts to broaden Marshall's scope. Among these concepts was the idea of a "spent-stage laboratory," which envisioned repurposing an empty rocket stage into a functional living and working space in orbit. This idea had been under consideration for some time, with a conceptual design study commencing at Marshall just two weeks before Mueller formally established the AAP office at NASA Headquarters.

The spent-stage laboratory concept soon gained traction, becoming the centerpiece of the Apollo Applications Program. Every orbiting spacecraft typically includes the final stage of its launch rocket. While these upper stages usually remain briefly in orbit, slightly adjusting their fuel-burning trajectory could stabilize their orbits for extended periods. This concept originated in the speculative work of von Braun and his colleagues during the Peenemünde days. In 1959, the Army Ballistic Missile Agency proposed a similar idea in a study called Project Horizon. This report, prepared by Heinz Koelle and Frank Williams, was the Army's last attempt to secure a role in crewed spaceflight and proposed establishing an armed lunar outpost.

Project Horizon suggested that by 1965, the United States should have a permanent space station in Earth orbit, serving as a launch base for lunar missions. If such a station was not in place, the report recommended a minimal orbital shelter for the crew tasked with refueling the lunar-bound rockets. The basic structure for this shelter was the empty third stage of the rocket that launched the crew's spacecraft. The crew would dock with this stage, remove the residual hydrogen, and fit it out for habitation with equipment from their spacecraft. As more payloads were launched, additional empty stages could be added, providing storage and shielding against meteoroids and cosmic radiation. The report even included sketches of a larger station assembled from 22 empty stages, resembling the familiar wheel-shaped design.

Although Project Horizon never advanced beyond the report stage and the Army's lunar outpost proposal was ultimately shelved, utilizing spent rocket stages did not fade away. When the Douglas Aircraft Company, known for building the Saturn S-IV stage, revisited the idea, it reignited interest. Douglas, a key player in the rocket industry since the end of World War II, had previously achieved significant success with the Thor missile system. Thor, the first intermediate-range missile deployed by a Western power, was succeeded by the intercontinental Atlas and Titan missiles and continued its legacy by launching satellites and space probes into the 1970s.

Thus, the spent-stage laboratory concept, refined and championed by Marshall Space Flight Center, became a cornerstone of the Apollo Applications Program, embodying a vision that bridged the gap between past speculations and future possibilities in space exploration.

In 1960, the S-IV stage contract was put out for bids, and Douglas Aircraft Company emerged victorious. This contract marked a significant milestone as the S-IV was the first major rocket stage to utilize cryogenic propellants—liquid hydrogen and liquid oxygen. Douglas's achievement in this area represented a substantial technological advance. The S-IV contract was managed by Marshall Space Flight Center, which facilitated a close and collaborative relationship between Douglas and Marshall. Engineers from both organizations worked closely together to resolve design and production challenges, reinforcing the interdependence of their efforts. When issues arose with the Saturn rockets, they were viewed as mutual problems, fostering a strong bond between the teams at Douglas and Marshall.

Similarly, at the Manned Spacecraft Center (MSC) in Houston, engineers and managers established close working relationships with their counterparts at McDonnell Aircraft Corporation, the primary contractor for both the Mercury and Gemini spacecraft. This spirit of collaboration was critical for the success of these programs.

Douglas struggled to break into the crewed spacecraft sector despite successfully developing launch vehicles. The company had bid on the Mercury spacecraft, the Apollo command module (as part of a consortium of four companies), and the lunar module, but without success. Determined to change this trajectory, Douglas management initiated a future studies program in the early 1960s to explore potential opportunities in crewed spaceflight. They identified small space stations and orbiting laboratories as promising areas for expansion.

By late 1963, Douglas had secured several study contracts with NASA and was actively competing for the Air Force's Manned Orbiting Laboratory project. However, Douglas entered the Apollo Applications Program through a different avenue. As the S-IVB replaced the S-IV stage, Douglas was awarded the contract for the new stage, rendering the S-IV obsolete. In 1962, Douglas's chief engineer for the Saturn program led a study group to explore potential uses for the S-IV. The group proposed converting the S-IV into a small orbiting laboratory.

The exact transition of the spent-stage concept from Marshall to Douglas remains somewhat unclear, though there were numerous opportunities for collaboration. Wernher von Braun, a key figure at Marshall, traced the origins of Skylab to the initial S-IV study, attributing its development to concurrent thinking among Marshall engineers. Heinz Koelle, who was involved in discussions about the spent-stage idea, recalled conversations with von Braun in 1960 and believed von Braun may have shared the concept with Douglas engineers, likely in 1961.

Douglas's group analyzed NASA's existing programs and future goals of space exploration and recognized a gap that could be filled. Their work, along with the collaborative spirit between Marshall and Douglas, eventually paved the way to developing Skylab, an endeavor that built upon these early ideas and technological advancements.

Though groundbreaking, the Gemini and Apollo programs were narrowly focused on specific goals. They did not fully address the broader questions of orbital operations or long-duration human habitation in zero gravity. The scope of these programs left gaps in understanding how both humans and their equipment would perform over extended periods in space. Recognizing the need for more comprehensive research, experts proposed an innovative solution: an orbiting laboratory.

The vision for this space laboratory involved repurposing the S-IV stage of a rocket, which, with minimal modifications, could be adapted for human habitation. This approach promised a cost-effective way to gather crucial data on long-term spaceflight. The plan called for outfitting the S-IV with a meteoroid shield to protect it from space debris and installing a storage module on top to house equipment that could not endure the harsh conditions of liquid hydrogen.

By 1965, the laboratory was projected to be operational, costing approximately $220 million. The design envisaged a mission in which two astronauts would dock their Gemini spacecraft with the S-IV's storage module. They would then transfer equipment into the stage, emptying its fuel and pressuring it with a nitrogen-oxygen atmosphere, thus converting it into a habitable environment.

During the planned hundred-day mission, the crew would conduct over seventy experiments focused on physiology, space technology, and orbital operations. The S-IV laboratory was equipped with advanced medical monitoring tools, including a one-man centrifuge designed to simulate artificial gravity. This device would allow scientists to study the effects of weightlessness on the human circulatory system and provide a means of reconditioning the astronauts if any serious health issues arose.

The idea of such a space laboratory was not entirely new. It had been conceptualized within Douglas Aircraft Company three years earlier. During this period, as the first Thor missile squadron was deployed in England, the London Daily Mail sought to capitalize on the public's fascination with space. For its annual Ideal Home Exhibition in 1960, the paper chose "A Home in Space" as its theme and invited American aerospace contractors to submit proposals. Douglas's advanced design team in Santa Monica responded with a full-scale model and promotional materials, showcasing their vision for a space laboratory.

Douglas Aircraft's proposal and subsequent studies were submitted to the Future Projects Office at Marshall Space Flight Center as unsolicited ideas, later published in professional journals. Over the following years, Douglas continued to explore and advocate for innovative uses of its rocket hardware, driven by a commitment to expanding the frontiers of space exploration.

Using spent rocket stages as space laboratories captured the imagination of many engineers at Douglas Aircraft Company. Driven by their enthusiasm for the project, the engineers dedicated far more time to the design than their budget would have allowed. Their proposal envisioned a space laboratory built into the upper stage of a hypothetical launch vehicle, capable of housing four crew for up to thirty days in Earth orbit. This laboratory would facilitate astronomical observations from above the atmosphere.

In March 1960, Douglas showcased a mockup of this space laboratory at London's Olympia Exhibition Hall. The display attracted significant public attention, with estimates suggesting over a million visitors, around 150,000 toured the exhibit. Despite this impressive public interest, the project garnered little attention back in the United States and was eventually filed away and forgotten until Skylab's launch years later.

At that time, NASA's Marshall Space Flight Center, which had limited involvement in space station studies compared to the Manned Spacecraft Center (MSC) or Langley Research Center, did not respond immediately to Douglas's proposals. By 1963, the future direction of space stations remained unclear. Both Wernher von Braun and his colleague Hermann Koelle recognized that the likelihood of a large-scale space station as the next step in space exploration was diminishing.

Financial constraints suggested that only a small station would be feasible within NASA's budget.

This shifting perspective was reflected in various studies. MSC's contract with North American Aviation to extend the Apollo mission's duration in orbit, and Langley's Manned Orbital Research Laboratory studies initiated in June 1963 with contracts awarded to Douglas and Boeing, highlighted this trend.

George Mueller, who sought to maintain the momentum of the crewed spaceflight program, recognized the need for a parallel program that could continue if Apollo faced delays or succeeded ahead of schedule. Public opinion on Apollo was also shifting, with growing criticism of the program's cost. Scientists and other critics argued that the vast sums spent on the moon mission could be better allocated to uncrewed satellite projects. Mueller's idea of using Apollo hardware for scientific investigations in space resonated with the evolving needs and criticisms.

The Saturn program was progressing smoothly at Marshall, but no new vehicles were planned. Concerns about the future of the Saturn IB—a vehicle well-suited for Earth orbital operations but with a limited future—prompted Marshall management to seek new uses for this hardware. This concern was exacerbated when the Air Force selected the Titan III for its Manned Orbiting Laboratory, leading Koelle to explore potential uses for the Saturn IB.

In response to Mueller's interest in alternative uses for existing hardware, Marshall resurrected the idea of utilizing spent rocket stages. A nine-month, $100,000 contract was awarded to North American Aviation to study the feasibility of using spent stages in NASA's space programs, including potential applications for orbital operations and laboratories. North American's study explored several concepts, such as refueling S-IVB stages in orbit for lunar or planetary missions, converting S-II stages into orbital hangars, and assembling empty S-IVBs into large stations.

Following a midterm review in January 1964, Marshall added a new concept to the study: the "Apollo Support Module," which proposed using an empty S-IVB fuel tank as a working space. The final report, completed in April 1965, recommended this concept for further development, highlighting the practical value of the tank's large volume for extensive experimentation in orbital operations.

Koelle's office had previously considered adapting Saturn hardware for space laboratories, including using the oxygen tank of the Saturn V's first stage as a structural shell for a laboratory. With the North American report in hand, the Future Projects Office began to develop a proposal for utilizing spent stages, incorporating both new ideas and those previously considered by Marshall.

Initially, the proposal involved a simple mission where an Apollo spacecraft would dock with a spent S-IVB stage, allowing astronauts to experiment with extravehicular mobility techniques in a controlled environment without major modifications to the S-IVB. However, the concept evolved to include pressurizing the tank and using its 281-cubic-meter volume as living quarters. This approach aimed to establish a continuing program, with each mission building on the results of previous ones.

Marshall saw significant potential in using spent stages for extended Apollo missions and regarded them as logical candidates for further development. Frank Williams was tasked with advancing this proposal, marking a crucial step in the evolution of space station concepts and NASA's future in orbital research.

The Management Task

The management of Skylab was as pioneering as its technological innovations. Coordinating the program's multifaceted activities required a sophisticated and effective management approach. Skylab's management challenge encompassed the collaboration of multiple NASA centers, numerous major contractors, various scientific and educational institutions, other government agencies, and contributions from international partners, all contributing experimental hardware and expertise. The management strategy needed to balance fiscal and schedule constraints to maximize the scientific output from a diverse array of experiments while ensuring Skylab's safe, reliable, and efficient operation.

To address this complex challenge, NASA leveraged the capabilities of its three Manned Space Flight centers. The Office of Manned Space Flight at NASA Headquarters in Washington, D.C., provided overall program direction, ensuring alignment with the broader objectives of the space program.

The George C. Marshall Space Flight Center (MSFC) in Huntsville, Alabama, was tasked with developing the hardware for the orbiting space station and overseeing overall systems engineering and integration. MSFC's role was crucial in ensuring that all mission hardware was compatible and functioned seamlessly during each flight.

The Lyndon B. Johnson Space Center (JSC) in Houston, Texas, was responsible for mission control, flight operations, and adapting the Apollo command and service modules for Skylab. JSC played a key role in the planning and executing these operations, ensuring that the mission's technical and operational aspects were meticulously managed.

The John F. Kennedy Space Center (KSC) in Florida handled the planning and execution of launch operations. KSC's expertise was vital in preparing and launching the Skylab missions, ensuring they were executed smoothly and efficiently.

These centers played a prominent role in the highly integrated technical and management team that implemented the Skylab program. Skylab's success depended on continuous, close communication and the implementation of carefully planned technical and management reviews. These reviews integrated the expertise of the various participants, coordinating their efforts to achieve the program's objectives and ensure its successful completion.

Skylab management responsibilities.

In March 1970, key figures from NASA and their staff convened at the McDonnell Douglas Astronautics Company facility in St. Louis to review the progress of the Skylab program. This visit was an important milestone, bringing together many prominent individuals involved in the project. Among those present were Christopher C. Kraft, Jr., a leading figure in NASA's mission control; Thomas W. Morgan; and Faget, who contributed significantly to developing the space station. Ludie G. Richard was also in attendance, positioned behind Faget, alongside Walter Burke and Sigurd A. Sjoberg. F. Brooks Moore, Eberhard Rees, Kenneth S. Kleinknecht, Lee B. James, T.J. Lee, Leland F. Belew, Floyd M. Drummond, and Fred A. Speer rounded out the group.

This assembly of experts was crucial for assessing the project's advancement and addressing any emerging challenges, ensuring that Skylab remained on track for its ambitious goals.

At the end of June 1965, a notable transition occurred in the American space program. Dr. Koelle concluded his decade-long involvement with NASA and assumed a professorship at the Technical University of Berlin. In his stead, Williams, who had been serving as Wernher von Braun's special assistant for advanced programs since late 1963, returned to his previous role as the director of the Future Projects Office. This shift also marked a rebranding of the office, now renamed the Advanced Systems Office.

Visitors to the neutral buoyancy simulator of the Marshall Space Flight Center, in November 1967, included top-level personnel from NASA Headquarters and several of its field centers. Left to right were Edgar M. Cortright, William E. Stoney, Robert F. Thompson, Kurt H. Debus, Edwin A. Weaver, Paul H. Schuerer, Charles W. Mathews, William R. Lucas, Wernher von Braun, George Hage, Robert R. Gilruth, Barney E. Evans, Hermann Weidner, George B. Hardy, Julian West, George S. Trimble, Thomas W. Morgan, Maxime A. Faget, and Robert F. Heiser.

Inspecting launch facilities at the Kennedy Space Center early in the Skylab program are, left to right: von Braun, George E. Mueller, James C. Elms, Rep. Olin E. Teague (D-Texas), and Robert C. Seamans, Jr., facing away from the camera.

Williams's immediate task was to finalize the material on the spent-stage proposal for submission to Headquarters. On July 20, 1965, Williams and von Braun presented their proposal to the Manned Space Flight Management Council. Their plan called for the initiation of a conceptual design study to detail the project further. Mueller, a key supporter of the initiative, allocated $150,000 for a four-month study to advance the proposal. By August 10, Williams had presented the proposal to Marshall's Future Planning Policy Board, and on August 20, he convened the first meeting of the conceptual design study group.

Members of the Congressional Subcommittee on Manned Space Flight received a briefing on the Skylab program during a visit to the Marshall Space Flight Center in March 1969. Left to right: von Braun, Rep. Joe D. Waggoner, Jr. (D-Louisiana), Rep. Earle Cabell (D-Texas), Representative Teague (D-Texas), Rep. James G. Fulton (R-Pennsylvania), and Ernst Stuhlinger.

The organizational meeting on August 25 was a crucial step in this endeavor. The primary objective was to orient the group with the project and review the plan previously presented to the Management Council. The study aimed to explore three potential configurations for the orbital workshop, the official term for the spent stage. Initially, the project was considered a "wet workshop," which would incorporate the original concept of using the spent stage as a research facility. Over time, this concept evolved to include a ground-equipped variant, known as the "dry workshop," launched by a Saturn V rocket. The dry workshop, which would ultimately be the model constructed, would not contain any fuel.

Three configurations were under consideration:

The "minimum configuration," consisting of the empty tank equipped only with a docking port, lacking power or life-support systems.

The "intermediate configuration," which included an airlock, power, and oxygen supply, though it would not have carbon dioxide removal capabilities. This setup allowed the crew to work without pressure suits.

The "baseline configuration," featuring a complete environmental control system, space for a broad range of experiments, a power system adequate to support these experiments, and positive attitude control.

The minimum and intermediate configurations were suited for missions lasting between 3 and 14 days, with only a limited number of experiments. In contrast, the baseline configuration could support missions of 14 to 28 days and accommodate a more extensive experimental program.

As the study progressed, numerous technical challenges emerged, necessitating extensive analysis and resolution. Questions arose regarding power supply, the readiness of experiments for the first flight, their weight, power requirements, and the fuel needed for attitude control. The study also had to address the removal of excess propellant from the tank, sealing tank openings, and mitigating risks from micrometeoroids. Constraints imposed by the launch vehicle, such as orbital altitude and inclination, along with various "ground rules," further complicate the process. The solutions had to be negotiated through various trade-offs involving Marshall's Saturn Program Office, Houston's Apollo Program Office, and Douglas Aircraft Company.

Douglas Aircraft Company had long been at the forefront of space station development, driven by its commitment to advancing space technology. By 1963, Douglas had secured significant contracts that set the stage for its future achievements. The first contract was with Langley Research Center for the Manned Orbital Research Laboratory (MORL) study, while the second, from the Manned Spacecraft Center (MSC), focused on a study of a 24-man "Saturn V Class" laboratory. These early engagements marked Douglas's entry into the burgeoning field of space station design.

By May 1965, Douglas had achieved a notable milestone: the design, construction, and testing of a flight-weight airlock, completed under contract with Langley. This was a crucial step in validating the technologies required for space habitation. That same year, in August, Douglas was appointed as the prime contractor for the Air Force Manned Orbiting Laboratory (MOL), further solidifying its role in space station technology. The company's Saturn Payload Applications Group continued to drive innovation, contributing significantly to the evolution of space stations.

During this period, Douglas closely watched the Extended Apollo program, meticulously compiling published information. This vigilance was justified when a document emerged from the group detailing the potential use of an empty S-IVB stage for mobility and maneuvering experiments. This proposal closely resembled the three workshop configurations under study at Marshall Space Flight Center. Among these, the most advanced Douglas design was a pressurized stage closely mirrored Marshall's most sophisticated workshop concepts.

In September 1965, the renowned space engineer Wernher von Braun visited Douglas, offering an unofficial briefing on the orbital workshop concept. This presentation was reminiscent of Frank Williams's presentation to the Management Council in July. During this visit, it became apparent to both Douglas and Marshall that their respective ideas were closely aligned, an exciting revelation that underscored the collaborative potential in their shared field of interest.

On October 20, 1965, Williams and a delegation from Marshall traveled to Houston to update MSC on their progress and findings. The briefing highlighted the advanced studies conducted on extravehicular activity, which left the MSC team visibly surprised by the depth of data and the energetic approach of Marshall's efforts. Despite this positive interaction, MSC maintained a strong preference for including a minimum of 0.1 g of artificial gravity in the workshop, which would be achieved by rotating the station on a radius of 20 to 30 meters. This requirement posed a significant challenge and became a persistent issue for the workshop study, eventually contributing to delays and complications.

As the various labs grappled with the workshop mission's complexities, Williams and von Braun grew increasingly confident that they were on the verge of developing a substantial and impactful space program. Despite the hurdles and differing viewpoints, Douglas and Marshall's combined efforts set the groundwork for future advancements in space station technology, reflecting a period of intense innovation and collaboration in the exploration of space.

At the November Management Council meeting, NASA's Associate Administrator, George Mueller, urged Wernher von Braun to continue his work on the space station study. By the end of the month, Marshall Space Flight Center had an opportunity to showcase their program when Mueller, accompanied by Deputy Administrator Robert Seamans, visited Huntsville. During this visit, Mueller and Seamans were briefed on the workshop concept and the results of the conceptual design study, which focused on a minimal configuration for the Saturn-Apollo mission 211, scheduled for August 1968.

Mueller's interest in the project was piqued. He requested a detailed presentation for the upcoming Management Council meeting three weeks later. He was particularly interested in seeing what Marshall could achieve with the Saturn-Apollo mission SA-209 and how soon a pressurized workshop could be prepared. Ideally, Mueller wanted the pressurized version ready for SA-209, six months before mission 211. Recognizing this as a significant opportunity, von Braun committed to preparing the presentation. He proposed advancing the spent-stage study into the project definition phase, a suggestion Mueller supported.

The newfound urgency led to immediate action. The following day, Williams announced that the workshop needed to be ready for SA-209, although there would be minimal financial support from Headquarters. This constraint meant that developing essential components, such as an airlock, would need to be managed in-house and funded through existing budgets. Williams demanded a half-day presentation, providing an honest assessment of Marshall's capabilities, to be prepared for von Braun within two weeks.

To meet these demands, Williams's office issued new guidelines for the study. The primary goal was to design, fabricate, and test a flight-qualified pressurized version of the S-IVB workshop for Saturn IB flights 209 or 211. Simultaneously, an effort would be made to develop an unpressurized version. Essential components, including an airlock, environmental control system, experiments, ground-support equipment, mockups, and training hardware, were to be designed and developed to support the mission. Marshall was tasked with developing the pressurized version, while Douglas handled the unpressurized version. The target was to have the pressurized version ready for flight 209 unless unforeseen costs, production delays, or technological issues necessitated a shift.

The guidelines emphasized maximizing using flight-qualified hardware, minimizing modifications to the S-IVB stage, and keeping costs low. The environmental control system was to be designed for a 14-day operation, although a shorter lifespan of 2-6 days would be acceptable if it was necessary to adhere to the schedule. A pure oxygen atmosphere would be used instead of a mixture of oxygen and nitrogen to simplify hardware requirements. Connections between the Apollo spacecraft and the spent stage were to be minimized, with the S-IVB maintaining its attitude until docking. Once the Apollo spacecraft was docked, its systems would take over control.

In determining the priority of experiment categories for the workshop, the following sequence was established: first, basic maneuvering experiments and biomedical observations; second, tasks involving maintenance, repair, and inspection of spacecraft systems, as well as rescue and cargo transfer operations; and third, prepackaged experiments where astronauts would primarily act as monitors.

Williams emphasized the critical importance of crafting a proposal that Marshall could confidently execute. The center's reputation was on the line, making it imperative that the proposed program was valuable and feasible. The labs responded with impressive speed. By December 21, Williams, having reviewed the proposal with von Braun, presented it to the Management Council. The proposal had evolved to include an additional piece of hardware: the "Spent Stage Experiment Support Module," which would serve as both an airlock and a carrier for certain equipment and expendables. Marshall planned to leverage Douglas's experience with the Langley airlock in constructing this module, though this collaboration was not highlighted during the presentation.

Following the presentation, Mueller suggested that Marshall explore the potential use of Gemini components in the airlock to save both time and money. With the Gemini program nearing its end, Mueller believed adapting existing hardware could be advantageous. Williams promptly contacted Charles W. Mathews, Gemini program manager at MSC, requesting assistance coordinating discussions with McDonnell Aircraft Corporation. The objective was to facilitate a technical briefing between Marshall and McDonnell to discuss potential collaboration.

Chapter 2 - Development and Design

The meeting in Houston, on January 4-5, 1966, set out the ground rules for the proposals that Marshall and McDonnell were to present to Headquarters. Key requirements included using Gemini-qualified environmental control and electrical power systems and ensuring the airlock remained functionally independent from the S-IVB instrument unit and the Apollo command module. The expectation was that McDonnell would provide the qualified systems while Marshall would handle the fabrication, leveraging Huntsville's civil service personnel and aligning with Mueller's zero-cost goal.

The involvement of McDonnell introduced unexpected complexities. While Marshall was familiar with Douglas's airlock design and testing, the existing airlock, though flight-weight and extensively tested, was not fully flight-qualified. Upgrading it to meet flight standards would significantly increase costs. In contrast, McDonnell could use components already qualified from the Gemini program, offering potential cost savings. However, the unfamiliarity between McDonnell and Marshall contrasted with MSC's long-standing relationship with McDonnell, dating back to the Mercury program. Due to its deep expertise in crew systems, this situation presented a strong argument for MSC to manage the airlock project.

Leveraging its experience, MSC made a compelling case to assume responsibility for the airlock project. By early February, the situation needed resolution. On February 11, during various meetings in Houston, von Braun passionately argued that NASA's interests would be best served by assigning the project to Huntsville. However, von Braun's case was complicated by his additional plans for Marshall, including future projects involving the assembly of large structures in orbit, which inadvertently strained the argument.

Despite von Braun's efforts, MSC's depth of experience and established relationship with McDonnell positioned them strongly to take over the airlock project, leading to a complex negotiation over its management and direction.

The management of the airlock project revealed underlying tensions between NASA's field centers, illustrating a broader challenge in the evolving landscape of crewed spaceflight. The division of responsibilities had always been clear-cut, with each center specializing in distinct areas of space exploration. However, converting an empty rocket stage into a crewed workshop blurred these traditional boundaries. The project posed a new problem: defining the roles and missions of the two field centers, Marshall Space Flight Center (MSFC) and the Manned Spacecraft Center (MSC) and negotiating their relationship with each other and with Headquarters.

MSC's reluctance to share responsibility for training further strained the situation. Sensing that the dispute over the airlock project could escalate into a disruptive confrontation, von Braun ultimately conceded the airlock management to MSC. This decision was a strategic move to avoid further conflict despite Marshall's initial interest in overseeing the project.

As Marshall returned from managing the airlock, Douglas Aircraft Company remained eager to participate. On March 9, 1966, a Douglas delegation traveled to Washington to present their proposal. Mueller showed interest in their presentation, particularly impressed by Douglas's prior experience with the Langley airlock and their proposal to build two new units for $4 million. Despite the project being under consideration, Mueller encouraged Douglas to submit formal proposals.

In contrast, on March 11, MSC's planners proposed a procurement plan for the airlock, favoring sole-source procurement from McDonnell Aircraft Corporation. However, following Douglas's presentation and their competitive pricing, Headquarters could not justify this exclusive procurement approach. Consequently, at the Management Council meeting on March 22, both centers were instructed to conduct studies to define the airlock's requirements and to establish cost and schedule projections for its construction.

Marshall, recognizing the high stakes, feared that the airlock's cost might jeopardize the workshop project. Marshall defined a minimal airlock concept to safeguard the project, which was just sufficient to support their intermediate-configuration workshop. To explore viable options, MSC granted Douglas, McDonnell, and Grumman three 60-day study contracts, each worth $50,000. Douglas focused on adapting the Langley airlock, McDonnell proposed modifications based on Gemini systems, and Grumman explored the possibility of repurposing the lunar module for airlock functions.

By late June, a source evaluation board chaired by Kenneth S. Kleinknecht, deputy program manager for Gemini at MSC, began reviewing the proposals from these three contractors. The board's evaluation would play a crucial role in determining the final design and selection of the airlock, reflecting the ongoing negotiations and shifting responsibilities within NASA's complex structure of crewed spaceflight operations.

On August 19, 1966, NASA Headquarters announced the selection of McDonnell Aircraft Corporation to negotiate the airlock contract. Following extensive negotiations, the final contract price was set at $10,509,000. This decision rendered Marshall's precautionary study redundant, and it was promptly terminated.

While the airlock negotiations were ongoing, the Technical Working Group focused on modifying the S-IVB stage to prepare it for use as a crewed workshop. Before astronauts could enter the stage's hydrogen tank, it had to undergo a rigorous "passivation" process. This involved venting hydrogen and oxygen tanks, emptying high-pressure helium bottles, and deactivating the stage's pyrotechnic systems. Additionally, all potential hazards inside the tank had to be eliminated. The internal insulation must be painted uniformly to ensure a consistent photographic background. Equipment mounting provisions, restraints, and mobility aids were also required.

By December 15, 1965, a list of necessary stage modifications was compiled. Marshall requested a quick-response cost estimate from Douglas for these changes, which came in at $1.5 million for modifications to stage 209. This cost was more than Marshall could afford, prompting further negotiations. A revised estimate, which included changes to five S-IVBs, was $4.5 million. Marshall forwarded the figure to George Mueller on January 20, 1966, finding this estimate the most feasible option in the limited time available. Mueller found the cost unsatisfactory and deferred any decisions for a month due to the lack of a budget for the Apollo Applications program. He instructed Marshall to explore the possibility of making the modifications in-house and requested a list of no-cost experiments to be prepared by mid-February.

Despite Frank Williams' earlier assertion that the spent-stage project was "in high gear," by late January, progress was hindered by unresolved funding issues. The S-IVB modifications identified before Christmas were advancing slowly. On February 21, Williams was informed that no additional funds could be allocated for changes until the workshop was officially approved. Immediate approval was necessary for the workshop to be included in the SA-209 mission.

Fortunately, a significant modification to the S-IVB had already been completed. Von Braun had initially identified that the "manhole," a 71-centimeter circular opening in the forward dome, was too small for a suited astronaut to pass through. Although funds were not available to address this issue, a July 1965 discovery by Douglas engineers of cracks in the welds around the manhole on stage 203 revealed a structural weakness in the dome design. Enlarging the manhole was identified as a viable solution, and this modification was applied to stages 211, 507, and all subsequent S-IVBs. This change, costing $600,000, was financed from Saturn program funds.

Additionally, von Braun advocated for redesigning the manhole to improve accessibility, further demonstrating the ongoing adjustments and adaptations required to meet the evolving needs of the crewed spaceflight program.

In the mid-1960s, the Skylab Space Station project faced various technical and financial challenges as it advanced toward becoming America's first space station. One of the initial hurdles was the design of the manhole cover on the S-IVB, the upper stage of the Saturn V rocket, to serve as Skylab's core. Wernher von Braun, a leading figure in the space program, had proposed replacing the cumbersome manhole cover—

secured by 72 bolts—with a more efficient quick-opening hatch. However, the cost estimate for this upgrade was $400,000, a sum deemed too high to justify. Despite this setback, the issue resurfaced a year later when operational concerns from Houston prompted the adoption of the quick-opening hatch, illustrating the shift in priorities as mission efficiency became a critical concern.

Another significant challenge was addressing the risk posed by micrometeoroids—tiny particles no larger than a few millimeters traveling at extremely high velocities through space. To better understand this risk, three Pegasus satellites, payloads on the final Saturn I flight, were placed into Earth orbit to measure the frequency and impact of these particles. Their data indicated a non-negligible risk that micrometeoroids might strike the S-IVB.

In response, the Marshall Space Flight Center arranged for tests at the Air Force's Arnold Engineering Development Center in Tullahoma, Tennessee, which specialized in producing high-speed projectiles. Preliminary results from these tests, reported in February 1966, were concerning. They showed that micrometeoroids could penetrate the metal skin of the S-IVB and ignite the polyurethane insulation, presenting a fire hazard. To mitigate this, engineers considered two solutions: adding an external shield to reduce impact velocity or applying a coating to the insulation to slow down combustion. Initial estimates of the cost for an external shield led engineers to pursue a coating solution instead, initiating tests to continue throughout the year.

By late November 1966, various approaches were being tested. Douglas Aircraft Company evaluated one coating, while the Manned Spacecraft Center (MSC) recommended another. Their findings varied significantly as Marshall and its contractor re-evaluated the micrometeoroid risk. Douglas's calculations suggested that the probability of micrometeoroid penetration during a 30-day mission could range from 1 in 3 to as low as 1 in 40, while Marshall's estimate was 1 in 50. Considering the potential benefits, Douglas engineers started exploring the feasibility of an external shield, which could lower the risk to 1 in 200.

The fire hazard was also closely tied to the atmospheric pressure within the workshop. A December 1966 medical staff paper prepared for NASA's Associate Administrator, George E. Mueller, recommended an atmosphere of 69% oxygen and 31% nitrogen at a pressure one-third of sea level for long-duration missions, though other compositions were also acceptable. In factoring in this atmospheric composition, Marshall engineers found that using a mixture with lower oxygen content at half the sea level pressure and an external shield would be advantageous. However, a definitive resolution on these measures remained elusive for some time.

In September 1966, the MSC raised concerns about the operational plan for activating the workshop. The plan at that point required two astronauts in spacesuits to remove the 72 bolts from the forward tank dome cover—a task that, based on experience from three Gemini missions, was deemed excessively time-consuming. This highlighted the importance of Von Braun's early caution regarding the ease of access. As a result, Douglas was tasked with designing and manufacturing a full-scale model of the quick-opening hatch for evaluation. Despite this progress, financial constraints remained a significant barrier. NASA's fiscal 1967 budget request had been severely cut, leaving the Apollo Applications Program with only $42 million—barely sufficient to sustain the program at a basic level.

By the end of 1966 and into 1967, developing Skylab, which had progressed slowly, faced additional complications. Meanwhile, the Office of Space Science and Applications (OSSA) was working on a major scientific project known as the Apollo Telescope Mount, which would eventually become the first astronomical facility to be launched into space. This new project marked a pivotal shift in the Apollo Applications Program, setting the stage for future advancements in space science.

In parallel with the efforts to define the Skylab orbital workshop, the Office of Space Science and Applications (OSSA) was advancing a significant scientific initiative to transform the Apollo Applications Program eventually. This initiative was the Apollo Telescope Mount (ATM), envisioned as the first astronomical facility to utilize human operators as in-orbit observers. The ATM's primary objective was to explore the effectiveness of human control over sophisticated instruments in space, with the goal of gaining valuable solar data during a period of anticipated peak solar activity around 1969-1970.

Homer Newell, the head of OSSA, discussed integrating the Apollo Telescope Mount into the Apollo Applications Program with George E. Mueller in early 1966. Both Newell and Mueller recognized the project's potential, yet they diverged on its implementation. OSSA proposed placing the ATM in the experiments bay of the Apollo service module, while the Office of Manned Space Flight (OMSF) favored a modified lunar module to accommodate such large and cumbersome experiments.

By the end of August 1966, NASA's Deputy Administrator Robert Seamans approved the project and endorsed Mueller's plan to use the lunar module for the telescope mount. This decision, however, introduced a range of technical and management challenges. An elaborate stabilization system was deemed necessary to counteract the inevitable movement of the crew, which could disrupt the alignment of the instruments. Additionally, the potential for contamination from spacecraft emissions posed a risk to the telescopes' optical surfaces, potentially impairing the quality of observations.

The Lunar Module, initially designed for lunar landings, was not ideally suited for prolonged operations in Earth orbit. Concerns arose about its ability to sustain itself as a free-flying spacecraft for extended periods, particularly for missions lasting up to 28 days. Moreover, the Lunar Module lacked a heat shield, making reentry impossible and complicating any potential rescue operations if it became disabled.

In response to these concerns, Mueller proposed a novel solution: operating the telescope mount while tethered to the Skylab workshop, with power, coolant, and oxygen supplied through an umbilical connection. This concept was presented to the Management Council on October 7, 1966, following a feasibility study conducted by Marshall Space Flight Center. Despite its potential, the tethered observatory module idea faced numerous challenges and did not garner significant support. Nonetheless, Mueller considered this backup concept for several months as the challenges associated with using the Lunar Module became more apparent.

By March 1966, mission plans began to include solar astronomy flights as standalone missions and as components of longer-duration Skylab missions. As the technical issues related to the Lunar Module became more evident, it became clear that operating the telescope mount from the Skylab workshop might offer a more viable solution. This shift in approach reflected the ongoing adaptation and refinement of NASA's plans as they worked to integrate the Apollo Telescope Mount into their broader space exploration goals.

In the aftermath of the crucial October meeting, it became evident that a new solution was necessary to facilitate the telescope module's docking with the Skylab workshop. Earlier, in May, while studies were underway to develop an airlock, NASA's decision-makers, including Mueller and the Headquarters staff, had firmly resolved to keep the airlock design straightforward and cost-effective. They had specifically excluded the possibility of implementing double docking, a design approach that, in hindsight, now seemed indispensable.

The Manned Orbiting Laboratory (MOL) was a pivotal yet ultimately unrealized component of the United States Air Force's (USAF) human spaceflight program during the 1960s. Emerging from the USAF's early concepts of crewed space stations designed for reconnaissance, the MOL was intended as a successor to the canceled Boeing X-20 Dyna-Soar military reconnaissance spaceplane. The project's vision evolved into a single-use laboratory, where crews would be launched on 30-day missions and return to Earth aboard a modified Gemini B spacecraft. This spacecraft, derived from NASA's Gemini program, was paired with the laboratory for a comprehensive mission profile.

Publicly announced on December 10, 1963, the MOL was presented as an inhabited platform to demonstrate the military utility of placing humans in space. However, its primary mission as a reconnaissance satellite was a highly classified black project. Seventeen astronauts were selected for this groundbreaking program, including Major Robert H. Lawrence Jr., the first African-American astronaut. The spacecraft was contracted to the McDonnell Aircraft Corporation, while the Douglas Aircraft Company was tasked with constructing the laboratory. Although the Gemini B strongly resembled NASA's Gemini spacecraft, it featured significant modifications, including a circular hatch through the heat shield to facilitate crew passage between the spacecraft and the laboratory. Launches into polar orbit were planned from the newly developed Vandenberg Space Launch Complex 6 (SLC-6).

As the decade progressed, the escalating Vietnam War increasingly diverted funding away from the MOL program, resulting in repeated delays to its first operational flight. Concurrently, advancements in automated systems began to diminish the perceived advantages of a crewed space platform over an unmanned one. The MOL program achieved a significant milestone with the uncrewed test flight of the Gemini B spacecraft on November 3, 1966. Despite this progress, the program was canceled in June 1969 without any crewed missions being flown.

After the MOL's cancellation, seven of its astronauts transferred to NASA in August 1969, joining NASA Astronaut Group 7. These astronauts would go on to contribute significantly to space exploration, all eventually flying on Space Shuttle missions between 1981 and 1985. The Titan IIIM rocket, explicitly developed for the MOL, never saw flight; however, its UA1207 solid rocket boosters found a second life as part of the Titan IV, and the materials, processes, and designs pioneered for these boosters informed developing the Space Shuttle's Solid Rocket Boosters.

Several elements of the MOL program had enduring impacts on NASA and other space initiatives. For instance, the spacesuits initially developed for MOL influenced subsequent NASA designs. Moreover, MOL's waste management system was later utilized aboard Skylab, and other MOL equipment was found to apply to NASA Earth Science missions. Although SLC-6 was refurbished to support military Space Shuttle launches, these plans were ultimately abandoned

following the Space Shuttle Challenger disaster in January 1986.

The airlock contract had, however, made it too late to amend the design without incurring significant delays and potentially initiating a new round of competitive bidding. Consequently, the only viable alternative was to devise an additional piece of hardware: a new docking adapter. This component, while seemingly simple in its design, was critical. It consisted of a cylindrical extension to attach to the existing airlock structure. Its primary function was to provide additional docking ports, accommodating the telescope module and other components without performing any active systems operations.

The new piece of hardware, officially named the Multiple Docking Adapter, was initially conceived as a straightforward cylindrical shell. It was designed to endure the stresses of launch and docking, while its sole function was to offer multiple docking ports. This adapter would be mounted above the airlock, featuring four radial tunnels at its upper end. Each of these tunnels and the main structure would be equipped with Apollo docking mechanisms. Notably, the adapter itself would not possess any active systems; instead, it would rely on power supplied by the airlock for its docking ports.

Introducing the Multiple Docking Adapter transformed the Skylab program. The "orbital cluster" concept emerged, presenting a vision of a core structure capable of supporting a diverse range of missions. Attaching specialized mission modules and performing resupply operations for extended missions marked a significant advancement. This configuration allowed crew rotations without necessitating a workshop shutdown and enabled the possibility of rescuing disabled spacecraft.

Thus, what had initially been conceived as a mere experiment utilizing empty rocket stages began to evolve into the semblance of a small space station, expanding the horizons of space exploration and mission capabilities.

By November, George Mueller had resolved to present the orbital cluster as the cornerstone of OMSF's post-Apollo strategy. When discussing future programs with James E. Webb and the Director of the Budget, Mueller was confident in his well-defined, coherent program and saw an opportunity to advocate strongly for funding. Despite this progress, a critical component—the multiple docking adapter—still required design work.

Initially, the plan was to have Marshall Space Flight Center fabricate the docking adapter. However, given the importance of the module's interface with the airlock, it was decided that McDonnell and the Manned Spacecraft Center (MSC) should also be considered for the task. McDonnell proposed a $9-million design, which Marshall deemed "rather sophisticated [and] 'unsellable'." Consequently, Marshall revised its design to include space for experimental equipment, allowing experiments to be carried into orbit rather than confined to the workshop. Despite these adjustments, the final design from Houston bore similarities to Huntsville's, though it required significant modifications to the airlock. These changes would increase the cost to $21.8 million and delay delivery by six months. By the end of November, Headquarters had reassigned responsibility for the multiple docking adapter to Marshall.

With these adjustments, earlier mission plans had become outdated. At the close of 1966, the Apollo Applications Program Office issued Program Directive 3A, which outlined the first four missions based on the newly adopted cluster concept. The initial two launches aimed to establish the orbital cluster, evaluate the workshop concept, and extend human spaceflight duration to 28 days. The subsequent missions would revisit the workshop and deploy the Apollo Telescope Mount to collect solar data. Although a first launch in 1968 was still under consideration, the schedule had slipped by three months. Nonetheless, after a year and a half of intensive planning and development, George Mueller had secured a viable program and a potential respite to refine further and advance the project.

As von Braun's engineers worked to resolve the technical issues of the S-IVB workshop, George Mueller and his Headquarters staff turned their focus to planning and funding concerns. The summer of 1965 brought a wave of optimism and encouragement. In June, Edward White's historic spacewalk during the second crewed Gemini mission captured the public's imagination and reignited interest in space exploration. The Apollo program also made impressive strides: the final three Saturn I rockets were launched within less than six months, and preparations were well underway for the first Apollo-Saturn IB flight, scheduled for early 1966. Even the Soviet Union's advancements, including the Voskhod 1 mission with its three-man crew in October and Aleksei Leonov's spacewalk outside Voskhod 2 in March, had a positive side. These achievements helped NASA's budget navigate through the executive branch and Congress with only minor reductions, ensuring that the Gemini and Apollo programs remained on track.

However, the landscape was not without its challenges. America's escalating involvement in Vietnam in 1965 brought a significant increase in troop commitments, rising from 23,000 to 184,000 as the U.S. took on a more active combat role. Domestically, the Watts riot in August exposed deep-seated social unrest among urban Black communities. Amid these issues, President Lyndon B. Johnson announced his intention to adhere to a $100 billion budget while simultaneously funding new initiatives such as Medicare and the War on Poverty. Johnson's Great Society programs placed substantial budgetary pressure on existing programs, including NASA's ambitious space efforts.

Apollo Applications Program (AAP) became an early casualty of these budgetary constraints. The White House did not support it adequately in the fiscal 1967 budget. While a lack of funds was a significant factor, critics within the administration also questioned the ambition and feasibility of the AAP. Despite these setbacks, Mueller remained resolute in his vision. His determination seemed justified when President Johnson later expressed strong support for a substantial post-Apollo program in the fiscal 1968 budget, offering a glimmer of hope for the future of space exploration.

The Apollo Applications Program Office embarked on its journey with a comprehensive planning guideline for the crewed spaceflight centers. The initial flight schedule, one of hundreds to be developed over the ensuing four years, outlined an ambitious plan involving 13 Saturn IB and 16 Saturn V launches. Four missions were designated to utilize surplus hardware from the lunar landing program, while the remaining 25 would require new Saturn-Apollo configurations.

The missions were categorized into four types: earth orbital, synchronous, lunar orbital, and lunar surface. They were further divided into two phases. The first phase consisted of eight missions using a standard Apollo command service module for 14-day flights. The subsequent missions, utilizing an extended Apollo module, were designed for longer-duration flights of up to 45 days.

One of the significant challenges facing the program was integrating experiment payloads. This process involved ensuring that the spacecraft and experiment hardware were compatible regarding mechanical, electrical, and other operational aspects. Additionally, a meticulous grouping of experiments was required to prevent their interference. To manage this complexity, the payload integration responsibilities were divided between the two crewed spaceflight centers: Houston and Huntsville.

Houston was tasked with handling experiments related to the Apollo command service module, including biomedical and behavioral experiments about the astronauts and those focused on advanced spacecraft subsystems. In contrast, Huntsville was responsible for integrating payloads related to lunar surface research, astronomy and astrophysics, and physical sciences.

The flight schedule allocated primary responsibility for 17 missions to Houston, including the inaugural flight that concentrated on earth resources. Meanwhile, the Marshall Space Flight Center in Huntsville was assigned to integrate payloads for 12 missions, including the final two flights planned for lunar exploration. This division of labor ensured a systematic and specialized approach to integrating the diverse array of experiments and objectives planned for the Apollo Applications Program.

As the planning guidelines were being dispatched to the field centers, the Headquarters staff focused on preparing briefings for the congressional space committees. There had been growing concern among several congressmen the previous year regarding the future direction of America's space programs. Recognizing the potential for a lull in NASA's activities following the lunar landing, the space committees, led by Senator Clinton Anderson and Representative George Miller, supported NASA's initiatives. They understood the urgency of initiating a new crewed program in 1966 to maintain momentum and avoid a period of inactivity. The chairmen were instrumental in publicizing and advocating for NASA's plans, aligning their support with the agency's needs.

George Mueller, grappling with a challenging fight to secure the Apollo Applications Program (AAP) budget from the Johnson administration, sought to consolidate as much congressional support as possible. On August 23, Mueller presented a comprehensive overview of the program's objectives, experiments, and proposed flight hardware to Anderson's committee. This testimony marked a notable shift in emphasis. Whereas previous NASA communications had emphasized the technological advancements of earth-orbital operations, the AAP focused more on space science.

The AAP office had compiled a list of 150 experiments categorized by their general field of interest and specific area of investigation. Among these, nearly half were classified under "space science/applications," with a significant portion dedicated to medical experiments designed to assess the physiological impacts of extended space missions. Additionally, 35 experiments were planned for lunar surface exploration, reflecting the scientific community's keen interest in lunar research. Until December 1967, when a dedicated Lunar Explorations office was established under Apollo, lunar exploration responsibilities remained with the AAP, underscoring the program's integral role in advancing space exploration.

The hearings underscored notable differences among NASA's leadership regarding the scope and direction of the Apollo Applications Program (AAP). George Mueller, driven by a bold vision, proposed an ambitious plan involving 29 flights between 1968 and 1971. His strategy included an annual launch schedule of six Saturn IB rockets, six Saturn V rockets, and eight Apollo missions. In contrast, NASA Administrator James E. Webb and Deputy Administrator Robert C. Seamans viewed AAP as an interim program for the early 1970s and were less focused on precise numbers.

Mueller believed that these differences reflected an attempt by his superiors to delay the costs associated with AAP. Some within NASA suggested that Mueller's enthusiasm for the program far surpassed that of his bosses. Despite well-organized presentations to Congress, AAP clearly required further refinement. NASA's traditional method of extending flight durations—doubling the previous longest flight to test endurance—was not systematically applied to AAP. Instead, the planned 14- and 45-day missions were determined more by hardware limitations than by medical considerations.

The AAP's experimental package resembled a lengthy shopping list, with only three experiments formally assigned to the program. The House staff report criticized NASA for failing to provide a comprehensive cost estimate, define program parameters, or clearly outline objectives. However, this criticism was somewhat misplaced, as Mueller had informed the Senate committee that the parameters and costs for AAP would be clarified during the project's definition phase. Although Mueller hoped for a prompt decision, AAP was not yet an "approved" program.

In the Fiscal Year 1967 budget request, NASA's preliminary estimates for a full-scale AAP program amounted to $450 million, with an additional $1 billion projected for the following year. Bureau of the Budget officials, who had anticipated only $100 million for AAP in FY 1967, were taken aback by the request. Nevertheless, they agreed to consider a compromise figure of $250 million. Mueller, however, deemed this amount inadequate and sought to increase it.

To persuade NASA's top administrators to support his funding request, Mueller presented five key arguments:

While the Gemini program offered a marginal advantage in the space race, it could lose its edge if funding for AAP was insufficient.

The scientific and technological communities identified several hundred experiments for AAP in collaboration with NASA. However, a $450 million program would only cover half of the 150 most promising experiments.

A slowdown in the program would affect the morale of NASA's crewed spaceflight team, which is known for its strong esprit de corps.

Mueller's efforts aimed to secure the necessary funding and momentum for AAP, ensuring that NASA could continue its pioneering work in space exploration.

The Bureau of the Budget's proposal for Apollo Applications Program (AAP) funding was criticized for its economic and political shortcomings. The suggested budget was seen as poor economic policy because it would lead to significant unemployment and result in America's substantial Apollo investment being underutilized from 1968 through 1971. Furthermore, potential unemployment and inefficient use of Apollo assets could become a political issue in the 1968 elections, compounding the program's challenges.

Despite these arguments, they failed to sway Robert C. Seamans, who recommended a $250 million budget on September 15, with James E. Webb's subsequent approval. However, this decision only marked the beginning of a more challenging phase: securing the $250 million request through the Bureau of the Budget.

NASA's attempts to navigate the budgetary process in the following weeks proved largely futile. By November, NASA presented two funding levels for Fiscal Year 1967: a desired $5.76 billion and a minimum of $5.25 billion. The Bureau of the Budget responded with a counteroffer of $5.1 billion, establishing slightly below the previous year's appropriation. This reduction included a $222 million cut to crewed spaceflight's share, with the Apollo Applications Program absorbing the brunt of this decrease. Webb's appeal to President Johnson at his ranch in December did not alter the decision. Consequently, the final budget request for NASA was set at $5.012 billion, including only $42 million for AAP—barely enough to maintain some program options.

The Senate Space Committee shared the Budget Bureau's lack of enthusiasm for AAP. On January 27, 1966, Senator Clinton Anderson conveyed the committee's preference for a post-Apollo program centered around a primary new goal rather than "loosely related scientific experiments." Some were concerned that continued use of Apollo hardware might impede developing new technologies. Given that the Department of Defense's Manned Orbiting Laboratory could achieve many of AAP's goals sooner, the committee advised close coordination with the Air Force. While the committee supported initial planning and experimentation work, it would not fund additional launch vehicles or spacecraft without further justification and specific recommendations.

AAP's early funding situation was complex. Initially, it fell under the purview of Advanced Missions and later Apollo Mission Support, with a dedicated AAP line item only appearing in Fiscal Year 1968. In FY 1966, AAP received $26 million in appropriations, which rose to $71 million in FY 1967. Experiment funds were also drawn from other sources, such as OSSA and OART. NASA's operating budget for FY 1966 allocated $51.2 million to AAP, including $40 million for experiments and $8.5 million for space vehicles. By FY 1967, AAP's operating budget had increased to $80 million, with $38.6 million dedicated to vehicle hardware and $35.6 million to experiments.

Anticipating criticism of the Apollo Applications Program (AAP), George Mueller sought the expertise of senior managers from major Apollo contractors to scrutinize its objectives. He posed five key goals for evaluation:

To explore and utilize Earth's resources for the benefit of humanity.

To define and develop the operational capabilities for future space vehicles beyond the Saturn-Apollo systems.

To expand human knowledge of the near-Earth and lunar environments.

To enhance national security through space operations.

To develop a sustainable, usable environment in space where humans could operate effectively for up to one year.

While the executives generally supported the first goal due to its public appeal, they expressed significant concerns about the feasibility of such an ambitious program. One executive remarked that although the goal was commendable, it was unclear how it could be achieved, suggesting that it might be beyond the scope of the current space program. Goal four, in particular, was seen as potentially confusing to the American public regarding NASA's role compared to that of the Department of Defense. Despite varying levels of support for the other objectives, no unified consensus emerged among the executives. Mueller concluded that, much like the "average U.S. citizen," there seemed to be no singular "average Apollo executive."

In the February 1966 issue of Astronautics and Aeronautics, columnist Henry Simmons compared the troubled Apollo Applications Program to "Wednesday's child," which was "full of woe." Simmons acknowledged the valid reasons for an ambitious program: maintaining the Apollo organization and securing a substantial return on the Apollo investment. However, the budget cuts for FY 1967 suggested that NASA might have to accept a scaled version of AAP, confined to using leftover hardware from the lunar landing missions. Developing sophisticated experiments and acquiring additional Saturn rockets and Apollo spacecraft seemed increasingly unlikely.

Simmons noted that space scientists were particularly dissatisfied with AAP, viewing many of its proposed experiments as "make-work." He suggested that deferring AAP funding had likely averted an "outright rebellion" within the scientific community and possibly prevented significant internal strife within NASA. Simmons criticized AAP on two fronts: NASA's failure to assess the value of crewed versus uncrewed space science and the agency's reluctance to utilize crewed flights for Earth-orbital missions on the Manned Orbiting Laboratory if they were deemed "cost-effective." Despite these criticisms, Simmons concluded that NASA had no choice but to continue with AAP in some capacity to prevent the disbandment of its Apollo team.

Inside NASA, the program faced significant reservations. While Simmons's suggestion of an "internal explosion" might have been an exaggeration, there was notable resentment toward AAP, particularly within the Office of Space Science and Applications. Additionally, strong opposition from the crewed spaceflight community, especially from the Manned

Spacecraft Center (MSC) in Houston, had surfaced. MSC officials had questioned fundamental aspects of AAP since its inception and continued to voice their objections throughout the winter of 1965-1966.

In March 1966, Robert Gilruth, Director of the Manned Spacecraft Center (MSC), formally expressed his center's opposition to the Advanced Apollo Program (AAP) in an eight-page letter addressed to William R. Mueller. Although Gilruth agreed with the fundamental purpose of the AAP—continuing the utilization of Apollo technology for scientific endeavors in Earth orbit and on the Moon—he criticized the program for its lack of clear, future-oriented goals for crewed spaceflight. According to Gilruth, NASA had failed to establish a definitive direction for the future of human space exploration, establishing crucial for guiding the program's development.

One major concern highlighted by Gilruth was the overly ambitious launch rate proposed by AAP. This accelerated schedule, he argued, forced the selection of missions and experiments that could be completed within the limited timeframe rather than those that were most scientifically valuable. This constraint hindered advancements in space technology as the focus shifted from pursuing meaningful research to merely meeting deadlines.

The proposed modification of Apollo hardware, including the lunar module, was particularly contentious. Engineers at MSC were deeply unsettled by the suggested changes to the lunar module, which they viewed as unsuitable for either a laboratory or a lunar habitat. Converting the lunar module into a space laboratory would require the removal of many existing subsystems and the addition of new ones for which the module had not been originally designed. Gilruth criticized this approach, arguing that the modifications would transform a highly specialized and costly vehicle into something less appropriate for its intended role. He suggested that the lunar module was not well-suited for the proposed uses, and that a different module would be more effective.

Gilruth also raised concerns about the potential negative impact of the AAP on the Apollo program itself. The proposed high launch rate would necessitate additional resources, such as trainers, simulators, and operational equipment. With AAP funding already constrained, there was a real risk that Apollo's budget could be compromised to support AAP's ambitious schedule. The constant revisions to AAP plans, driven largely by financial limitations, had diverted management's focus and threatened the overall integrity of the Apollo program.

In response to these issues, Gilruth proposed an alternative strategy. He recommended that NASA define its goals for crewed spaceflight over the next two decades, focusing on developing a permanent, crewed orbital station and a spacecraft for planetary exploration. By aligning the AAP with these long-term objectives, NASA could ensure that Apollo hardware was used for tasks that did not require extensive redesigns. This approach would better match the available funding and maintain the momentum of the crewed spaceflight program.

Gilruth's letter underscored the dilemma faced by NASA in early 1966: a critical mismatch existed between the current planning for AAP, the significant opportunities for human space exploration, and the available resources. Without a major program to drive innovation and significant research and development, the AAP risked becoming merely a continuation of Apollo's production and flight rates, without advancing the broader goals of space exploration.

Mueller's response to Robert Gilruth's objections to the Advanced Apollo Program (AAP) was not recorded, but subsequent developments suggest that Gilruth's concerns had little impact on the program's course. The AAP office continued to pursue a trajectory largely contrary to MSC's recommendations, and Houston's objections persisted in later stages.

A significant challenge to AAP's progress came from the competing Manned Orbiting Laboratory (MOL) program. Despite NASA's official stance that AAP and MOL were unrelated, they were seen as competitors by Congress and the executive branch. Both NASA and the Air Force supported each other on technical fronts, yet they were vying for the same limited pool of political and financial support. Their interactions were multifaceted: Houston provided support for the

Gemini component of MOL, and personnel from NASA and the Air Force collaborated on joint panels and coordinated experiments of mutual interest. This exchange included lending key officials between agencies.

As NASA formulated its post-Apollo plans, it frequently evaluated the merits of MOL, with figures such as James E. Webb showing unexpected objectivity toward the Air Force's program. However, the Office of Manned Space Flight struggled to view MOL favorably, given that AAP and MOL were competing for scarce space funds. The presence of MOL created a financially detrimental environment for AAP, with the uncertainty of whether both programs could be sustained.

The approval of MOL by President Lyndon B. Johnson in August 1965 did not immediately alleviate its financial constraints. The Air Force's Space Systems Division continued to face budgetary shortfalls. By late 1965, the choice of a Titan IIIC rocket with strap-on solid-fuel boosters for MOL's launch vehicle had been finalized. Concurrently, a new launch complex was being constructed at the Western Test Range in California. In November 1965, the Air Force prepared a position paper advocating for an expansion of the Satellite Control Facility in Sunnyvale, California, a proposal met with resistance from congressional critics who favored using NASA's mission control center.

In early 1966, construction began at Vandenberg Air Force Base for the new launch facility, with lone lead items for the launch vehicle ordered by June. Before MOL's approval, NASA and the Department of Defense (DoD) addressed mutual interests through informal groups or the Aeronautics and Astronautics Coordinating Board and its panels. The approval of MOL necessitated more formal arrangements to manage the increased coordination. By mid-October 1965, Mueller and General Bernard Schriever, head of Air Force Systems Command, had signed the first agreement concerning joint experiments. Over the next year, various agreements defined the working relationships between the agencies, culminating in the creation of the Manned Space Flight Policy Committee in January 1966. This committee, which included Mueller, NASA's Associate Administrator Robert Seamans, and key DoD counterparts, ensured high-level coordination.

Throughout this period, program officials maintained a unified front in congressional testimony, asserting that MOL and AAP were independent programs serving distinct but valuable objectives.

The debate over the future of American space exploration programs reached a critical juncture in early 1966, marked by significant congressional scrutiny and internal NASA deliberations. On January 27, 1966, Senator Anderson addressed a letter to NASA Administrator James Webb, suggesting the integration of the Manned Orbiting Laboratory (MOL) into the space program. This correspondence came amid intense discussions on Capitol Hill, where both the House Military Operations Subcommittee and critics from the Johnson administration voiced concerns about the overlapping objectives of NASA's Apollo Applications Program (AAP) and the MOL.

The Military Operations Subcommittee, concluding three days of hearings on Missile Ground Operations, delivered a scathing report criticizing what they perceived as "unwarranted duplication" and an unapproved program that could potentially incur annual costs between $1 to $2 billion. They highlighted the backing of prominent space scientists for a joint program and argued that merging NASA's and the Department of Defense's (DoD) efforts could result in significant financial savings. The Budget Bureau had been questioning the necessity of maintaining separate Earth-orbiting laboratories for several years. By the time the discussions turned to the fiscal year 1968 budget, bureau officials advocated for a unified approach, proposing that NASA conduct its experiments on MOL missions or, alternatively, utilize the more cost-effective Titan III rocket.

In September 1966, the President's Scientific Advisory Committee joined the growing chorus of critics. The committee expressed dissatisfaction with the spent stage concept proposed for the AAP, which involved substantial construction at the mission's outset and could potentially skew medical results. Their report urged NASA to closely examine the MOL option before committing significant funds to the AAP.

In response to these critiques, NASA undertook a two-pronged approach. The agency commissioned Douglas Aircraft to assess the potential of the MOL in fulfilling early AAP objectives and initiated an in-depth internal comparison of the two programs. The Office of Manned Space Flight (OMSF) first considered using the Titan III rocket for the AAP. Although NASA officials acknowledged that the Saturn IB launch vehicle was economically inefficient—costing approximately twice as much per launch as the Titan III—utilizing the Air Force rocket could save around $15 million per mission. Nonetheless, OMSF found that integrating the Titan with the Apollo spacecraft would entail significant costs and delays. Systems integration, launch facility modifications, additional checkout equipment, and two qualification flights would require at least 3.5 years and approximately $250 million. Such a delay would push the first AAP mission back by two years and necessitate 17 launches before the savings from Titan integration offset the initial conversion costs.

Furthermore, the OMSF found that the basic MOL configuration was insufficient for AAP objectives. A proposed larger MOL would require four additional years of development and an extra $480 million in facility modifications, ultimately proving more costly annually than continuing with the Saturn IB and Apollo systems. With these findings, NASA officials advocated maintaining an independent AAP during congressional hearings.

Throughout NASA's history, the roles of its various centers had been well-defined. Von Braun's team in Huntsville managed launch vehicles, while Robert Gilruth's engineers in Houston handled spacecraft development. The Mercury-Redstone flights marked their first collaboration, and subsequent programs like Gemini and Apollo saw each center specializing in distinct areas. Gemini, for instance, was managed mainly by the Manned Spacecraft Center (MSC), with the Air Force supplying the Titan launch vehicle. At the same time, Apollo divided responsibilities among various centers, including Saturn development, spacecraft design, launch operations, and communications.

With the AAP, however, the potential for conflict increased. The program promised benefits for all involved, but for Huntsville, which faced a future without a successor to the Saturn V and the imminent phase-out of the Saturn IB, it offered a glimmer of hope. Marshall Space Flight Center (MSFC) seized upon the wet workshop concept as a cost-effective means for long-duration flights, enhancing its prospects. Yet, this decision was resisted by Houston officials, who viewed space stations as their domain. The introduction of the wet workshop fostered competition between MSC and MSFC, complicating their previously collaborative relationship.

NASA's field centers enjoyed a degree of autonomy, a legacy from the National Advisory Committee for Aeronautics (NACA) era, which had pursued work independently. Despite efforts from NASA Administrator James E. Webb to centralize control, the centers retained considerable independence. By August 1965, AAP's planning guidelines designated specific integration tasks for the centers as per their Apollo responsibilities. However, after informal consultations, Webb revised these assignments in September. The MSC was tasked with spacecraft development, astronaut training, mission control, and flight operations, while Marshall would oversee the integration of experiments into the lunar module. This decision marked a significant shift in responsibilities and further intensified the center competition.

In summary, the early years of the Skylab program were characterized by complex interactions between NASA, the Department of Defense, and Congress and internal rivalries within NASA itself. Debates over cost, efficiency marked the period, and the strategic direction of the U.S. space program, setting the stage for the eventual realization of Skylab as a pioneering space station.

On October 14, 1965, the Houston Post reported a potential shift in NASA's organizational structure that sparked significant discussion. The article "Marshall May Take 2nd Apollo Control" suggested that the Marshall Space Flight Center in Huntsville might be tasked with integrating the Apollo Applications Program (AAP) payloads, while NASA Headquarters would likely oversee the program. Despite this possible reorganization, the article confirmed that mission control and astronaut training would remain at the Johnson Space Center in Houston.

This news stirred a notable reaction, particularly from Representative Olin Teague, the Texas Democrat who chaired the subcommittee overseeing NASA. Teague initiated a review of the proposed changes, causing NASA officials at the Office of Manned Space Flight (OMSF) to address the situation with caution and discretion until the matter was clarified.

The division of roles and missions was still considered tentative at this stage. Before finalizing these responsibilities, including the allocation of duties related to the lunar module (LM), NASA's key figures—Administrator James E. Webb and Associate Administrator for Manned Space Flight, Dr. George E. Mueller—needed to be convinced of the practicality of these proposals.

Dr. Mueller argued that centralizing the responsibility for payload integration at a single center would be overly burdensome. Specifically, splitting the lunar module integration between Houston and Huntsville would strain the personnel resources at the Manned Spacecraft Center (MSC) in Houston, exceeding its 1968 staffing limit while leaving Marshall with a surplus of personnel. Additionally, this approach would lead to unnecessary duplication of mock-ups and support equipment.

Mueller proposed that consolidating the LM payload integration responsibilities at Huntsville would keep both centers within their personnel ceilings and ensure more efficient use of resources. He assured his superiors that Marshall possessed the necessary engineering expertise to handle this task. Webb agreed to the division of responsibilities, with the condition that Huntsville's program office be renamed to "LM Applications" or "LM Integration Office" rather than "Apollo Applications." Webb's stipulation was intended to clarify that the focus of NASA's crewed flight program would not shift but that all available resources would be utilized more effectively.

Huntsville seized this opportunity, establishing the Experiments and Applications Office in mid-December. By March 1966, Leland F. Belew, the former manager of Saturn engines at Marshall Space Flight Center, was appointed as the director of this new office. Belew, born in Salem, Missouri in 1925, had earned a Bachelor of Science in mechanical engineering from the University of Missouri-Rolla in 1950. He began his career at Redstone Arsenal and joined NASA with the Army Ballistic Missile Agency Development Operations Division in 1960. His subsequent career included becoming deputy director of the Science and Engineering Directorate at MSFC in 1975, underscoring his extensive experience and expertise in the field.

In the early 1960s, the burgeoning space program faced a pivotal challenge as the Marshall Space Flight Center in Huntsville, Alabama, undertook the ambitious task of integrating payloads for the Apollo Applications Program (AAP). This initiative, guided by the vision of expanding human spaceflight capabilities, became a significant focus for engineers like Belew and his team. Their eight-year endeavor began with payload integration—a crucial step in ensuring the successful execution of AAP missions.

By May 1966, Marshall had initiated a critical phase of this process, awarding parallel contracts worth $1 million each to Lockheed and the Martin Company of Denver. These contracts tasked the contractors with a comprehensive evaluation of the experimental hardware, installation, and integration of equipment, crew and launch facility requirements, tracking, and mission analysis. In September of that year, both companies independently reviewed the Office of Manned Space Flight (OMSF) plans for the initial four flights of the AAP.

As the program progressed, Belew enlisted Martin's expertise for detailed planning of the spent-stage mission, while Lockheed focused on the Apollo Telescope Mount missions. Initially, Marshall had considered handling payload integration in-house, but the extensive work performed by Lockheed and Martin highlighted the necessity for their involvement. By November 1966, Marshall began preparing a formal work statement for an integration contract, marking a significant step in the program's development.

Amidst these tensions, Gilruth appointed his deputy, George Low, as Houston's point of contact for the AAP in April 1966. The AAP office at MSC officially opened on July 6, 1966. However, Low's involvement was limited due to other pressing duties, leaving his deputy to handle many AAP matters.

Concerns about maintaining authority over various aspects of the program persisted at MSC. Officials worried that Marshall's extensive role in payload integration might challenge MSC's control over mission planning and astronaut training. To address these concerns, an agreement was reached: astronauts would train with specific experiments during integration work at Huntsville, but Marshall would not establish an "Astronaut Training Center."

In early 1966, some progress was made in delineating roles and responsibilities. OMSF and the two centers divided tasks for the spent-stage mission. Huntsville was responsible for designing the workshop and implementing an experiment program incorporating elements from MSC and other sources. Meanwhile, Houston's Gemini office would oversee developing the airlock module. This agreement, however, only covered one mission, and disputes over other AAP roles and responsibilities continued to arise as the program evolved.

In August 1966, to address mounting disagreements regarding the Apollo Applications Program (AAP), Mueller convened a critical session of OMSF's Management Council at Lake Logan, North Carolina. This three-day retreat brought together the deputy directors from the three crewed spaceflight centers: George Low from Houston, Eberhard Rees from Marshall, and Albert Siepert from Kennedy. The meeting was intended to reconcile differences and streamline the planning process for the AAP, with a focus on developing a coherent space station model to serve as a bridge between early AAP missions and future planetary explorations.

The council began with a shared understanding that a space station was a logical intermediary step in human spaceflight. They envisioned a modular design consisting of a command post, a mission module, and one or more experiment modules. The command module, responsible for guidance, navigation, control, and communications, would be developed by the Manned Spacecraft Center (MSC) in Houston. The mission module, where the crew would live, sleep, and conduct experiments, would fall under the Marshall Space Flight Center (MSFC) purview. Both centers were tasked with contributing to designing and implementing experiment modules.

The "Lake Logan Accord," as it came to be known, formalized this approach, applying the space-station model to the AAP. It designated the Apollo command-service module and the airlock module as the command post, the orbital workshop as the mission module, and the Apollo telescope mount as an experiment module, all under Marshall's direction. This agreement delineated Huntsville's primary role in the early AAP launch while reaffirming Houston's responsibilities for flight operations, astronaut activities, life-support systems, and medical research.

Despite the signing of the Lake Logan agreement by key figures such as Gilruth and von Braun in late August, not all was settled. The Houston Post continued to express dissatisfaction with the arrangement. On October 10, Jim Maloney's front-page article, "Von Braun a Persuasive Voice—Some MSC Tasks Being Moved," highlighted the ongoing tensions. While acknowledging Huntsville's achievements in rocket development, Maloney criticized the shifting responsibilities, particularly regarding payload integration and the Apollo telescope mount, as encroachments on Houston's traditional spacecraft role. He questioned the wisdom of fragmenting the team that had successfully developed Mercury, Gemini, and Apollo spacecraft, suggesting that such moves could undermine future missions to Mars and beyond.

Maloney's article sparked renewed congressional scrutiny. Mueller's response, emphasizing the August agreement and its provisions, appeased NASA's congressional committees but did little to quell Maloney's concerns. Subsequent articles by Maloney criticized the spent-stage mission and alleged that MSC leaders were capitulating to headquarters pressures. Although Maloney's portrayal may have exaggerated some issues, his concerns resonated with certain engineers within the space program.

The Lake Logan agreement, while a pragmatic solution at the time, did not entirely resolve the competition and friction between the centers. The collaboration it aimed to foster was tempered by ongoing rivalries and disputes over post-Apollo priorities, reflecting the complexities and challenges inherent in managing such a pioneering space program.

In 1965, the Johnson administration deferred a decision on the Apollo Applications Program (AAP), hoping for more favorable circumstances in the following year. However, the situation only deteriorated. The U.S. troop presence in Vietnam surged from 184,000 to 385,000, and the cost of the war skyrocketed from $6 billion to $20 billion. President Lyndon B. Johnson maintained that it was possible to protect American interests in Southeast Asia without compromising his domestic Great Society programs—a notion that critics derided as an attempt to have both "guns and butter." This belief was increasingly questioned, as evidenced by the landslide Republican victories in the 1966 elections, which reflected widespread dissatisfaction with Johnson's policies.

The repercussions of Johnson's troubles extended to NASA, a reality keenly understood by James E. Webb, NASA's Administrator. Shortly after the elections, Webb addressed the challenging climate during a management review. He noted that the space programs were facing mounting scrutiny, with critics focusing on the enormity of the Apollo program and the potential for expansive post-Apollo initiatives. The Bureau of the Budget was pressuring Webb to eliminate the remaining five Saturn V rockets from the Apollo program, reflecting a lack of enthusiasm for AAP. Webb recognized that the administration and Congress were unlikely to support the program until NASA could present clear and tangible goals for it.

Webb urged his management team to shift their focus away from promoting the Apollo-Saturn hardware and instead highlight the national needs that could be addressed by Apollo's capabilities. He advised against emphasizing internal NASA concerns, such as the desire to keep the Apollo team employed, and warned against center parochialism, which could jeopardize future programs. Webb stressed that the internal divisions within NASA could significantly undermine the prospects for post-Apollo projects.

By mid-December, there was a noticeable improvement in the perception of AAP, particularly from the perspective of George Mueller, Associate Administrator of the Office of Manned Space Flight. In a meeting with staff from the Office of Manned Space Flight and various NASA centers, Mueller acknowledged that, just a few months earlier, AAP had been regarded by many outsiders as little more than an assortment of ambitious but impractical ideas. There had been serious skepticism about the value of human involvement in space science. During briefings in August, neither the Bureau of the Budget nor the President's scientific advisers had sh much interest in a post-Apollo program. However, Webb's strategic focus on utilizing the workshop cluster concept as a cost-effective means of enabling long-duration spaceflight and advancing scientific research—particularly in solar astronomy through the telescope mount—had significantly enhanced AAP's standing with both the administration and the broader scientific community.

The evolving status of the Apollo Applications Program (AAP) was most clearly reflected in NASA's Fiscal Year 1968 budget proposal. Despite persistent concerns from the Budget Bureau regarding the program's lack of clear objectives, potential overlap with the Air Force's Manned Orbiting Laboratory, the debate over crewed versus uncrewed missions for space science, and the timing of AAP and Apollo missions, the administration remained committed to the future of crewed spaceflight. The original request for AAP funding had been reduced from $626 million to $454 million, a cut that represented a slow down but nonetheless marked the allocation of a significant sum toward the program. More importantly, this decision underscored President Lyndon B. Johnson's formal commitment to maintaining America's crewed space capability, as highlighted in his budget message, emphasizing that abandoning this capability was not an option.

In the mid-1960s, AAP was often described as a transitional phase between the Apollo program and NASA's next major crewed initiative.

Johnson's approval, despite the severe funding constraints of the time, positioned AAP as a crucial bridge over these troubled waters. For 18 months, the AAP office had struggled to gain recognition. Initially deferred and then scaled back, the program's viability had been in question by August 1966. Johnson's approval revitalized its prospects, with renewed optimism for its success. George Mueller, who remained hopeful, noted that AAP's 1966 schedule anticipated 37 flights through 1973 at an estimated cost of $7 billion. However, firm public and congressional support was still essential.

A significant opportunity to secure this support came with the release of the budget message in January 1967. At NASA's FY 1968 budget briefing on January 23, Robert Seamans outlined the AAP funding requirements: $263.7 million for additional Saturn-Apollo hardware, including four Saturn IBs and four Saturn Vs per year; $140.7 million for experiments; and $50.3 million for mission support. The allocation for mission support highlighted the urgency as the first AAP mission was scheduled for June 1968. During the briefing's question-and-answer session, Mueller provided further details on NASA's plans. The program aimed to launch its orbital workshop in mid-1968, followed by a solar observatory equipped with a telescope mount six months later. Subsequent missions to the workshop were planned for 1969.

Administrator James Webb emphasized the program's evolution from a reliance on one-time use of spacecraft to a model where large systems would be stationed in orbit and utilized repeatedly. This shift in focus was central to AAP's appeal to the president. On January 26, Mueller presented a more detailed overview of AAP to the press, showcasing its advancement since the earlier, more fragmented schedule of August 1965. Over the intervening 18 months, the orbital cluster had matured into a central component of the program, marking a significant evolution from its initial conception.

During a briefing, Dr. George Mueller described the Skylab Space Station project as an embryonic form of a space station, emphasizing its potential as a significant step in space exploration.

The initial plan focused on four Apollo Applications Program (AAP) missions, each contributing crucial components to developing Skylab.

The first mission was a complex endeavor involving two launches. The first launch would deploy an Apollo Command/Service Module, followed by the launch of Skylab itself, which included the workshop, an airlock, and a multiple docking adapter. Skylab, designed to remain in orbit at an altitude of approximately 510 kilometers, was planned to function for a minimum of three years. Upon docking, astronauts would inhabit the station for 28 days, doubling the duration of the longest Gemini flight to that point.

Construction of the two-story workshop within the spent S-IVB stage was expected to take four days. The lower level was designed as living quarters, featuring fabric curtains to create separate areas for sleeping, food preparation, waste management, and exercise. The upper level would house workstations, similarly partitioned for efficiency. The airlock, developed by McDonnell Corporation, was a critical component, providing the necessary oxygen, nitrogen for a shirtsleeve environment, electrical power, and other consumables required for the 28-day mission.

The size of Skylab impressed journalists, who contrasted it with the cramped interiors of the Gemini and Apollo modules. One reporter inquired whether the workshop was comparable in size to an average ranch house. Mueller responded that it was indeed similar, though he noted it was comparable to a "small ranch house" – the kind he himself could afford.

The first mission's primary focus was on medical experiments, addressing the physiological challenges of extended spaceflight. Experiments would include vectorcardiograms, studies of metabolic activity, bone and muscle changes, and vestibular function assessments. Additionally, 18 engineering and technology experiments were planned, such as testing "jet shoes" – devices resembling skates with gas jets developed at Langley Research Center. These would be evaluated for their potential as maneuvering aids during extravehicular activities. The mission would also assess crew comfort through experiments on sleeping arrangements, suit management, and overall habitability.

The second mission, scheduled three to six months after the first, would extend to a 56-day stay in orbit. This mission required a Saturn IB rocket to carry a crewed Apollo spacecraft and a supply module, while another rocket would launch a telescope mount. The solar telescope apparatus was complex, measuring two meters in diameter and nearly four meters in length, weighing one ton, and housing a dozen delicate instruments. The telescope was designed to be mounted on Skylab for most observations but could be tethered at a short distance under specific conditions. While normal operations would require only one astronaut to operate the telescope, the remaining crew would manage other tasks such as eating, sleeping, and conducting additional experiments. Mueller described the telescope array as "the most comprehensive array of instruments ever assembled for observing the Sun," with hopes to have it operational by early 1969, coinciding with a peak in sunspot activity.

Skylab permitted man to take a long look at the Sun from a vantage point never before enjoyed. (Naval Research Laboratory)

Looking beyond the first two missions, Mueller outlined a broad vision for the Skylab program. Four more crews were scheduled to visit Skylab in 1969 for new experiments and further solar observations. Although specific experiments for these later flights were not yet detailed, anticipated payloads included Earth-resource cameras and weather instruments. In 1970, NASA planned to launch a second Saturn IB workshop, followed by another telescope mount in January 1971. Through resupply and crew-transfer flights, NASA aimed to achieve a year-long mission by 1971. Additionally, a biomedical laboratory was scheduled for launch in 1970 to monitor the effects of space on the human body. The first lunar-mapping flight was set for December 1969, with subsequent two-week lunar missions planned for 1971. Two Saturn V launches were anticipated for each extended lunar mission, and equipment developed for Apollo lunar exploration, such as the lunar rover, was under consideration for the AAP program. By late 1971, NASA planned to launch the first of two Saturn V workshops, with four Apollo flights programmed to visit each laboratory. As a NASA official remarked, it was "quite an ambitious program," underscoring the scale and scope of the Skylab project.

During the fiscal years 1967 and 1968, the Apollo Applications Program (AAP) Office set ambitious goals to advance space exploration. The office anticipated launching seven major initiatives: developing an airlock and workshop, a telescope mount, a lunar mapping and survey system, modifications to Apollo spacecraft for extended missions, a lunar shelter derived from the lunar module, experiment payloads, and a land-landing capability for the Apollo spacecraft. This final project was particularly notable as it aimed to enable the reuse of Apollo spacecraft, contributing to cost savings and supporting the program's long-term viability. The land-landing capability was envisioned to not only facilitate the reuse of spacecraft but also to accommodate up to six astronauts. This would be achieved through a soft landing mechanism that reduced the required space for shock absorption by the astronauts' couches. Despite the promising outlook, Houston had previously dismissed the practicality of land recovery for Apollo due to the challenges associated with braking the descent of the spacecraft. Nevertheless, the AAP Office persisted in believing that the benefits of a land-landing capability justified its development.

In January 1967, during a press briefing, George Mueller outlined an ambitious scientific agenda for the Apollo Applications Program. The program, he explained, would encompass a significant scientific component, although much of this work had only been outlined in the preceding year. By late 1965, it had become clear that the list of scientific experiments ready for early AAP flights was disappointingly short. The space science director at Marshall Space Flight Center had noted to Wernher von Braun that the scientific experiment options were limited at that time. Despite NASA's efforts to compile an extensive list of scientific studies for an Earth-orbiting laboratory, only a handful of experiments were actively under development. This scarcity of prepared experiments was unsurprising, given the nascent state of the AAP and the rapid pace of hardware development.

In December 1965, Mueller had called for accelerating the orbital workshop project, highlighting the urgent need to advance scientific experiments. As 1966 progressed, NASA Headquarters began seeking out sufficiently developed experiments to be included in early missions with minimal additional cost. The challenge was compounded by the fact that crewed spaceflight experiments were still relatively novel, with the Office of Manned Space Flight and its associated field centers preoccupied with Saturn and Apollo projects. The Office of Space Science, newly engaged in crewed spaceflight, faced coordinating with the Manned Space Flight Office. This collaboration was necessary but challenging, given the two offices' different backgrounds, objectives, and approaches.

Scientific research in space in the United States originated in the postwar V-2 rocket flights and evolved with the launch of orbiting satellites. The first of these satellites marked the beginning of a new era in space exploration, setting the stage for the Apollo Applications Program's scientific endeavors.

In the United States, the quest for scientific understanding in space began with the postwar V-2 rocket flights, an early venture into space exploration that laid the groundwork for future research. This pioneering effort evolved with the launch of orbiting satellites, marking a significant milestone when Explorer 1 entered orbit on January 31, 1958—under the guidance of NASA's Office of Space Sciences (OSS), astronomy and space physics researchers embarked on a journey to gather extensive data and develop increasingly advanced instruments to expand the boundaries of human knowledge.

The OSS played a crucial role in advancing scientific inquiry by supporting developing sophisticated space instrumentation and conducting pioneering experiments. Concurrently, NASA's Office of Applications focused on the practical applications of space technology, including communications, navigation, and weather satellites. Between 1963 and 1971, these two offices were merged into the Office of Space Science and Applications (OSSA). By 1965, OSSA had established a well-organized program that included launch vehicles, a tracking and data-acquisition network, and a dedicated center for science and applications programs, located at the Goddard Space Flight Center in Greenbelt, Maryland. The office served a diverse clientele of scientists, supported university research programs, and provided research fellowships for graduate students. By the fiscal year 1965, this support had grown to $46 million—a substantial figure that, while modest compared to some agencies' research budgets, was significant to the academic community.

Homer E. Newell, who led OSS and later OSSA, was instrumental in shaping the direction of NASA's scientific endeavors. Joining NASA in 1958 from the U.S. Naval Research Laboratory, where he coordinated the science program for the Vanguard satellite project, Newell brought a wealth of experience to his role. With a Ph.D. in mathematics, his background included investigations into radio propagation and upper-air phenomena before he delved into satellite work. Appointed as director of the Office of Space Sciences in 1961, Newell was responsible for overseeing NASA's science programs. His role required him to balance the scientific community's demand for research support against the constraints imposed by a practical-minded Congress. Despite its early start and significant achievements, space research often struggled to capture the public's imagination compared to the more glamorous and immediately impactful crewed space programs.

Newell's tenure highlighted the challenges faced by space science. While uncrewed satellites, equipped with miniaturized electronics, quietly transmitted valuable data from orbit, they failed to generate the same public excitement as the crewed missions. Though remarkable, the dramatic images of Earth from space or the Moon's surface could not compete with the human stories and technological feats of crewed spaceflight. Consequently, NASA's research and development funds were heavily skewed toward the Apollo program and other crewed missions. In fiscal year 1960, before the first crewed Mercury flight, the Office of Manned Space Flight (OMSF) received 45.5% of NASA's funds compared to OSS's 34.6%. By 1964, the disparity had widened significantly, with OMSF securing 69.7% of the budget and OSS receiving only 17.6%. This shift led to frequent complaints from scientists about the perceived imbalance in funding distribution.

Convincing key stakeholders of Apollo's significance was a challenging endeavor. The Apollo program was not solely driven by its scientific merit but rather by a broader national goal. Unlike the Office of Scientific Research (OSS), established and dedicated to a wide-ranging scientific agenda, the Office of Manned Space Flight (OMSF) focused more narrowly on achieving specific engineering objectives. This divergence in focus stemmed from the nature of their respective missions: while scientific research was inherently open-ended, encompassing areas such as astronomy and space physics with limited objectives within a larger framework, engineering programs were characterized by clear, time-bound goals.

The OSS's approach to scientific research was marked by a deliberate pace and a tolerance for less immediate results. Constrained by budgetary limits rather than stringent deadlines, OSS scientists could afford a more measured approach, accepting lower reliability in their launch vehicles if it meant yielding greater scientific returns over time. In contrast, the OMSF's mission was driven by the imperative of meeting tight deadlines, particularly the challenge of landing a crewed mission on the Moon. This urgency necessitated a focus on engineering solutions and immediate problem-solving, with a staff predominantly composed of engineers, including prominent figures like George Mueller.

The OMSF's engineering-centric approach had clear priorities. It was paramount to ensure the safety and reliability of every component of crewed spacecraft and their boosters. This led to rigorous testing programs and meticulous documentation of every test and inspection. The stakes were high; a failure in a Delta booster leading to losing an astronomy satellite was a setback, but a failure involving a Titan rocket with astronauts aboard was a critical issue with potentially grave consequences. The survival of astronauts and the successful completion of mission objectives were non-negotiable priorities.

Project Mercury, America's inaugural crewed space program, exemplified this engineering and operational focus. Initially, Mercury had little emphasis on scientific experimentation, with resources primarily allocated toward achieving the lunar landing goals. However, as the program progressed and it became evident that scientific experiments could be conducted in orbit, the need to formalize the process for integrating scientific research into missions became apparent. To address this, MSC Director Robert Gilruth established the Mercury Scientific Experiments Panel, later known as the MSC In-Flight Experiments Panel. This panel, consisting of representatives from eleven MSC divisions and program offices, along with an ex officio member from OSS's Manned Space Sciences Division, was tasked with ensuring that experiments were well-conceived and seamlessly incorporated into mission plans. This shift marked a significant evolution in the approach to integrating scientific objectives within the framework of crewed space missions.

The Mercury Scientific Experiments Panel was crucial in evaluating proposed experiments for their scientific merit, relevance to crewed spaceflight, potential impact on the spacecraft, and operational feasibility. Despite this directive encouraging developing valuable investigations, the panel and, by extension, the Manned Spacecraft Center (MSC) earned a reputation for being indifferent or even antagonistic toward scientific experiments. Scientists often voiced frustrations about the excessive paperwork required to demonstrate the safety and reliability of their experiments, which they felt rendered them prohibitively expensive. They also believed that engineers lacked an understanding of the nuances of scientific research. Conversely, engineers criticized scientists for being lax regarding schedules, changes, and the operational impacts of their experiments. Nevertheless, despite these tensions, the two groups collaborated on a few basic visual and photographic observations during Mercury flights. Although these experiments were relatively minor, they demonstrated that valuable observations could be made from orbit.

In 1963, Homer Newell established the Manned Space Sciences Division to improve the integration of scientific and engineering objectives. This new division was coordinated between the Space Sciences Steering Committee, which established the OSSA review board for experiments, scientific researchers, and MSC's experiment coordinators. The aim was to align NASA's scientific and engineering goals better. The director of this division reported to both Newell and Brainerd Holmes, his counterpart at OMSF. Despite these changes, OMSF did not initially adjust its organizational structure for experiment management; this responsibility remained with the In-Flight Experiments Panel in Houston.

Following the Apollo decision in 1961, the organizational structure assigned OSSA responsibility for all science programs while OMSF controlled crewed flights. This created a somewhat awkward arrangement in which OMSF managed the budget for experiments, but OSSA was expected to oversee them. Newell highlighted the difficulties of this arrangement to Associate Administrator Robert Seamans, but Seamans declined to make changes or reallocate funds after Congress had approved the budget.

In mid-1963, an agreement was signed by deputies representing Newell and Holmes to clarify the roles and responsibilities in this complex arrangement. According to the agreement, OSSA solicited, evaluated, and selected experiments for flight and developed experiment hardware to the "breadboard" stage. A breadboard experiment was a prototype containing all the components of the final flight model but not permanently assembled, allowing for easy

modifications during design studies. OMSF would then select a center to develop the flight hardware, contract with experimenters and equipment developers, and oversee the testing and development of the experiments. OSSA would also plan and develop the science training program for astronauts, but OMSF would conduct the training.

In the early 1960s, the management and coordination of scientific experiments for NASA's Gemini program were evolving, reflecting the ambitions and complexities of early human spaceflight. In 1965, the arrangement between NASA's Office of Space Science and Applications (OSSA) and the Office of Manned Space Flight (OMSF) had become workable, albeit not optimal. This arrangement was briefly renewed with only minor adjustments, reflecting the ongoing adjustments in managing the Gemini missions.

The Gemini program began to take shape in August 1963 when Dr. William Newell formally launched its science program. He reached out to the scientific community by sending 600 letters, outlining the capabilities of the Gemini spacecraft and soliciting proposals for potential experiments. This initiative aimed to harness the scientific potential of the Gemini missions, focusing on those that could be performed during the spacecraft's planned flights.

To manage the influx of ideas, a Panel on In-Flight Scientific Experiments was established to screen the submitted proposals. This panel carefully evaluated approximately 100 proposals, discarding those lacking scientific merit or deemed infeasible for spaceflight. The remaining proposals were forwarded to the Manned Space Science Division of the Space Sciences Steering Committee for further review. After rigorous evaluation by specialized disciplinary subcommittees within OSSA, the steering committee recommended 12 experiments for flight on the Gemini missions to the OMSF.

During this period, Brainerd Holmes faced challenges related to NASA's commitment to include Air Force experiments on the Gemini spacecraft whenever possible. The involvement of the Department of Defense (DoD) added layers of complexity and sensitivity to the program. To address these issues, Holmes assigned several

OMSF observers to participate in the review of experiments conducted in Houston. The In-Flight Experiments Panel was responsible for reporting its recommendations to NASA Headquarters, while the joint NASA-Air Force Gemini Program Planning Board allocated experiments to specific flights. Holmes also established a priority system for Gemini experiments: experiments directly supporting Apollo objectives, including medical studies, were given top priority, followed by DoD experiments and then other experiments. This prioritization led to a perception among many scientists that their experiments, often falling into the third category, were being given a lower level of encouragement.

When George Mueller took over from Holmes in the fall of 1963, he sought to streamline and centralize the management of spaceflight experiments. Mueller proposed the creation of a new board to oversee all crewed spaceflight experiments and to consolidate control within NASA Headquarters. Despite some resistance from OSSA, which felt that the proposed board might encroach on its responsibilities, Mueller's proposal was finalized. On January 14, 1964, he issued a directive establishing the Manned Space Flight Experiments Board.

The newly formed board, staffed with an executive secretary and a full-time team based in Washington, coordinated the various experiments. It was organized into four categories: scientific, technological, medical, and DoD experiments. The process involved several steps: each sponsoring office solicited and evaluated proposals before forwarding them to the experiments board. The board then assessed the feasibility of each experiment in collaboration with OMSF program offices and approved or rejected them based on flight feasibility and priority. In cases where the board could not reach a unanimous decision, Mueller had the final say. Each approved experiment was prioritized and assigned to a development center, ensuring a structured approach to integrating scientific research into the Gemini missions.

In March 1964, under the direction of Robert R. Gilruth, NASA's Manned Space Flight Center (MSC) undertook a significant reorganization to streamline the process of reviewing and managing experiments conducted in space. The In-Flight Experiments Panel, previously responsible for evaluating experiments, was effectively replaced by the Manned Space Flight Experiments Board. Despite this change, the MSC retained a crucial role in assessing the feasibility of experiments through its newly established Experiments Coordinating Office, establishing part of the Engineering and Development Directorate. This office collaborated closely with several departments, including Flight Operations, Flight Crew Operations, the Gemini and Apollo program offices, and center medical programs, all of which contributed to evaluating technical and operational feasibility.

The Experiments Coordinating Office was tasked with overseeing the development and execution of each experiment, assigning a technical monitor from MSC to liaise with the principal investigator and guide the hardware through its development phase. By the time the second program of the Office of Manned Space Flight (OMSF) began, this organizational framework was well-established to solicit, evaluate, and develop experiments. Despite the efficient system, scientists found the bureaucratic procedures cumbersome, especially the extensive documentation required. Nonetheless, NASA's rigorous approach was necessary to ensure that experiments were scientifically valuable, operable in flight, manageable by the crew and did not compromise mission safety or astronaut well-being. The system proved effective and maintained its core elements throughout the crewed spaceflight program.

The Gemini program, establishing instrumental in integrating scientific research into crewed spaceflight, became a testing ground for this experimental framework. During the ten Gemini missions, 111 experiments were conducted, encompassing 17 scientific, 12 technological, eight medical, and 15 Department of Defense (DoD) experiments, with 36 investigators from 24 organizations participating. The official assessment of these results acknowledged the invaluable knowledge gained, which would benefit future crewed spaceflight endeavors. However, many involved participants privately regarded the scientific outcomes as relatively modest. They recognized that the primary value of the experiments lay in refining the process of managing experiments in space rather than in groundbreaking scientific discoveries.

The task of integrating experiments into spacecraft was far from straightforward. Scientists, astronauts, flight planners, and spacecraft engineers had to navigate a complex learning curve. Over time, and through trial and error, they learned how to design reliable and flightworthy experiment hardware, integrate it into the spacecraft, manage it within the mission timeline, and train crews to operate it effectively. Tensions between scientists and engineers were evident, with scientists often frustrated by the engineers' stringent requirements and engineers finding scientists 'disregard for schedules and last-minute changes problematic. Astronauts displayed varied attitudes toward the experiments; while some took them seriously, others viewed them as inconvenient.

Ultimately, as the Gemini program concluded, scientists and astronauts gained a greater mutual understanding, and most expressed satisfaction with the overall results. The experience gained during this period laid a foundation for future space missions, demonstrating the importance of collaboration and adaptation in pursuing scientific and exploratory goals in space.

The Gemini program marked a significant advancement for medical researchers, offering them their first substantial opportunity to address critical questions about the effects of spaceflight on the human body—questions that had emerged from the earlier Mercury missions. During Mercury, medical personnel primarily focused on supporting flight operations rather than conducting in-depth medical research. While Mercury had resolved many of the concerns voiced in the 1950s, it also revealed potential weightlessness-related issues, particularly concerning the circulatory and skeletal systems.

Gemini's longer-duration missions provided a valuable platform for more comprehensive physiological monitoring and in-flight medical experiments. The primary concern for Gemini managers remained the medical certification of weightlessness for short missions to ensure safety for at least the eight-day duration anticipated for lunar landing missions. Medical researchers, however, were eager to gather as much data as possible to enhance the statistical reliability of their findings. They sought to understand the immediate effects of weightlessness and the broader implications for human physiology.

The eight-day flight of Gemini 5 in August 1965 intensified the pressure to reduce medical studies, as these investigations were perceived to encroach upon valuable training and flight time. Consequently, Gemini 7, which flew from December 4 to 18, 1965, became the final mission to include multiple medical experiments. While the Gemini missions effectively alleviated major concerns about weightlessness during shorter flights, they also highlighted potential issues that could become significant during extended missions. The unresolved questions at the end of Gemini provided a compelling rationale for the Apollo Applications Program (AAP) medical program, focusing on understanding how the body adapts to prolonged weightlessness, the duration of these changes, and the potential effectiveness of countermeasures.

In parallel to these developments in human spaceflight, the early scientific satellites represented a different phase of space exploration. These small, relatively inexpensive satellites were designed to be launched on available boosters and proved remarkably successful. Despite their limitations, the low cost of these missions allowed more frequent launches, establishing a crucial factor in advancing space science. President Eisenhower's science advisers, recognizing the high cost of crewed spaceflight and its uncertain scientific return, viewed the pursuit of crewed missions skeptically. They argued that the scientific benefits of crewed spaceflight were minimal compared to its expense and saw no urgent need to rush into such endeavors.

The President's Science Advisory Committee (PSAC) dismissed the significance of crewed spaceflight, emphasizing that the United States should not be drawn into a competitive race with the Soviets, seeking prestige through their space achievements. Instead, the PSAC strongly supported the continuation of the space science program, establishing it as a more valuable and strategically advantageous area of space exploration. The United States' leadership in this field and its potential for maintaining that lead was seen as a more effective approach to advancing space exploration.

John F. Kennedy recognized the broader significance of crewed spaceflight and aimed to position the United States at the forefront of space exploration. Despite his visionary outlook, his science adviser, Jerome Wiesner, who had been a member of the President's Science Advisory Committee (PSAC) since its inception during the Eisenhower administration, sought to persuade Kennedy to reconsider. On January 12, 1961, Wiesner's task force presented a report suggesting that the focus on Project Mercury was misplaced. The report argued that NASA should de-emphasize Mercury's significance and instead promote the cultural, public service, and military benefits of space activities beyond this project's scope. Wiesner contended that Project Mercury, plagued by numerous challenges, offered limited prospects for the United States to outpace the Soviets in the immediate future.

At this juncture, NASA was also guided by the recommendations of its external advisors, the Space Science Board of the National Academy of Sciences, which had been established in 1958. Unlike Wiesner's perspective, the Space Science Board advocated for ambitious plans that included human participation in space exploration. A month after Wiesner's report, the board emphasized that the long-term strategy for NASA should anticipate crewed missions to the moon and other planets. According to the board, scientific exploration would benefit profoundly from human involvement, as there was no mechanical substitute for the nuanced judgment that trained human beings could provide.

Kennedy's decision to endorse the Apollo program marked a pivotal shift in NASA's focus, prioritizing engineering achievements over scientific exploration. This move triggered immediate criticism from the scientific community, which felt that the program's engineering emphasis overshadowed the potential for scientific discovery. Many scientists lamented that the public and Congress, captivated by the technological spectacle of the Apollo missions, did not fully appreciate the scientific advancements that the program could foster. The excitement surrounding "Man on the Moon" overshadowed the scientific discourse, turning the space program into an engineering spectacle.

The Space Science Board adopted a pragmatic approach in response to the Apollo initiative. While it accepted the Apollo mission as the central objective, it sought to maximize the scientific value achievable within this framework. By 1962, the board had already endorsed crewed lunar and planetary exploration and reaffirmed this position in a summer study that year. This study, involving ninety-two academic and industrial scientists, evaluated the state of the uncrewed program and plans for NASA. A notable aspect of the study was its emphasis on "Man as a Scientist in Space," highlighting the unique capabilities of human judgment and analysis. The report underscored that scientifically trained individuals were indispensable for thoroughly exploring the moon and planets. It recommended that Ph.D. scientists be recruited and trained as astronauts as soon as possible, ideally in time to participate in the initial lunar landing missions. Furthermore, it advocated for enhanced scientific training for existing astronauts to ensure their readiness for the challenges of space exploration.

In the months preceding developing Skylab, a crucial working group examined the evolving role of scientists in the Apollo missions, guided by responses to a questionnaire dispatched by the Space Science Board. This survey sought the input of space scientists on how they perceived the integration of scientific expertise into the Apollo program. The feedback underscored a prevailing consensus: that each Apollo mission should ideally include at least one crew member who was primarily a scientist rather than an astronaut. The rationale behind this perspective was clear—those who ventured to the Moon should possess specialized knowledge, enabling them to collect samples efficiently and precisely.

The respondents advocated for a model where scientist-astronauts could continue their professional, scientific work without the overwhelming demands of astronaut training. They suggested that astronaut training should not dominate their time, ideally taking up only a fraction of each year. While this proposal may have seemed somewhat optimistic regarding the rigorous demands of space missions, it reflected a broader desire to balance scientific expertise and space travel's physical and technical requirements. This viewpoint was met with skepticism by some within NASA. A NASA official succinctly captured this sentiment by remarking that "it was easier to teach an astronaut to pick up rocks than to teach geologists to land on the moon," highlighting the challenges of reconciling these differing perspectives.

As the Apollo program progressed, the summer study briefly considered the future of space exploration beyond lunar missions, focusing on Earth-orbiting laboratories. The report acknowledged the need for further investigation into the design and timing of such laboratories but was cautious in its evaluation. It identified the primary role of an orbiting laboratory as being in biological research while also noting its potential utility for modifying, maintaining, and repairing orbiting satellites. However, there was significant skepticism about mounting telescopes on crewed orbiting stations, as astronomers were concerned that the movement of the crew might disrupt the alignment of the instruments.

Despite the clear acknowledgment of the Apollo program's scientific value, the report revealed underlying discontent among some scientists regarding the program's focus. There was considerable debate about the mission's justification and the balance between its technological achievements and scientific outcomes. The report's authors urged NASA to justify the enormous costs of the Apollo missions by emphasizing the long-term scientific benefits beyond the immediate technological milestones. They also called upon the scientific community to recognize that the Apollo program's goals were shaped by a blend of scientific and nonscientific considerations, an acknowledgment that both scientists and NASA needed to accept as part of the evolving landscape of space exploration.

Within a year of the Apollo program's ambitious launch, concerns about its financial demands became alarmingly justified. By the fall of 1962, the preliminary review of NASA's fiscal year 1964 budget nearly jeopardized the continuation of uncrewed science programs. Only a compelling defense from NASA Administrator James Webb persuaded the President to preserve these programs.

The situation became more critical in the spring of 1963 when NASA sought a substantial $5.71 billion from Congress. This request sparked widespread discussions about potential budget cuts. Webb and his team maintained that the ambitious goal of landing on the Moon by 1970 could not be achieved with a reduced budget. The sharp increase in NASA's budget request and a growing awareness of the Apollo program's total projected cost—estimated at over $20 billion—prompted critics to question national priorities. They dismissed the lunar program as a costly technological stunt with dubious value.

During the spring and summer of 1963, this critical sentiment gained traction among respected scientists, many of whom were not directly involved with the space program. On April 19, 1963, Philip H. Abelson, editor of the prestigious journal Science, ignited a wave of dissent with a scathing editorial. Abelson criticized the justifications for Apollo, arguing that the program's propaganda value was exaggerated, the potential for military advantage was minimal, and the so-called "technological fallout" would not offset the project's massive costs. He also contended that the scientific returns from the mission were negligible, especially since no scientist was slated to be among the first crew members. Abelson suggested that uncrewed probes, costing a fraction of an Apollo mission, could provide more valuable data and assist in designing a crewed landing vehicle. In essence, he questioned the high priority afforded to Apollo.

Abelson's editorial opened the floodgates for further criticism. Headlines like "Scientist Blasts Moon Project" became common for several weeks. Prominent figures, including Nobel Prize winners, voiced their disapproval, sparking a vigorous debate. Defenders of the Apollo program responded with counterarguments. NASA Deputy Administrator Hugh L. Dryden accused the critics of misrepresenting the program's objectives, asserting that NASA had never claimed the program's value was based solely on scientific returns. An aerospace magazine suggested that the critics were part of a "specialized segment of the scientific community" who resented the successful pursuit of goals they deemed outside their domain. In response, eight distinguished scientists, including three Nobel laureates, defended Apollo, asserting that the program significantly contributed to the future welfare and security of the United States.

The growing debate over the Apollo program's scientific value did not go unnoticed by Congress. In June 1964, before resuming deliberations on NASA's budget, the Senate Aeronautical and Space Sciences Committee convened two hearings to examine the matter more thoroughly. They invited ten esteemed scientists to provide their testimonies, hoping to gain insights beyond what had been reported in the press.

During the hearings, senators encountered little new information. The scientific community shared many of the same concerns as the general public and lobbyists. Scientists questioned whether the $20 billion earmarked for moon landings might be better spent on pressing domestic needs such as education, social programs, medical research, environmental protection, urban improvement, and other scientific fields. Despite these reservations, Harry H. Hess, the Space Science Board chairman, and Lloyd V. Berkner, its first chairman, defended NASA's mission. They conceded that, under certain circumstances, the presence of a scientifically trained observer on the moon could justify the expense.

As summer progressed, criticism from the scientific community gradually subsided. While figures like Abelson continued to express skepticism about the Apollo program, by early 1965, his enthusiasm for opposition had waned. Abelson acknowledged that it was acceptable if the public wanted lunar exploration, provided it was recognized as a venture primarily for enjoyment and adventure rather than a major scientific breakthrough. The tide of scientific dissent had receded significantly. As Daniel Greenberg noted, with President Lyndon Johnson's staunch support for the lunar mission and much of the capital investment already secured, it would take more than a few dissenting voices to prompt Congress to reconsider the Apollo program.

The opposition's impact on the Apollo mission was minimal, but it did influence NASA's long-term policies. With Apollo's apparent success, lunar scientists and engineers at the Manned Spacecraft Center (MSC) began to appreciate the mutual benefits of collaboration. By Apollo 72, their relationship improved significantly. The space science community recognized the need for a balanced approach that integrated scientific exploration with engineering advancements.

In 1964, the Space Science Board demonstrated its attentiveness to the ongoing debate by reassessing the nation's space objectives. When President Johnson requested a review of space goals, NASA tasked the board with revisiting its 1961 recommendations and contemplating future directions post-Apollo. Although the board had traditionally supported crewed space exploration, its updated October 30, 1964 report adjusted its stance. The board affirmed the importance of continuing lunar and planetary exploration, placing crewed exploration in a secondary position. It advocated for intensive uncrewed exploration of Mars alongside a gradual approach to solving biomedical issues to prepare for potential crewed missions to Mars by 1985. The board recommended that the space science program continue and even expand in certain areas, emphasizing a balanced and adaptable research strategy to leverage Apollo's technological advancements and ensure ongoing scientific benefits from space exploration, regardless of potential delays in the Apollo program.

Lunar exploration and crewed orbital stations were recognized as significant programs, but they were deemed secondary in terms of scientific importance compared to other research priorities. The primary focus was on operational techniques rather than on direct scientific outcomes. An Earth-orbiting station, while important, was seen more as a means to develop operational procedures than as a center for scientific discovery.

In late 1964, NASA sought further guidance from the Space Science Board, requesting a new study to evaluate future directions in space research beyond the Apollo program. This summer study, conducted at Woods Hole, Massachusetts, in June and July of 1965, aimed to assess priorities in planetary exploration, astronomy, and crewed space science. The report from this study broadly concurred with the board's previous recommendations from October 1964 but placed greater emphasis on the role of human participation in space research.

The report suggested a more balanced approach to space research, challenging the artificial separation between crewed and uncrewed missions. It advocated for selecting the mode of investigation—crewed or uncrewed—based on which would yield the best scientific results. The need for a greater number of scientist-astronauts was underscored, with the argument that scientific expertise would become increasingly crucial as the crewed space program advanced.

Among the notable recommendations were the endorsement of a solar telescope mount for the Apollo service module and a strong call for an Earth-orbiting laboratory to study human responses to the space environment. These suggestions were particularly significant for the Advanced Apollo Program (AAP) planners.

When George Mueller and Homer Newell presented their findings to Congress in the spring of 1966, they made the case for incorporating these recommendations into NASA's plans. Mueller highlighted the orbital workshop as a crucial step toward a future space station, proposing it as a cost-effective way to gain valuable experience before embarking on more extensive missions with a six- to eight-man laboratory. Newell expressed satisfaction with the proposal to include a telescope on the Apollo spacecraft but noted that the report's more ambitious suggestions would exceed the Office of Space Science and Applications (OSSA) budget. Indeed, budget constraints had already led to the cancellation of the Advanced Orbiting Solar Observatory (AOSO), despite its enthusiastic endorsement by the Woods Hole study.

The loss of the AOSO was a significant blow, mainly because solar research was a primary focus for OSSA. The AOSO project, valued at $167.4 million, had been designed to enhance solar observations with improved stabilization, pointing accuracy, and data-storage capacity. It was intended to launch four observatories by 1971 to cover the expected solar maximum period in 1969. The cancellation was a setback for studying solar phenomena, which had already benefited from the data gathered by previous Orbiting Solar Observatories (OSOs). These earlier missions had provided valuable insights into solar radiation, particularly in ultraviolet and x-ray wavelengths that do not penetrate Earth's atmosphere.

In 1965, NASA faced significant financial constraints that profoundly affected its space science programs. The Office of Space Science and Applications (OSSA) experienced a substantial budget cut of 16%, reducing its allocation from $783.2 million in fiscal year 1966 to $661.4 million for fiscal year 1967. This reduction necessitated difficult decisions, leading to the cancellation of several high-profile projects, including the Advanced Orbiting Solar Observatory (AOSO). The AOSO was particularly affected due to its high funding requirements, and its cancellation left a notable gap in solar observation capabilities. Although other observatories (OSOs) would continue to operate and partially fill the void, they could not match the quality or quantity of data the AOSO was designed to collect.

As 1966 began, Homer Newell, then Associate Administrator for OSSA, was deeply concerned about the survival of space science programs. The memory of the close call space science had faced just four years earlier was still fresh, and Newell saw similar pressures emerging once more. With the nation heavily invested in the Apollo program and resources increasingly diverted due to the escalating demands of the Vietnam War, Newell feared the allure of crewed spaceflight would overshadow that space science. He recognized that unless scientists advocated strongly for their research needs, Congress might view space science as a lower priority.

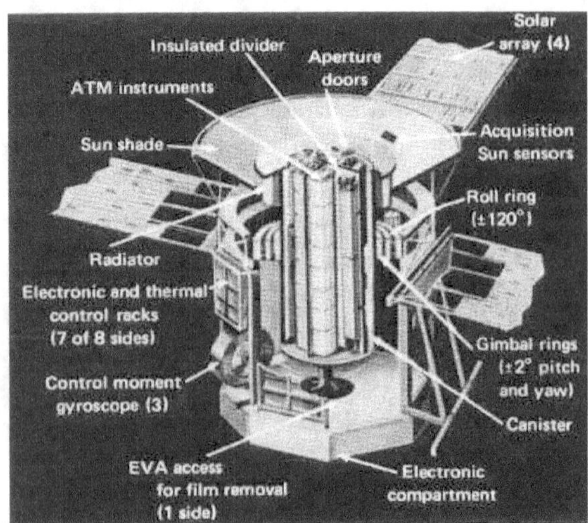

In early 1966, Newell reached out to Gordon MacDonald from UCLA, who had experience with the President's Science Advisory Committee and the Space Science Board and was a staunch supporter of OSSA's initiatives. Newell expressed his concern that the focus on crewed missions might again overshadow the achievements of space science. He urged MacDonald to speak out, to testify before congressional committees, and to rally fellow scientists to support high-quality research proposals.

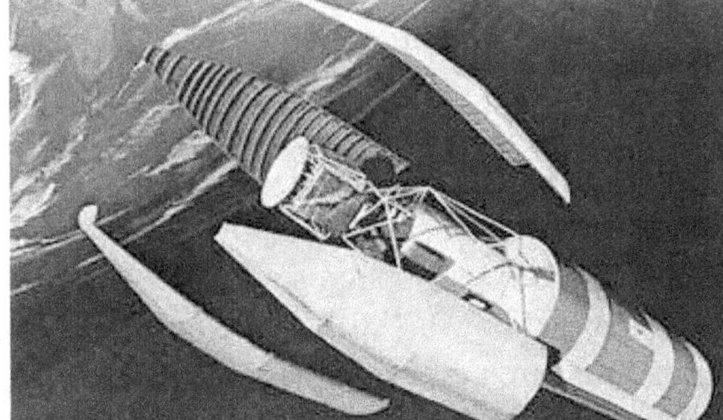

Solar Observatory.

Amid these concerns, OSSA continued its efforts to advance space science. In September 1965, the Physics and Astronomy Section took a significant step toward integrating astronomy with crewed spaceflight. A contract was awarded to Ball Brothers Research Corporation of Boulder, Colorado, to study the feasibility of installing a telescope mount in the Apollo service module. This study evaluated whether astronomical instruments could be stabilized adequately for data collection in space and whether human operators would enhance observational capabilities. Initially referred to as the Apollo Telescope Orientation Mount, or ATM, this device was designed to be controlled and adjusted by astronauts, using photographic film to record data.

While the ATM could not replace the AOSO, which was intended for a nine-month solar observation mission, it emerged as the sole option for high-resolution solar observations during the maximum solar period. With AOSO no longer possible, the ATM represented a critical scientific project that needed a suitable vehicle. Early discussions between Newell and Robert R. Mueller of the Office of Manned Space Flight (OMSF) centered on integrating the ATM into the Apollo Applications Program.

As the project advanced, several logistical challenges arose. Although the Goddard Space Flight Center, known for its expertise in astronomy programs, had directed the Ball Brothers study, the OMSF centers had more experience in integrating experiments into crewed spacecraft. In late January, Newell's office solicited proposals from Langley Research Center, Marshall Space Flight Center, Goddard, and the Manned Spacecraft Center (MSC) for managing the ATM project. Langley could not allocate resources for the project, and after reviewing the proposals, OSSA decided to retain management of the ATM project at Goddard.

With the management issue resolved, Newell sought approval from Deputy Administrator Robert Seamans to proceed with the project, emphasizing the urgency of starting immediately, given the impending solar maximum in 1969. The project approval documentation highlighted the instruments' compatibility with various spacecraft locations, with a specific mention of the service module's experiments sector. However, Mueller advocated for mounting the ATM on the lunar module, presenting a further point of contention as the program offices debated the optimal setup for this critical scientific endeavor.

On April 8, 1966, a pivotal status review meeting for the Apollo Applications Program (AAP) involved key figures such as Homer Newell, Robert Mueller, and their technical teams. They assessed two competing proposals for the Apollo Telescope Mount (ATM) project, establishing crucial for advancing solar observation during the upcoming solar maximum.

OSSA's position was clear: mounting the ATM in the Apollo service module was both cost-effective and reliable. This option required fewer modifications to the spacecraft and would meet the critical 1968 launch deadline. However, this approach had a significant limitation—the service module would burn up upon reentry, allowing for only a single 14-day mission.

In contrast, the Office of Manned Space Flight (OMSF) argued in favor of the lunar module-ATM combination. They posited that this option would initially cost less, benefit from the lunar module's capability to remain in orbit and be reused and facilitate future lunar module laboratories due to the operational experience gained. Yet, the lunar module was still untested, and its production was behind schedule. This uncertainty left both proposals fraught with risks and benefits.

Seamans, recognizing the complexity of the decision, requested additional details before committing. He was cautious about approving the project immediately, concerned about the fiscal constraints for the ATM's FY 1967 requirements, and wary of initiating an industry competition without secure funding. Consequently, he approved two additional studies by Ball Brothers. One study would explore the feasibility of the ATM's automatic operation if it could remain in orbit post-mission. At the same time, the other would examine its adaptation for the lunar module.

Despite these deliberations, Mueller was determined to pursue a Marshall Space Flight Center (MSFC)-based solution. He faced internal dissent, notably a strongly worded letter from Robert Gilruth, Director of the Manned Spacecraft Center (MSC), who objected to using the lunar spacecraft as a laboratory and the integration responsibilities being assigned to Huntsville. Nevertheless, Mueller proceeded with his plan, deciding on May 18 that the ATM system, excluding the telescopes, would be designed, built, and integrated into the lunar module at Marshall.

By June 8, planners at Huntsville had initiated discussions with Grumman Aircraft Engineering Corporation, the prime contractor for the lunar module, to assess the ATM's compatibility with the module. Shortly after that, MSC authorized Grumman to study this integration. OSSA raised objections to a mission assignments document issued by the AAP office in June, noting discrepancies with OSSA's intended operational plans, orbital parameters, and launch dates.

Despite these objections, Mueller continued to press forward. In June and July, he and Newell met with Seamans multiple times, advocating for their competing project approval documents. On July 11, a consensus was reached to transfer the entire ATM project—including the experiments—to Marshall for development. This decision was influenced by the recognition that managing such a complex project was best handled without splitting responsibilities, as Goddard could not manage the full scope alone.

With the transfer to Marshall confirmed it became increasingly clear that Mueller's plan would be adopted. He continued to provide Seamans with technical data, including tradeoff studies comparing various ATM locations. During the Gemini 10 mission in July 1966, Mueller sought feedback from MSC officials on these studies. The response from Houston was clear: while all ATM work should indeed be assigned to Marshall, selecting the lunar module as the experiment carrier would forfeit the benefits associated with that choice. Instead, MSC recommended that Marshall design and build a specialized structure, or "rack," to carry the ATM and its supporting systems within the CSM-LM adapter.

Ultimately, this period of intensive discussion and negotiation set the stage for the ATM project's development, highlighting the intricate balancing act of budget constraints, technical feasibility, and strategic priorities in pursuing space science advancements.

The crew was tasked with operating the solar telescopes from the command module in the orbit of the Skylab Space Station. If conducted using an Extended Apollo spacecraft, the mission could potentially sustain up to 30 days of uninterrupted observation. However, this ambitious plan faced substantial challenges.

NASA's Manned Spacecraft Center (MSC) had estimated that modifying the lunar module, as Dr. Robert R. Mueller proposed, would incur an additional cost of at least $100 million over the existing budget for a standard mission. Moreover, the modification would likely extend the project timeline by two to three years. MSC's managers were also concerned about the operational logistics of Mueller's plan, which involved the Apollo spacecraft rendezvousing with a separately launched telescope mount. The procedure required two crew members to transfer to the solar observatory, which would then separate from the Command and Service Module (CSM). After conducting 14 days of solar observations, the Solar Observatory Module (ATM) was supposed to rejoin the Apollo craft, allowing the two crew members to return to the command module. However, if the second rendezvous failed, the ATM crew would have no means of returning home, presenting a significant and unacceptable risk.

Safety considerations were paramount, and Houston's MSC could not support the proposed approach due to the high risk of mission failure. The summer of 1966 had already been a challenging period for the lunar module project, with Grumman facing severe technical and management problems. MSC's program office struggled to resolve these issues, and the introduction of another complex project, as Mueller suggested, added to their difficulties.

Despite Mueller's insistence on pursuing the lunar module laboratory, he faced direct questioning from MSC's Apollo Spacecraft Program Manager. The manager queried whether Mueller's continued support was driven by political motives rather than technical feasibility. Mueller candidly acknowledged that his motives were indeed "completely political," driven by the necessity to maintain the Marshall Space Flight Center's cohesion and avoid the appearance of introducing a major new project.

In response to these challenges, on August 2, 1966, the Office of Manned Space Flight (OMSF) recommended that NASA Administrator James E. Seamans approve a modified version of MSC's suggestion. This revised approach involved contracting approximately $60 million of major components for the Lunar Module - Astronomical Telescope Module (LM-ATM) system and utilizing Marshall personnel for select development tasks. Mueller expressed confidence that the 1968 launch date could still be met with immediate approval and initiation of work.

However, optimism was tempered by concerns within the Office of Space Science and Applications (OSSA) about the feasibility of observing the sun during its peak activity.

On August 29, 1966, following the completion of congressional action on NASA's fiscal year 1967 appropriation, Seamans signed the approval document for Mueller's version of the ATM project. This document authorized developing one set of instruments for flight on the second Apollo Applications mission. Seamans noted that several key details remained undefined and requested to be informed of major decisions throughout the project's development phase. The subsequent three months were dedicated to finalizing the ATM design and operational mode, culminating in developing an orbital cluster based on the multiple docking adapter system.

After a prolonged four-month wait, during which experimenters were held back from advancing their work, they were finally cleared to construct their instruments. The delay had been particularly frustrating, as the design of the Apollo Observatory Solar Observatory (AOSO) instruments had stalled due to the project's termination. Nevertheless, the Office of Space Science and Applications (OSSA) managed to keep the project alive, hoping to find a way to utilize these instruments effectively. Meanwhile, the Goddard Space Flight Center's ATM team had maintained enthusiasm for the project by organizing a betting pool on the anticipated date when NASA Administrator James E. Seamans would sign the project approval document.

The complexity of the solar observatory canister was clear in this photograph.

With the approval now in place, the development schedule for integrating the instruments into the ATM was extremely compressed. Despite the urgency, the waiting period had not been entirely in vain. It provided valuable time to address some fundamental issues, particularly concerning the stabilization of the ATM. Achieving the superior resolution that film could deliver required an exceptionally stable mount, a technical challenge that had to be met to ensure the precision needed for astronomical observations.

The project specifications demanded that the telescopes' alignment be maintained within ±2.5 seconds of arc for 15 minutes. This level of precision was equivalent to keeping the ATM pointed at a specific point on a distant bridge—a kilometer away—without allowing the alignment to drift more than the width of a human pupil. Many experimenters were skeptical about whether such stringent requirements could be met. Conventional attitude-control thrusters, typically used for stabilization, were deemed insufficient.

At the May Apollo Applications Program (AAP) review, Mueller made the critical decision to employ gyroscopes as the primary means of stabilizing the ATM. Research conducted at Langley Research Center had produced prototypes of "control moment gyroscopes" with 90-centimeter rotors, which were deemed adequate for stabilizing a vehicle of the ATM's size. However, further development was necessary to ensure the long-term reliability of these gyroscopes in space conditions. Both Langley and Marshall Space Flight Center embarked on this crucial work to qualify the gyroscopes for space use, to achieve the precise stabilization required for the mission's success.

Three days after Dr. Robert Seamans approved the Skylab program's Advanced Technology Module (ATM) project, four research agencies were informed that their experiments had been selected for inclusion. On September 6, the contracts for the ATM were officially transferred from NASA's Goddard Space Flight Center to the Marshall Space Flight Center. By September 19, the basic ATM program received the endorsement of the Manned Space Flight Experiments Board, marking a significant milestone.

Marshall Space Flight Center presented its compatibility studies for the Lunar Module ATM hardware and mission at this juncture. Their findings detailed a design featuring an experiment canister with a diameter of 1.5 meters and a length of 3.3 meters. This canister, equipped with instruments mounted on a cruciform spar dividing it into quadrants, was engineered to attach securely to the lunar module's ascent stage. The total estimated weight of the canister fell well within the lift capacity of the Saturn IB rocket, allowing for a comfortable margin.

The ATM's five instruments were designed to capture the sun's spectrum across a broad range, from visible light to high-energy X-rays. This array of instruments represented a coordinated approach to solar research, unprecedented in its scope. Although the general public might not recognize these instruments as traditional telescopes, all but one could record images of the sun or specific regions of it on film.

The Naval Research Laboratory contributed two ultraviolet instruments that photograph the entire sun or smaller, targeted areas. These instruments utilized wavelengths that revealed detailed information about the sun's composition. Meanwhile, the Massachusetts-based American Science and Engineering was tasked with building an X-ray instrument to record detailed images of solar flares and monitor X-ray output from the sun. The High Altitude Observatory in Boulder designed a white-light coronagraph to photograph the sun's faint corona by blocking out the intense light from the sun's disk. Harvard College Observatory provided a complementary ultraviolet spectrometer and spectroheliometer. Unlike the other instruments, this one used photoelectric detector to telemeter readings to Earth, allowing it to be operated remotely when the ATM was uncrewed. However, this mode of operation lacked the fine-pointing control that a crew could provide.

Marshall's compatibility study did not uncover any issues to prevent scheduling the ATM for launch in the fourth quarter of 1968. Nevertheless, there were concerns about whether two instruments could be delivered six months before the launch. It was hoped that any scheduling conflicts would be resolved during contract negotiations. Power requirements for the module were still undecided, with planners considering an array of solar cells capable of generating up to three kilowatts of electricity.

With the approval of the ATM instruments, the Apollo Applications Program (AAP) embarked on its largest and most complex scientific endeavor. As Lee Belew of Marshall's AAP office noted, the project warranted significant attention. However, a growing challenge was finding experiments that required human participation and effectively utilized the workshop's extensive volume. Wernher von Braun highlighted this issue in May 1965, noting that an optimistic schedule could potentially place 970 metric tons of payload into a 225-kilometer orbit annually—an amount that could be achieved by a single Saturn V rocket, surpassing all previous NASA payloads combined.

By early 1966, NASA Administrator James E. Webb informed the centers of anticipated funding shortfalls for the experiment program. Experiment development could no longer rely on contractors as it had previously; instead, centers were encouraged to utilize commercially available components wherever feasible. Von Braun conveyed this directive to Marshall, emphasizing that the centers needed to fill the workshop with experiments.

After the February 1966 AAP review, Robert Seamans instructed the Office of Manned Space Flight (OMSF) to include experiments in periodic program evaluations. By March, three experiments were under development, the Manned Space Flight Experiments Board was considering ten, and another thirteen were nearing submission. Additionally, eleven experiments were in the definition phase, 108 were planned for definition studies, and 72 were awaiting the process to begin. Given that an experiment typically required 32 months from inception to readiness for flight, the timeline for a substantial program of experiments by 1968 appeared challenging.

Funding remained a critical issue, compounded by inadequate manpower and a complex division of responsibilities between the Office of Space Science and Applications (OSSA) and OMSF. Seventeen biomedical experiments had been identified, but work statements outlining center responsibilities were still pending. Despite its promising potential, the ATM project faced a financial shortfall of $19 million, with no identified funding source.

Within OMSF, early phases of the experiment program fell under the purview of E.2. Gray's Advanced Manned Missions Office. Mueller urged this office to accelerate its efforts, and Gray responded by appointing Douglas Lord as chief of the Experiments Division. Lord was tasked with assembling a coherent set of experiments for the Skylab workshop, setting the stage for the forthcoming scientific endeavors aboard the space station.

In mid-May, a critical call to action was issued after preliminary consultations with the experiments offices at Houston and Huntsville. Lord requested the centers submit a comprehensive list of experiments, complete with priorities, development funding plans, and schedules, for presentation to the experiments board at its meeting in July. However, when the deadline arrived, the response was disappointing; no proposals had been received.

A month later, when the proposals finally did come in, they fell short of expectations. On June 28, Gray expressed his dissatisfaction directly to von Braun and Gilruth, stating that "the proposed workshop experiments do not constitute a reasonable program." The proposed experiments were criticized for their lack of relevance. Notably, no experiments were designed to assess the habitability of the spent stage or to provide crucial design parameters for future space stations. Many of the proposed experiments did not require the specialized environment of the workshop, and others demanded minimal or no crew involvement. Gray's candid assessment was that the effort had not adequately addressed the problem of defining a useful set of experiments. Instead of developing experiments in-house that could be conducted in the workshop, the proposals seemed to bypass this essential step.

In response to this challenge, Lord assembled a team to visit the centers for viable experiments. He described their efforts as "beating the bushes" to uncover low-cost, practical experiments. Reflecting on the situation later, Lord acknowledged the task's difficulty, noting that insufficient funding had been allocated to define the experiments properly, leading to a scarcity of suitable proposals. Meanwhile, von Braun criticized the approval process as so cumbersome that it undermined the efforts to find and implement experiments, rendering it ineffective for the impending late 1968 flight. Despite their diligent efforts over the next six months to press the centers for viable experiments and to evaluate them, the team faced persistent obstacles in their quest to meet the program's needs.

In August 1966, various reviews for proposed experiments were conducted at Houston and Huntsville. During these evaluations, twenty-four experiments, predominantly focused on engineering tasks, were meticulously scrutinized. The outcome of these reviews was mixed: eight experiments were rejected, thirteen were approved, and three were either withdrawn or merged with other proposals. Among these, a "Habitability/Crew Quarters" experiment was prioritized, involving participation from both centers. Other notable experiments included assessing the astronauts' ability to repair and maintain equipment, investigating the flammability of materials in a zero-gravity environment, and evaluating the effectiveness of spacesuits and extravehicular mobility aids.

By the September experiments board meeting, eleven of the proposed experiments had been approved, contingent upon the availability of development funding. The board emphasized cost-efficiency, urging the centers to utilize in-house facilities and manpower wherever possible. Despite this, the initiation of medical experiments was surprisingly sluggish. Medical research, which had been a primary justification for the workshop-type missions, faced delays. Houston's Medical Research and Operations Directorate was already stretched thin with the ongoing evaluation of Gemini medical results and developing sixteen medical experiments for Earth-orbiting Apollo missions. This burden on the Directorate was compounded by the demanding tasks of planning Apollo experiments and conducting ground-based research.

Similarly, the Crew Systems Division at Houston, crucial to developing medical experiments, was operating at full capacity on life-support and environmental-control systems. Nevertheless, the essential medical studies for the upcoming 28-day mission were outlined. At the September board meeting, OMSF's Office of Space Medicine presented three key proposals. The first two, Metabolic Activities and Cardiovascular Assessment were designed to measure the effects of zero gravity on the muscular and circulatory systems using telemetry. The third proposal, Bone and Muscle Changes, continued the Gemini M-7 experiment. This study involved pre- and post-flight bone calcium measurements and collecting urine samples in flight for subsequent analysis.

The board approved these medical experiments with the provision that detailed plans would be submitted later. Additionally, the board supported the recommendation to include a physician-astronaut in the crew for the first workshop mission, acknowledging the importance of having medical expertise onboard for the pioneering flight.

In November, the board meeting was a hive of activity, primarily focused on the experiments planned for the Apollo Applications Program (AAP). During this session, the board approved a diverse set of experiments, comprising two medical, four technological, and six scientific studies. However, these approvals were contingent on securing the necessary funding. At this stage, the first workshop mission started feeling somewhat congested, raising concerns about the crew's ability to complete all the approved experiments due to time constraints.

One significant challenge was the integration of a proposed artificial gravity experiment. The available maneuvering fuel was inadequate to spin the experiment cluster while retaining a reserve for deorbiting the command module if required. This issue underscored the complexities of balancing various experimental needs with the mission's operational requirements. The integration team at Marshall Space Flight Center faced considerable difficulties, exacerbated by the fact that experimental parameters were evolving every two months and the spacecraft configuration was still under development.

The board also addressed the issue of experiment prioritization. While sponsoring agencies set initial priorities for their respective experiments, it was up to the board to create a cohesive and integrated list. In November, habitability studies were prioritized, followed by biomedical research and experiments focused on crew mobility and work capability. The artificial gravity experiment was assigned the lowest priority. These priorities were flexible and subject to change as the list of experiments expanded. The board would continue to grapple with this prioritization issue for the next year.

By the end of 1966, the experimental program had seen some stabilization. Only two experiments had been firmly assigned to specific missions, while thirty-one, including the Apollo Telescope Mount (ATM) experiments and medical studies, were approved and tentatively scheduled. Nineteen additional experiments were approved but awaited mission assignment. Adopting the cluster concept and finalizing the first four launches by December marked a significant step forward in solidifying the experiment program.

By February 1967, the experimental assignments had become more definitive, with eight additional experiments scheduled and several new ones proposed and approved. George Mueller's press briefing on January 26, 1967, highlighted the substantial progress in crewed orbital science. The medical experiments planned for the first mission aimed to assess human performance and endurance in zero gravity. The ATM experiments were expected to advance knowledge about the role of humans as scientists in space and gather unprecedented solar data. Smaller experiments were anticipated to contribute valuable insights into space technology and operations.

Despite these advancements, the program was not without its challenges. While the Office of Manned Space Flight (OMSF) worked through the intricacies of its first post-Apollo project, the President's Science Advisory Committee (PSAC) was evaluating the future direction of the U.S. space program. Throughout 1966, the 24-member panels on space science and space technology scrutinized the nation's space efforts. Their report, which focused on broad policy recommendations, also included specific criticisms of the AAP. These criticisms challenged Mueller as he prepared to present the Fiscal Year 1968 budget to Congress.

On February 11, 1967, the President's Science Advisory Committee (PSAC) published a report that resonated with the Space Science Board's 1965 Woods Hole study, advocating for lunar and planetary exploration as the most fruitful areas for immediate space research. However, PSAC took a distinct stance on the goals for the 1970s. Rather than endorsing a major endeavor with a strict deadline, PSAC emphasized the need to focus on how to exploit space capabilities best to advance national interests. Their perspective shifted from identifying a singular major project to exploring the most effective ways to utilize space technology for broader national objectives.

PSAC proposed a balanced program that anticipated eventual crewed planetary exploration. This approach required a robust planetary program, extensive lunar exploration, qualifications for long-duration space missions, and advancements in space technology across all domains. Additionally, they recommended leveraging Earth-orbital operations to advance scientific understanding, particularly in astronomy. The fundamental questions PSAC aimed to address included extraterrestrial life, the universe's origin, and the solar system's evolution.

The committee's broad approach aligned with the Apollo Applications Program (AAP) goals but suggested a shift in emphasis. They recommended that any Apollo Saturn hardware not utilized for the initial lunar landings be repurposed for comprehensive lunar exploration rather than AAP projects. PSAC also advocated for limiting Saturn production to four Saturn V rockets annually and an unspecified number of Saturn IB rockets. They critiqued the Saturn IB, comparing it unfavorably with the Titan rocket, which offered comparable payload capacity despite being half as costly.

The report reignited discussions on the need for a permanent Earth-orbiting space station, which PSAC viewed as essential for qualifying humans for extended space missions. Such a station would facilitate studies on the effects of zero gravity on various life forms and support research across multiple scientific disciplines and space technologies. PSAC recommended launching the first module of a permanent space station in the 1970s. While acknowledging the AAP orbital workshop as a valuable precursor, they raised concerns about the extensive construction efforts required by the wet workshop scheme, suggesting that it might compromise the biomedical data crucial for early mission phases. They proposed that NASA consider funding the Manned Orbiting Laboratory (MOL) to accelerate the acquisition of biomedical information. They urged the Air Force to focus more on biomedical research within the MOL program.

In astronomy, PSAC identified this field as particularly ripe for post-Apollo exploration. However, their Apollo Telescope Mount (ATM) plan review revealed significant flaws. They argued that the current approach was inadequate for a man-supported astronomy project in Earth orbit. PSAC recommended alternative solutions, such as using a microwave control link between the Apollo spacecraft and a free-flying ATM or even a global communications network to allow ground-based operators to manage the instruments. They criticized the notion that humans needed to be near the instruments, suggesting that electromechanical systems or ground-based operations would be more efficient. The committee noted that the ATM's design did not allow for necessary repairs and adjustments, further questioning the practicality of the current plan for crewed involvement in the project.

In the late 1960s, the Office of Manned Space Flight (OMSF) faced mounting pressure to meet a highly ambitious timeline for the launch of the Apollo Telescope Mount (ATM), a key component of the Skylab Space Station program. The original goal was to launch the ATM in 1968, a target set to address the scientific community's eagerness for solar observations. However, this aggressive schedule placed immense pressure on developing critical instruments, leading to significant concerns and criticism.

By 1970, the period of maximum solar activity, characterized by frequent solar flares, was expected to continue, potentially justifying the mission's objectives. Nonetheless, the rushed timeline put undue strain on the ATM's instrumentation and threatened to compromise the design and operational effectiveness of the entire system. This situation was recognized in a report acknowledging the ATM's shortcomings but deemed its cost reasonable. For astronomers eager to deploy high-resolution instruments, the ATM represented a valuable opportunity, albeit imperfect.

The Panel on Science and Applications (PSAC) recommended delaying the launch by one year. They argued that this extension would allow for a much-needed redesign of the ATM, addressing its fundamental flaws and alleviating the pressures on the instrument developers. The panel's report emphasized that the proposed mission did not align well with the envisioned future of Earth orbital astronomy. It warned that the constraints imposed on the astronauts and the demanding operational schedule might undermine physical and mental performance during the mission. Thus, the report advocated optimizing all mission parameters to maximize scientific returns.

Additionally, discussions with seasoned astronauts raised concerns about the complexity of mission operations. The scientists were apprehensive about the communication protocol, which required all communications with the orbiting spacecraft to pass through the Capsule Communicator (CapCom). In their view, this setup might be inadequate for managing an astronomy mission effectively.

The report, establishing somewhat ambivalent about the workshop mission and critical of specific ATM details, nonetheless recommended proceeding with both. Despite its nuanced stance, the report drew minimal public attention. However, when Homer Newell presented the findings to the House Subcommittee on Science and Applications, Chairman Joseph Karth had thoroughly reviewed and annotated the document. The ensuing debate between Karth and Newell centered on whether the PSAC report supported the solar astronomy mission. While Karth was skeptical, Newell, citing clarifying letters from panel members, interpreted the report as a conditional endorsement.

In response, George Mueller, aiming to address the criticisms preemptively, presented Congress with evidence of the Apollo Applications Program (AAP) actively working to meet the report's recommendations. He countered the PSAC's critiques of the ATM in written responses to congressional inquiries.

Just days before the PSAC report was published, NASA was rocked by a tragic fire during an Apollo spacecraft test at Kennedy Space Center. This disaster overshadowed the report's impact and introduced unprecedented challenges for the AAP. Over the next 18 months, the program faced pressures far exceeding the scientific criticisms it had previously encountered, shifting the focus from the ATM's development to addressing the broader implications of the Apollo fire.

1967 to 1969 were marked by profound uncertainty and adversity for the United States and its space program. Under President Lyndon B. Johnson, the nation's focus was heavily skewed toward the escalating conflict in Vietnam. This war effort consumed vast resources, with 535,000 American troops deployed, 24,000 lives lost, and costs soaring to $2 billion per month. The relentless toll of the war was compounded by civil unrest and high-profile assassinations, contributing to a pervasive sense of malaise that overshadowed public optimism.

The early enthusiasm for space exploration, characterized by the 1960s, began to wane. By 1968, the country faced a staggering deficit of $25 billion, prompting President Johnson to agree to significant cuts in non-defense spending. Although the Apollo program continued to receive support as a symbol of national commitment, the future of post-Apollo programs, including the Apollo Applications Program (AAP), was far less certain. These budgetary constraints deeply affected the AAP, designed to extend the benefits of the Apollo missions into new scientific and technological realms.

The catastrophic spacecraft fire at Kennedy Space Center further tarnished NASA's reputation, raising fundamental questions about the agency's competence. In the aftermath, NASA focused primarily on ensuring the success of the lunar landing missions, which left AAP planners grappling with an uncertain environment. They faced mounting difficulties, including fluctuating funding and a reliance on the uncertain performance of the Apollo missions.

As budget cuts continued to slice through the AAP, the program was forced to retreat from the ambitious goals set in 1966. Projected flights were scaled back, and launch dates were repeatedly postponed. The cluster missions, in particular, seemed perpetually stalled, becoming a running joke within NASA.

By the late summer of 1968, the AAP's most ambitious project, the Apollo Telescope Mount (ATM), faced severe threats of cancellation. Rising costs and persistent technical problems plagued the project. The general election exacerbated the situation, which ushered in a new administration with no clear space policy. However, various successes, notably Apollo 8's historic mission orbiting the moon during Christmas 1968, boosted the nation's spirits and the space program.

A change in leadership at NASA also contributed to a shift in momentum. James Webb, who had carefully shielded the lunar landing from any potential interference, was succeeded by Thomas O. Paine. Eager to leave his mark on the space program, Paine worked to promote ambitious plans for NASA's future. Although his proposals were not ultimately adopted, they helped reinvigorate the Apollo Applications Program, offering a glimmer of hope amidst the era's challenges.

In the spring of 1969, NASA faced a pivotal moment in the Apollo Applications Program (AAP) as they sought effective solutions to numerous technical challenges. The Saturn V rocket emerged as a promising candidate for launching a ground-equipped ("dry") S-IVB workshop, which significantly improved over the previous plan to use a Saturn IB rocket. The Saturn V, combined with a dry workshop, promised to overcome the limitations of earlier designs and the complex task of converting a fuel tank into functional living and working quarters in space. This development marked a critical shift in the program's approach, allowing for more ambitious and feasible mission objectives.

Canceling the Air Force's Manned Orbiting Laboratory in June 1969 provided an additional boost to the AAP. This unexpected turn of events allowed NASA to view the AAP in a new, more favorable light. After Apollo 17's successful lunar landing, NASA made a significant announcement: the AAP would now be launched using a Saturn V rocket with a dry workshop. This decision was a substantial advancement from previous plans and was expected to allow the program to make significant progress for the first time.

However, the road to this breakthrough was fraught with challenges. The day before the tragic fire at Kennedy Space Center, George Mueller, who was leading NASA's efforts, had referred to the first four AAP flights as a firm program. Despite his optimistic outlook, several crucial issues remained unresolved. Notably, there was ongoing resistance from the Houston center regarding the plan to carry solar telescopes on a modified lunar module. Although Houston had temporarily accepted this concept, it was clear that further study was needed.

The center program offices had already identified the mid-1968 launch date as unrealistic. Charles W. Mathews, the new director of AAP, had established a committee to clarify tasks and set a more realistic timeline. This committee, composed of Mathews and three center program managers, had initially baselined the first four flights in February 1967. They had agreed on the essential mission features, including payload allowances, orbital parameters, and operational modes. However, including a solar-cell array on the Apollo Telescope Mount and identifying numerous required tasks for the center complicated matters.

In March 1968, the center program offices reviewed schedules and test programs. By the end of the month, it was confirmed that the June 1968 launch date was unachievable. Consequently, a revised schedule was established, postponing the first launch to December 1968, with the solar astronomy mission to follow six months later.

Despite this revised timeline, two significant problems persisted. Developing the solar telescopes was behind schedule. While two of the five experimenters were optimistic about meeting a mid-1969 launch date, the other three— representing the High Altitude Observatory, the Naval Research Laboratory, and Harvard College Observatory—required additional time.

Furthermore, the fire's aftermath raised doubts about the availability of command and service modules for AAP missions. North American Aviation was still defining the necessary modifications to the spacecraft for the applications missions. Determined to ensure Apollo's recovery and progress, James Webb proposed a different approach to address these emerging challenges, reflecting his commitment to overcoming the obstacles facing NASA's space endeavors.

After a significant setback, the Skylab Space Station program experienced a tumultuous period marked by intense planning and frequent revisions. The uncertainty surrounding the introduction of a new contractor led to hesitations in making necessary modifications, resulting in minimal progress.

The fire that occurred at the headquarters of the Apollo Applications Program (AAP) initiated a frenzied phase of reorganization and strategic adjustments. Although not fully grasped initially, the immediate repercussions of the fire made it clear that there would be considerable delays. This reality prompted a cascade of revisions to both Apollo and AAP schedules. The AAP, established to offer NASA flexibility in its space endeavors, faced the challenge of accommodating numerous potential scenarios. The objective was to ensure that any unforeseen delays or advances in the Apollo program could be managed effectively, with AAP missions as a buffer.

In the aftermath of the fire, the AAP office undertook an extraordinary amount of work. According to James Mathews, a key figure in the program, the six months following the fire were particularly grueling. During this period, the office developed an astonishing number of program plans, with estimates ranging from 55 to 57 separate plans in just one month. This figure highlights the substantial workload, and the complexity of the adjustments required.

The accident's impact was deeply felt at the Houston headquarters, where the focus shifted to managing the crisis. Max Faget, a prominent figure in the space program, recalled that discussions were monopolized by the aftermath of the fire, overshadowing other concerns. Webb appointed George Low to oversee the spacecraft recovery efforts in response to the incident, leaving Robert Thompson to manage the AAP office under challenging conditions. The office grappled with ongoing uncertainties regarding reliability and quality standards, further complicating their efforts.

As Mathews settled into his new role, he thoroughly reviewed AAP plans across various centers. During his visit to Huntsville, he encountered Wernher von Braun and his AAP manager, Leland F. Belew, who were primarily focused on the pressing deadlines for solar instruments. The urgency of meeting these deadlines underscored the broader challenges faced by the program in the wake of the fire.

This period of intense recalibration and adjustment was crucial in navigating the complexities introduced by the fire and ensuring the continued advancement of the Skylab Space Station program.

In Houston, concerns about North American's ability to deliver spacecraft on schedule were mounting. Both Gilruth and Thompson questioned whether a new contractor could meet the demanding modifications required for the Apollo Applications Program (AAP). Their apprehensions were compounded by the ambitious flight schedule proposed for 1969, which called for a total of ten crewed launches—six Apollo missions and four AAP flights. The Flight Operations Directorate at the Manned Spacecraft Center (MSC) was only prepared to handle six crewed missions per year, raising doubts about the feasibility of the proposed schedule.

James Mathews acknowledged the improbability of all scheduled missions proceeding as planned and noted that some were included merely to provide the Office of Manned Space Flight (OMSF) with flexibility to manage program contingencies. However, MSC officials and the AAP manager at the Kennedy Space Center were less impressed with this rationale, favoring a more realistic schedule to allow for firm planning.

The debate over the 1969 schedule was moot that summer as Congress made significant cuts to the AAP budget. The fire that had impacted NASA's reputation led to diminished confidence among legislators, with some, including Don Fuqua, who would later chair the House Space Sciences Subcommittee, noting its considerable effect on previously indifferent or critical members of Congress. NASA might have weathered the setback during more prosperous times without severe repercussions. However, by mid-1967, the growing federal deficit, with NASA contributing to the fiscal strain, became a significant concern on Capitol Hill.

In a conference committee, the Senate and House agreed to trim $107 million from AAP's initial $454 million budget request. Consequently, the AAP office was forced to develop a revised schedule, which included postponing the first missions by five months and eliminating using refurbished spacecraft. By August, the appropriation bill set AAP funding at $300 million, nearly $50 million below the authorization level, with President Johnson choosing not to oppose this reduction.

Each congressional cut necessitated a flurry of adjustments by the AAP team. One plan sought to avoid further delays by significantly reducing funds allocated for experiment definition and payload integration. Another approach proposed extending the schedule by three more months, allowing for an October 1969 launch of the wet workshop while providing additional funding for experiments and integration. Faced with the likelihood of further reductions, the AAP office devised a third plan based on a $250 million budget. This strategy included a further three-month delay, which would enable the postponement of launch vehicle deliveries and a substantial reduction in hardware purchases.

Based on the situation, George Mueller considered $250 million the minimum acceptable funding level. He warned that anything less would delay the program's "real start" by another year and result in long-term waste. Drawing on lessons from earlier programs like the Air Force's Skybolt and Dyna-Soar, Mueller highlighted the inefficiency of maintaining high levels of design activity without advancing to hardware development. He concluded that such practices typically increased costs in subsequent years and often prevented program elements from reaching a logical conclusion.

In September 1967, NASA Administrator James E. Webb made a significant budgetary decision to impact the future of the Apollo Applications Program (AAP). He ordered a transfer of funds from the AAP to NASA's Office of Tracking and Data Acquisition, leaving only $253 million allocated for the Apollo Applications in NASA's fiscal year 1968 operating plan. This financial adjustment effectively precluded the possibility of concurrent Apollo and AAP flights. Even if suitable launch vehicles and spacecraft became available, NASA would lack the funds to launch and track these missions. As a result, the first AAP mission was rescheduled to January 1970, with subsequent wet workshop and solar astronomy missions planned for later that year.

The revised schedule for October 1967 outlined a significant reduction in planned launches: 17 Saturn IBs and 7 Saturn Vs, a sharp decline from the 40 launches initially listed in May. These figures, however, appeared overly optimistic given that production of the Saturn IB was expected to cease after 16 vehicles.

The challenges faced by the AAP were highlighted by the fate of Mission AAP 1A, which symbolized the program's difficulties in 1967. The devastating Apollo spacecraft fire, which occurred earlier, led NASA to revise its priorities. Apollo missions were to carry only those experiments directly contributing to the lunar landing, leaving several scientists without flight opportunities for their planned experiments. Concurrently, AAP planners grappled with issues related to payload weights and crew workloads for the workshop mission.

In response to these setbacks, the Office of Manned Space Flight (OMSF) began planning a new mission to mark the beginning of AAP. This mission was envisioned as a two-week Command and Service Module (CSM) flight in late 1968. The primary objectives of this mission were to test the lunar mapping and survey system in Earth orbit and to conduct various Earth and space science experiments. The lunar mapping and survey system was originally designed to complement the Lunar Orbiter and Surveyor missions by assisting in selecting Apollo landing sites. However, by mid-1967, the data obtained from these two missions was deemed sufficient, rendering the lunar mapping and survey system redundant. Consequently, NASA Administrator William Seaborn canceled this mission component in August.

Despite eliminating its principal experiment, AAP 1A continued to advance rapidly, drawing substantial support from NASA's scientific community. For the Office of Space Science and Applications (OSSA), AAP 1A represented a pioneering effort in crewed space science. One OSSA project manager, reflecting on an August briefing, expressed skepticism regarding the mission's justification, noting that "the justification for the mission appears to be the experiments and not crewed spaceflight." He commented, "A 14-day flight does not seem to be a cost-effective way of obtaining space data for the experiments selected." Nevertheless, AAP 1A generated considerable enthusiasm within OSSA, prompting significant efforts to accelerate developing experiment hardware.

By the latter part of 1967, AAP 1A was evolving into an Earth-resources mission equipped with specialized instruments to gather valuable data about the planet. This mission was set to carry half a dozen specialized cameras and four infrared sensors. The intricate process of mission planning was underway, involving coordination between the Houston-based Mission Control Center and the Denver facility of Martin Marietta, the contractor responsible for integrating the payload.

On August 25, the Manned Spacecraft Center (MSC) published preliminary designs for an experiment carrier. This carrier was to be installed in the spacecraft-lunar module adapter, positioned between the service module and the S-IVB stage. The carrier was designed to provide a shirtsleeve environment, ensuring that one astronaut could operate the instruments comfortably. Martin Marietta engineers devised a flight plan that included six passes over the United States each day at an altitude of 260 kilometers and an orbital inclination of 50 degrees relative to the equator.

Aside from its scientific objectives, AAP 1A held significant value for NASA in terms of training. NASA's tradition of progressively increasing mission complexity meant that starting with a relatively simple flight would allow focus on management and operational relationships. AAP 1A was intended to allow Martin Marietta to collaborate closely with principal investigators, Apollo contractors, and various NASA centers. Furthermore, if NASA decided to switch contractors for spacecraft modification and refurbishment, AAP 1A would be a useful training ground for the new firm.

By late October 1967, planning for AAP 1A was in full swing. On October 27, the Flight Operations Planning Group convened for its first meeting. On November 10, MSC released a detailed project plan, and ten days later, the Manned Space Flight Experiments Board approved ten experiments focused on Earth resources and meteorology. In mid-December, engineers gathered in Denver for a presentation by Martin Marietta on the experiment carrier.

However, just as momentum was building, the mission was abruptly canceled. At the end of the year, NASA Administrator James E. Webb announced the termination of AAP 1A due to financial constraints. This decision was met with considerable dissatisfaction within the Office of Space Science and Applications (OSSA), where it was perceived as further evidence of NASA Administrator Robert F. Kennedy's lack of interest in scientific endeavors. At the AAP office, the termination felt like a significant setback, rendering much of the previous work seemingly futile.

Despite the cancellation of AAP 1A, the effort was not entirely in vain. Three years later, when NASA revisited Earth-resources experiments, several sensors from the AAP 1A mission found their way into the Skylab program, contributing to future scientific endeavors.

Problems With the Cluster Missions

As 1967 progressed, technical and financial challenges mounted, significantly impacting the development of the Skylab Space Station. The tragic Apollo fire had once again highlighted the critical issue of micrometeoroid hazards, prompting urgent revisions in spacecraft protection strategies. By early 1968, engineers at Douglas Aircraft Company, the contractor responsible for the S-IVB stage, opted for a highly cautious approach. They decided to enhance the stage's insulation by covering it with aluminum foil and adding an external shield. This shield, designed as a thin aluminum sheet, was initially flush with the S-IVB's surface at launch. Once in orbit, it would extend 13 centimeters away from the tank. The intent was to allow small particles to lose most of their energy upon impact with the shield, thereby minimizing potential damage to the tank.

On February 27, 1968, Wernher von Braun presided over a crucial meeting in Huntsville to address the meteoroid issue. The meeting brought together engineers from four contractors and two Marshall Space Flight Center (MSFC) centers. They reviewed data, examined test films, and discussed the Douglas shield design. It was evident that the aluminum foil would suppress flame propagation, and the shield would significantly reduce the likelihood of severe penetrations. Martin Marietta engineers noted that the shield's design would also facilitate temperature control inside the workshop. Despite the shield's weight—estimated at 320 kilograms—and its cost of approximately $250,000, these were not considered significant obstacles. It was agreed that the shield should be adopted, although further studies would continue to test the effect of liquid hydrogen on the foil lining and explore alternative insulating materials. Contingency plans were also developed to apply insulation to the S-IVB's exterior if necessary.

Payload weight continued to be a persistent issue throughout the year. By early January, the weight of the AAP 4 payload, which included the lunar module equipped with solar telescopes, was nearing the Saturn IB's lifting capacity. To address this, mission planners increased the orbital altitude for the AAP 4 mission, reducing the allowable payload. The Mission Control Center (MSC) also imposed higher power requirements, necessitating a larger solar array. By mid-year, the experiments canister for the Apollo Telescope Mount (ATM) required an active cooling system due to excessive heat generated by two instruments, which risked distorting their optical axes. Simultaneously, heavy shielding was needed for film storage vaults to prevent radiation-induced fogging.

By late April 1967, all payloads except AAP 1 were overweight, and design modifications in the workshop also exacerbated weight issues for this mission. The wet workshop had been reinforced with rigid metal floors and walls, and the increasing number of experiments necessitated additional power, achieved by adding two sets of solar panels. The Apollo command module, undergoing extensive redesign following the fire, also accumulated weight, exceeding its design limit by 900 kilograms by mid-year.

Mission planning faced another significant challenge: the expanding list of experiments demanded more crew time than was available. A compatibility analysis conducted in late 1966 revealed that the total time required for the assigned experiments exceeded the available 288 man-hours. The director of flight crew operations in Houston voiced concerns over the excessive training time required for the experiments. In response, in February 1967, the experiments board determined that weight, power, and crew-time constraints necessitated a reallocation of experiments among the four AAP missions. This task required establishing a system of experiment priorities. George Mueller assigned this responsibility to Douglas Lord, who reported in July with a prioritization scheme accepted by the board with minimal changes. In addition to practical factors like weight, space, crew time, power consumption, and hardware availability, Lord's criteria included more subjective considerations such as "the value of the experiment to the overall national space effort," allowing for some flexibility in prioritization. By the end of 1967, Lord and Bellcomm—a consulting systems engineering firm affiliated with American Telephone & Telegraph and tasked with independent analyses for the Office of Manned Space Flight (OMSF)—had determined the relative priority of all approved AAP experiments. This development streamlined the process of assigning experiments to missions.

In early 1967, the Apollo Telescope Mount (ATM) program encountered significant challenges, creating a substantial backlog for three of the five planned experiments. These delays, exacerbated by persistent technical issues, caused considerable concern among program officials. By May, adjustments to the schedule had pushed the ATM mission to mid-1969. While this postponement alleviated some of the pressure on one experiment, it did little to mitigate the delays affecting the other two.

On July 13, 1967, representatives from the Harvard College Observatory and the Naval Research Laboratory reported that their instruments would be delivered far too late to meet the launch schedule. Compounding the issue, thermal control problems threatened further delivery delays. During an American Astronautical Society (AAP) review on July 18, various solutions were proposed to address these issues. However, extending the launch date to accommodate the delayed experiments was not considered a viable option. Frustrated by the lack of flexibility, the scientists voiced their concerns to higher management.

At the July Management Council meeting, NASA Administrator James Webb acknowledged the scientists' dissatisfaction and emphasized the importance of maintaining a positive relationship with the scientific community in the context of the Apollo Applications Program (AAP). Despite this, the council debated only two potential solutions: either proceed with the experiments that could be delivered on time or modify the requirements for the delayed experiments to expedite their development.

On July 27, 1967, a crucial meeting was held in Washington, D.C., involving key figures such as Mueller, principal investigators, and program managers from the Office of Space Science and Applications (OSSA) and the Marshall Space Flight Center. Harvard proposed a reduction in the complexity of its instrument to overcome production hurdles, asserting that a simplified version could still provide the necessary data if the launch proceeded as scheduled. Although Richard Tousey from the Naval Research Laboratory was absent, his representative hesitated to approve changes. However, when Mueller announced that a second ATM mission would be launched approximately a year later, allowing the Naval Research Laboratory to use its original instruments, the laboratory's representative was more receptive.

Upon Tousey's return, another meeting was convened. OSSA expressed concern that the push for a timely launch might compromise the quality of the scientific data. Tousey, in particular, was reluctant to simplify the Naval Research Laboratory's experiments and preferred to focus on completing one instrument to the original specifications, even if the other might be excluded if not ready in time. After extensive discussions and reassurances from the Office of Manned Space Flight (OMSF) about the subsequent solar astronomy mission, the Naval Research Laboratory agreed to accept some reduction in performance to proceed with both experiments. The revised plan was endorsed by OSSA's Space Science and Applications Steering Committee on August 14, 1967. While the scientists reluctantly accepted the compromise, the series of meetings underscored their growing perception that OMSF prioritized mission schedules over the quality of scientific research.

Throughout 1967, medical experiments also faced considerable delays. In April, the difficulties with medical experiments mirrored the broader challenges of the ATM program, reflecting the pervasive issues impacting the Apollo Applications Program's ambitious goals.

The Uncertainty years

In the summer of 1967, amidst mounting delays and escalating frustrations, Mueller prodded the Manned Spacecraft Center (MSC) to accelerate experimental hardware development. This push resulted from the $1.46 million allocated over the previous six months. However, MSC's commitment fell short, with only $876,000 earmarked and a meager $8,000 spent. The irony of the situation was not lost on those observing the proceedings. An internal MSC summary lamented that four major medical experiments were still in various stages of preliminary work, with some only nearing final contract negotiations. The critique of the alleged schedule slippage seemed misplaced, given the self-evident need to adjust the Apollo Applications Program (AAP) timelines. Indeed, as things stood, delivering prototype hardware for training by mid-June 1967 was an exercise in futility.

Despite the mounting challenges, Mueller remained steadfast in his resolve to proceed with the basic workshop and Apollo Telescope Mount (ATM) missions as planned. Doubts lingered about whether the wet workshop would ever see the light of day, but if Mueller harbored any reservations, he kept them close to the vest. Challenges were anticipated, but none had yet proven insurmountable. This resolute attitude was mirrored by the managers and engineers at Marshall Space Flight Center (MSFC), who were determined to uphold the center's "can-do" reputation. Even if some had private doubts about the feasibility of the AAP missions, they quickly suppressed them and focused on finding solutions.

In stark contrast, the Houston officials, who had long been skeptical, did not hesitate to express their reservations. Since March 1966, when Gilruth had articulated MSC's criticisms to Mueller, the Houston center had reluctantly participated in workshop planning. Perhaps emboldened by Administrator Webb's tepid support for the AAP, Houston officials were quick to highlight perceived flaws in both the concept and execution of the program. From the outset of the AAP office's establishment at MSC, the Houston-Huntsville relationship had been tense. Houston viewed itself as the bastion of practical, level-headed critique, while Huntsville saw these criticisms as mere roadblocks rather than cooperative efforts to address genuine issues.

This dichotomy underscored the complexity of the AAP's internal dynamics, where earnest efforts to address genuine concerns often collided with entrenched positions and institutional pride. As the program grappled with its myriad problems, the interplay of ambition, skepticism, and the quest for scientific and engineering excellence painted a vivid, albeit turbulent, picture of the era.

In the mid-1960s, NASA faced significant challenges in developing and executing its Apollo Applications Program (AAP). While the program's overarching goals were broadly supported, disagreements emerged over the specific approaches and configurations chosen to achieve these objectives.

By mid-1967, Bob Thompson and his team at the Manned Spacecraft Center (MSC) were concerned about the trajectory of the AAP. The transition from various loosely connected Earth-orbital missions to a more integrated workshop concept—often referred to disparagingly as the "Kluge" by Thompson—resulted in a configuration many believed was less than ideal. Thompson later remarked, "Had we started with a clean sheet of paper, we would never have done it that way." This sentiment encapsulated the frustration within MSC regarding the workshop's design and mission planning.

Thompson, with the approval of MSC Director Chris Gilruth, began to formulate a comprehensive critique of the AAP missions. This critique was part of a broader reassessment of the program. At MSC's Engineering and Development Directorate, efforts were underway to devise an alternative to the initially proposed "wet workshop"—a spacecraft repurposed from a spent rocket stage.

The critique of the AAP was not confined to the Manned Spacecraft Center. The Associate Administrator for Advanced Research and Technology expressed a lack of confidence in the program during discussions with NASA Administrator James Webb in August 1967. Scientific advisory groups also voiced substantial concerns. On April 11, 1967, the Science and Technology Panel of the President's Science Advisory Committee visited the mockups at Huntsville and evaluated the experiment plans. Some panel members criticized the allocation of experiment time, arguing that priority should be given to medical experiments essential for assessing human adaptability to space. They contended that other experiments should be excluded if they risked compromising these critical medical objectives.

Similarly, after a briefing in late June, the Space Science Board's Life Sciences Committee criticized the tight scheduling of crew activities. They felt this approach undermined the key benefits of crewed experiments, which required time for reflection, judgment, and creative problem-solving.

In late June 1967, Robert Seamans toured the MSC centers to assess the recovery from the Apollo fire. During his visit to Houston, he became aware of the skepticism regarding the AAP. On July 26, he inquired with Associate Administrator George Mueller about the validity of the program plans presented to Congress in May and the extent of the centers' involvement in their preparation. Having heard various complaints about the AAP plans, Mueller took the initiative to defend the program. In a detailed seven-page response, he asserted that all significant program decisions had been thoroughly coordinated with the relevant organizational elements, though this did not always ensure complete agreement. He outlined the extensive planning and review processes that had been applied, concluding that the program had achieved a stable and realistic stance given the existing budget constraints.

On August 29, 1967, Bob Thompson submitted recommendations to Charles Mathews for consideration in the next phase of AAP planning. Thompson agreed with the program's broad objectives but proposed that the sequence of missions be guided by mission complexity rather than by predetermined priorities, which he found debatable. His recommendations included delaying the workshop missions by one year, separating the telescope mount from the workshop missions, postponing it by more than a year, and concluding with a ground-outfitted "dry" S-IVB workshop launched on a Saturn V rocket. As an initial step, MSC proposed developing a small experiment carrier to fit within the spacecraft-lunar module adapter, capable of accommodating a range of experimental payloads.

In 1969, a significant milestone in space exploration was anticipated with the launch of a mission to incorporate leftover Apollo experiments and new Earth-sensing instruments. As part of a broader plan between 1969 and 1971, this mission represented the beginning of various exploratory initiatives designed to extend human presence and capabilities in space.

The planned activities for 1970 included two key missions of varying durations. The first was a 28-day workshop mission, followed by a more extended 56-day mission. Both missions focused on biomedical and engineering experiments to test human and mechanical endurance in the space environment. These workshops were designed to provide valuable data on how humans and technology perform in prolonged space missions, laying the groundwork for future space station operations.

A particularly noteworthy component of the space program was scheduled for 1971. This mission was set to involve the deployment of a telescope mount, planned for two to four weeks. During this mission, a lunar module and telescopes would be docked with the command module throughout the entire period. This arrangement was intended to maximize the observational capabilities of the telescopes and provide a comprehensive view of celestial phenomena, further advancing scientific knowledge.

The mission plan, as envisioned by the Manned Spacecraft Center (MSC), presented several advantages. It was designed to start with relatively simple missions, gradually evolving into more complex and extended missions. This progressive approach aimed to build on the experiences gained from each flight, adapting to new challenges and opportunities. The initial missions were expected to yield early results from Earth-resource experiments, providing immediate scientific and practical benefits.

A key element of the plan was removing the solar telescopes from the workshop, effectively discarding the cluster concept. This decision was driven by the need to address several critical issues, including payload weight, crew workload, and the specialized skills required of the crew. By simplifying operations and focusing on more manageable objectives, the plan aimed to streamline the mission's execution and enhance its overall feasibility. Additionally, the proposed schedule aligned well with the anticipated availability of Saturn IB rockets, facilitating the planned launch timeline.

Despite these thoughtful preparations, the response to the Headquarters proposal was less favorable than anticipated. At the time, James E. Mueller, an influential figure in the space program, expressed skepticism about the proposed plan. He directed George Mathews to address each of Houston's objections and to proceed with the plan as outlined. This directive led to continued alternate design studies, with informal reviews held to refine the mission approach.

On November 18, a general review of the Office of Manned Space Flight's (OMSF) future Earth-orbiting missions was convened. The MSC team was prepared to argue against the prevailing concerns. The review began with assessments by Mueller, Mathews, and George E. Disher, who all acknowledged existing problems but maintained fundamental confidence in the Apollo Applications Program (AAP) and its mission plans. Robert Gilruth, representing MSC, introduced the center's presentation, highlighting the complexities and uncertainties associated with the cluster missions. While no single issue was deemed insurmountable, the cumulative technical and managerial challenges posed significant concerns.

Bob Thompson took the floor to present a revised proposal from the Manned Spacecraft Center (MSC), offering an alternative to the previously under consideration wet workshop and orbital cluster concept. His plan introduced a smaller workshop module designed to be built within the spacecraft-lunar module adapter. This module would serve as a living and working space for astronauts, providing a more streamlined approach than the larger S-IVB workshop.

Although less than half the size of the S-IVB workshop, the proposed module was designed to accommodate a variety of specialized experiment modules. These could include solar telescopes or Earth-sensing instruments, depending on mission requirements. Despite its reduced size, the module was adequately spacious to house critical experiments and control panels for major external experiment packages. It was equipped with solar cells capable of generating up to 3.7 kilowatts of electrical power, ensuring a robust power supply for the experiments.

Thompson's presentation highlighted MSC's primary concerns and strategic objectives. Their flight schedule proposed 10 Saturn IB launches between 1969 and 1972, which matched the existing Apollo Applications Program (AAP) schedule. However, with the new plan, the need for a double rendezvous for the Apollo Telescope Mount (ATM) mission and the requirement for the lunar module were eliminated. This simplification promised to streamline operations and reduce complexity.

The first mission under this new plan was intended to focus on essential medical experiments and Earth-resources data collection. A subsequent mission would utilize the smaller workshop for up to 112 days of crewed operation by the end of 1970, further establishing human capability to function effectively in a zero-gravity environment. In 1971, the solar astronomy mission was scheduled to proceed without the medical experiments, eliminating the overlap and competition for crew time. This approach would allow for up to 112 days of orbital experience and 84 days dedicated to solar observations.

For 1972, MSC envisioned two additional missions, though specific experiment plans for these flights had not yet been detailed. Thompson's presentation also addressed critical concerns with the current program. He described the issues with the S-IVB workshop and the lunar module solar observatory as significant "warning flags."

One major concern was the complexity involved in stowing experiments in the multiple docking adapter before launch and then transferring them to the workshop in orbit. This design required equipment to be compatible with both storage and operational modes, adding layers of complexity. Furthermore, some medical experiments needed to be conducted in both the adapter and the workshop, complicating the mission logistics. Thompson pointed out that the initial four days of the mission, devoted to outfitting the workshop, were particularly problematic. This setup phase introduced additional tasks and interfered with crucial early medical measurements during the astronauts' adjustment to weightlessness—a period for which insufficient data existed.

Thompson also raised concerns about the feasibility of preflight testing for the cluster components and the adequacy of contingency plans in case of mission failure. He questioned how the cluster could be effectively tested before launch and the implications if the crew could not open or utilize the workshop. In contrast, the proposed smaller workshop could be fully tested in its flight configuration before launch and did not require in-orbit assembly, thereby reducing potential risks and operational complexities.

In the early days of the Skylab Space Station program, the management structure established by the Apollo Applications Program (AAP) was deemed cumbersome and inefficient. The structure created numerous interfaces between NASA centers and contractors, leading to complexities in coordination and oversight. Thompson, a key figure in the program, criticized this approach, arguing that it convoluted program management. He proposed an alternative plan from the Manned Spacecraft Center (MSC), which promised to streamline operations by assigning each center responsibility for a single mission and overseeing one primary contractor. This adjustment aimed to simplify program management, enhance flexibility, and foster better relationships between NASA centers and contractors.

The modified lab design would more effectively balance program objectives and improve mission execution by integrating full preflight assembly, checkout, and testing. Thompson recommended conducting an early study to reassess the entire cluster concept.

The following day, the team traveled to Huntsville to review hardware mockups and address objections. During this visit, the group meticulously discussed the so-called "warning flags" associated with the project. While the MSC representatives found these concerns troubling, officials in Huntsville maintained that most issues could be resolved through diligent resource application. Gilruth, representing the MSC, reiterated that his center's proposal would simplify the program without compromising its goals, making it more achievable with existing resources.

Despite Gilruth's reassurances, no other stakeholders saw significant advantages in switching to new hardware. Von Braun, a prominent wet workshop and cluster mission advocate, defended their feasibility and desirability. He argued that adopting an alternative approach would effectively constitute the start of a new program, potentially causing a delay of at least a year and wasting both time and money already invested. Kurt Debus, Director of the Kennedy Space Center, shared this sentiment, noting that a change in direction would result in inefficiencies without offering substantial benefits.

Mueller, another key decision-maker, raised critical questions regarding the feasibility and merits of the proposed approach. He asked whether compelling technical reasons render the current approach infeasible, if there were reasons that the Office of Manned Space Flight (OMSF) would not be proud of the results, and if the present strategy could not meet program objectives. Responses from those present were unanimously negative, indicating no significant drawbacks to continuing the current approach.

Mueller concluded that the primary challenges lay in providing adequate manpower across the three centers and ensuring effective coordination of efforts. He tasked Charles Mathews with balancing the workload to adequately support the cluster program.

Upon returning to Washington, Mueller briefed Robert Seamans on the weekend's discussions. Following further consultations among NASA's senior managers over the next week, Webb initiated a comprehensive review. This review, held on December 6th, examined the launch schedule, orbital configurations, and the anticipated contributions of each mission to the overarching goals of crewed spaceflight.

Following the preliminary proposal from the Manned Spacecraft Center (MSC), a comprehensive review was conducted to assess the technical challenges associated with the workshop and solar astronomy missions proposed for Skylab. This review scrutinized twenty-one concerns from the Houston center, categorizing them into two primary groups: issues deemed resolved or straightforwardly addressable and major unresolved problems common to current or alternative approaches. Only four concerns were identified as specific to the current cluster configuration, presenting unique challenges.

The evaluation included a detailed examination of the Apollo Applications Program (AAP) experiments, mission plans, and hardware components. Concluding this thorough assessment, the Office of Manned Space Flight (OMSF) echoed the sentiments previously articulated by George Mueller at Huntsville. The review's final determination affirmed that the current approach to the AAP was feasible and should proceed as planned.

George Mueller had already reached his decision before Webb's scheduled briefing. On December 1, he distributed John Disher's notes from the November 18-19 discussions and a draft letter indicating no substantial reason to abandon the wet workshop concept. Mueller's stance was clear: "I have decided that we should continue with the present AAP approach and request that you proceed accordingly in implementing AAP requirements."

Initially, Wernher von Braun expressed considerable frustration with the sudden objections from Houston. He supported Mueller's decision but criticized MSC for its delayed objections, raising several concerns about mission AAP 1A and the Apollo spacecraft that Houston had overlooked. Von Braun's initial response was harsh, disparaging MSC's alternative proposal and its perceived risks. However, before his letter could reach Washington, it was recalled and replaced with a more conciliatory version. In this revised communication, von Braun acknowledged MSC's concerns as part of a logical progression from Gemini and lunar missions. He suggested that adjustments to long-term AAP plans could potentially reduce the costs of the first cluster mission below the 1967 estimates. Von Braun also encouraged Mueller to initiate studies for subsequent missions, noting that a dry Saturn V-launched workshop could be an impressive candidate for future endeavors. He enclosed summaries of Marshall's studies that addressed MSC's concerns point by point.

In his response to Mueller's letter, Robert Gilruth sought to reiterate MSC's core arguments. He disagreed with adhering to the current plans unless they could be unequivocally proven inadequate. Gilruth argued that the program should be optimized within the constraints of available Congressional funding, pointing out that Congress had not explicitly approved the cluster concept or the proposed missions. Given the reduction in AAP funding for fiscal year 1968, a thorough review was deemed necessary, even if it meant potential program content and schedule changes. MSC advocated for a complete program review rather than endorsing a specific alternative configuration.

Gilruth also submitted a separate letter containing eight pages of comments on Disher's notes, presenting the central question bluntly: why persist with a complex approach when simpler alternatives existed? He conceded that while fireproofing the S-IVB insulation was possible, it would be unnecessary if a ground-fitted vehicle were used. Acknowledging Marshall's admission of difficulties in making the S-IVB habitable due to its conversion from a liquid hydrogen tank to a living compartment, Gilruth underscored that many issues deemed solvable in the current approach were not required in the alternative approach—this lengthy critique distilled into a single theme: the preference for a simpler, more straightforward solution.

The briefing on December 6 appeared to reinforce Webb's long-standing reservations about the AAP, underscoring the ongoing debate surrounding the program's direction.

On January 6, 1968, James E. Webb requested Floyd L. Thompson, the director of Langley Research Center, to chair a review of alternative possibilities for post-Apollo crewed spaceflight. This committee comprised high-ranking officials from across the space agency, including directors from the three OMSF field centers—Langley, Lewis, and the AAP office at Headquarters. Thompson, a venerable figure in aeronautical and space research, was preparing to conclude a distinguished 42-year career with the National Advisory Committee for Aeronautics (NACA) and NASA. His deep experience included serving as associate director at Langley when the Space Task Group, led by Robert Gilruth, was established in 1958. Thompson had also recently led the investigation into the Apollo spacecraft fire, adding to his reputation for thoroughness and expertise.

Under Thompson's leadership, the Post-Apollo Advisory Group convened four times between late January and late March 1968, visiting each crewed spaceflight center. Their findings indicated a clear consensus on the future of crewed spaceflight: the next logical step was to enhance human participation in orbital science. Between 1971 and 1975, they proposed several objectives, including assessing human performance in zero gravity for 100 to 200 days, evaluating the necessity of artificial gravity, and advancing the technology required to support human life in space.

The committee found that the goals of the early Apollo Applications Program (AAP) missions were generally aligned with post-Apollo needs but critiqued the program for its broad and unfocused approach. They concluded that the wet workshop concept was only marginally sufficient for obtaining essential data on human adaptation to weightlessness. The report suggested that if unresolved difficulties persisted with the current near-term approach, it might be prudent to develop plans to assemble the workshop on the ground and launch it dry using the more expensive Saturn V rocket, even if this meant accepting schedule delays.

Thompson recognized the value of the committee's work in prompting a critical review of the AAP plans by both Houston and Huntsville centers, fostering a program that both centers could support. However, the report highlighted the growing discord between the centers over the wet workshop concept.

In addition to Thompson's committee, two other groups were established within the Office of Manned Space Flight to scrutinize specific aspects of the AAP. One group, led by George Mueller, focused on reviewing solar astronomy missions. At the same time, another, headed by Douglas R. Lord, deputy director of the Advanced Manned Missions Program, was tasked with defining two versions of a ground-outfitted "dry" S-IVB workshop. To be launched on a Saturn V rocket, this dry workshop was considered a potential follow-on mission after the wet workshop.

Lord's group, which included 150 individuals organized into six task teams, evaluated thirteen options ranging from a simple conversion of the wet workshop to a highly advanced workshop with extensive experiments. The cost estimates for these options varied significantly. The least expensive option offered minimal scientific return and advancement and was estimated to exceed current plans by $412 million. Conversely, the most advanced workshop, promising the highest scientific return, would exceed the launch date by more than a year and cost an additional $2.3 billion. The findings indicated that none of the proposed options were ideal, as they would compete with the cluster missions for funding. Additionally, the wet workshop's significant budgetary demands for fiscal years 1969 and 1970 complicated the financial planning for a potential dry workshop.

As a result, discussion of the dry workshop as a follow-on project diminished in official correspondence throughout the remainder of 1968. This shift in focus reflected the financial and logistical challenges of pursuing the dry workshop.

Meanwhile, Mueller's LM-ATM Evaluation Board critically examined the astronomy module and mission, addressing concerns raised by OMSF, particularly from Houston. Although Mueller was engaged in budget hearings and the board's meetings did not commence until early March, the evaluations revealed that the experiments were well-prepared. The primary concern was the escalating cost of adapting the lunar module for its new role. The modifications required to support the crew during rendezvous and docking, an area of recurring criticism from MSC, proved costly.

Ultimately, Mueller conceded to the necessity of changes in the mission approach. Manned double rendezvous and docking were replaced by automatic rendezvous and remote-controlled docking. This adjustment was acceptable to Houston, as it simplified operational complexity and reduced crew hazards, aligning with the preference for a less risky and more manageable technique.

As NASA grappled with the future of its space programs during the late 1960s, budgetary constraints became a defining factor in the direction of the Apollo Applications Program (AAP). The fiscal climate was increasingly harsh, with Congress tightening its grip on the federal purse strings. NASA's budget request for Fiscal Year 1969 was the smallest since 1963, reflecting a significant contraction in available funds.

In the early 1960s, NASA had anticipated that the decline in Apollo program costs—d nearly $1 billion over two years—might provide financial relief for the AAP. However, the tumultuous passage of the Fiscal Year 1968 budget had diminished those expectations. Consequently, the FY 1969 budget proposal for the AAP amounted to $439 million, a 16% reduction from the previous fall's submission and less than half of what had been anticipated in the earlier budget.

The proposed budget allocated nearly half of the $439 million to new launch vehicles and spacecraft modifications, with another 40% earmarked for experiments. Notably, the funding did not include any new spacecraft and the number of Saturn IB and Saturn V rockets had been scaled

back to just two of each—far short of the ambitious goals initially set by NASA's Associate Administrator for Manned Space Flight, George E. Mueller. The plans for additional missions were scrapped, such as a second wet workshop, a second solar observatory, and the earth-resources mission (AAP 1A). The envisioned lunar exploration missions were also abandoned.

During Congressional hearings, a potential new direction for AAP emerged, focusing on a Saturn V workshop. NASA Administrator James E. Webb testified that the wet workshop was envisioned as a preliminary step toward a larger Saturn V-based facility. Mueller echoed this sentiment, suggesting that the planned workshop aligned with the mid-1970s orbital station concept proposed by the President's Science Advisory Committee in February 1967. The AAP's revised budget included $70 million allocated for initial work on this Saturn V station—more than double the funds designated for the wet workshop or telescope mount and exceeding the budget for Saturn V production.

The Congressional response to AAP's budget request was tepid and fraught with skepticism. Representative William Ryan of New York, a vocal critic of NASA, questioned the overlap between AAP and the Air Force's Manned Orbiting Laboratory (MOL). Webb's assurances that there was no duplication did little to placate Ryan, who remained convinced of the redundancy. Ranking minority member James Fulton of Pennsylvania expressed concerns about the need for additional Saturn rockets, given the expected surplus from the Apollo program. Fulton also criticized the perceived lack of clarity in AAP's objectives. He questioned specific details of the wet workshop's design, including a $300,000 shower, suggesting a need for a comprehensive review of the entire program.

Despite these criticisms, most Congressional Space Committees recognized that further budget cuts could jeopardize NASA's operations. Representative Emilio Daddario of Connecticut warned that additional reductions could force NASA out of business. Supporters argued that the practical benefits of space exploration were not adequately communicated to the public. In a discussion with Wernher von Braun, Daddario highlighted the challenge of demonstrating the value of the technology developed through NASA's programs. Senate committee members, including Clinton Anderson and Margaret Chase Smith, echoed this sentiment, stressing the need to articulate each proposed project's benefits clearly.

As the hearings progressed, external events further eroded NASA's tenuous support in Congress. The Tet Offensive in February had intensified the Vietnam War, increasing military expenditures, while riots following the assassination of Martin Luther King Jr. spurred demands for greater domestic spending. With Congress preoccupied with contentious debates over a proposed income-tax surcharge and demands for substantial cuts in non-defense spending, the future of post-Apollo programs, including AAP, remained uncertain and vulnerable to further scrutiny.

By March 1968, the outlook for NASA's budget had become increasingly bleak, with uncertainties about the extent of the impending cuts. Amid this troubling atmosphere, the Management Council and Apollo Applications Program (AAP) managers convened at Kennedy Space Center on March 21 to review the program's status in light of recent studies. At that time, the AAP comprised three missions utilizing five Saturn IB rockets. These missions included two flights dedicated to the wet workshop, which aimed to establish a space-based laboratory and conduct various experiments. Initially slated for 1969, these launches were delayed until the latter half of 1970. The first of these missions, AAP 1, was to be followed by AAP 2, a 28-day mission focused on setting up the workshop and initiating various experiments. A second crew was scheduled to follow three months after the initial crew's return for a 56-day biomedical mission, known as AAP 3A.

The final two launches, anticipated for mid-1971, were planned to transport the Lunar Module-Atmospheric Test Module (LM-ATM) and its crew on missions AAP 3 and 4. These missions were primarily dedicated to solar observations. Despite the lack of specific details for additional flights, there was consensus at the meeting on the need to explore earth-oriented experiments as potential additions to the program. Furthermore, it was decided to construct a duplicate workshop as a backup measure. There remained some discussion of developing a Saturn V dry workshop as a follow-up to the cluster, but it was agreed that such a project would require insights gained from the wet workshop and that the AAP budget could not support both.

On March 21, Mathews addressed AAP officials and emphasized the need for a strategy to decelerate AAP activities. This strategy aimed to minimize current expenditures while deferring new commitments, all while maintaining the ability to proceed with the program. He warned that this adjustment period could last several months, as spacecraft modifications—an essential pacing item—could not commence until the fiscal year 1969 budget was finalized. In the interim, centers were advised to synchronize all AAP projects to a uniform stage of development.

In April, NASA Headquarters worked out a holding plan for the remainder of 1968. This plan imposed substantial reductions, causing several contracting issues but successfully reducing AAP expenditures by over 50%. The situation took a turn for the worse on May 3, when the House of Representatives sharply cut the AAP authorization. During a two-hour debate, Representative Olin Teague defended the agency's budget, including the $395 million his committee had recommended for AAP. However, Republican representatives, while supportive of space activities, favored reductions in certain programs. An amendment proposed by Representative James Fulton to cut AAP funding to $253.2 million—equivalent to the previous fiscal year's funding—passed by voice vote. Many representatives, including Donald Rumsfeld, expressed regret but deemed the cuts necessary in favor of higher-priority programs.

The Senate space committee deemed the House cuts excessive, recommending $350 million for Apollo Applications. However, a proposal by William Proxmire to slash NASA's authorization by $1 billion narrowly failed by five votes. Ultimately, Congress agreed on a revised authorization figure just above $4 billion, with

$253 million allocated to Apollo Applications—about 60% of the requested amount.

NASA's authorization was still subject to the Revenue and Expenditure Control Act, which mandated a $6 billion reduction in authorized funds. The impact on NASA and AAP remained uncertain. On June 20, the AAP office submitted a program plan based on $119 million in new funding. This financial constraint led to the shutting down of Saturn's production lines. Webb instructed his management team to refrain from finalizing any AAP contracts until it was clear that the program would proceed cohesively and consistently.

The Apollo Applications Program appeared to be disintegrating under the weight of budgetary constraints. Various cost-saving measures were considered, including eliminating continuous workshop occupation, reducing the number of experiments, and simplifying experimental equipment. Concerns from solar scientists about participating under these new conditions led to discussions about potentially canceling solar experiments altogether.

The schedule for the telescope mount had caused discontent among principal investigators from Harvard College Observatory and the Naval Research Laboratory, who could not meet the 1969 launch date. Mueller temporarily resolved the issue by promising a second solar mission and negotiating a simplified instrument design with the scientists. However, subsequent funding cuts further jeopardized the program's future.

On May 16, 1968, Leo Goldberg, director of the Harvard College Observatory, informed Harold Luskin, the newly appointed director of the Apollo Applications Program (AAP), of his intention to discontinue the development of instrument HCO-C, a scanning ultraviolet spectrometer designed primarily for studying large solar flares during the solar maximum. Goldberg proposed reinstating the original ultraviolet spectroheliometer, HCO-A, instead. In response, Luskin directed Goldberg to continue working on HCO-C, given that the Apollo Telescope Mount (ATM) was scheduled for launch in June 1971. Luskin's directive was based on the current plans to cease funding for HCO-A, although he promised that NASA would review Harvard's proposal in June.

Goldberg's frustration with the situation became apparent when he expressed his discontent in a letter to John Naugle, the head of the Office of Space Science and Applications (OSSA). Goldberg reminded Naugle that Harvard had agreed to fly the simplified HCO-C instrument as a favor to NASA, contingent upon two conditions: that Harvard would be permitted to fly its original HCO-A instruments on a second ATM mission and that the first ATM mission would launch in 1969. However, with the ATM launch delayed until 1971, two years beyond the solar maximum, Goldberg argued that the simpler HCO-C instrument had become less valuable. He suggested that with the project's schedule slipping further, the HCO-C might no longer be necessary, and the first mission might be better off without it. Goldberg noted that the ATM project consumed many of the observatory's resources, leaving little room for other scientific pursuits. He urged that astronomy and NASA would suffer if action were not taken to address these issues.

In response to mounting pressure and dissatisfaction among scientists, OSSA supported the scientific community's concerns more. After NASA administrator James E. Webb made several concessions to the ATM experimenters, Luskin agreed to halt work on HCO-C and allowed Goldberg to proceed with HCO-A.

Despite this resolution, the project continued to face challenges. In July 1968, Webb announced that NASA could no longer afford to fund the ATM project, leading to its removal from the Fiscal Year 1969 budget. Webb opened a review of the situation on August 5, citing NASA's strained financial situation and the uncertain status of the appropriations bill, which had yet to clear the Senate. Until the budget was finalized, Webb set a spending limit of $3.8 billion and warned of a "course of peril." During this period, various spokespeople defended the ATM project, emphasizing its merits and commitments to external groups. Naugle commended Marshall's project leadership, noting no major unresolved technical issues.

Floyd Thompson emphasized the technological significance of the Apollo Telescope

Mount (ATM) program, asserting that advancements such as the control moment gyroscopes made the mission valuable even if its scientific outcomes were less impactful. Edward G. Gibson, an astronaut and physicist, believed that the ATM provided a unique opportunity for observers to apply their judgment in enhancing the quality of space science. This perspective played a role in convincing NASA Administrator James E. Webb to continue supporting the ATM project, though Webb remained concerned about securing congressional backing.

In response to these concerns, the Fiscal Year 1969 operating budget was adjusted to allocate $50 million for the continued development of the ATM. The Apollo Applications Program (AAP) commenced the new fiscal year on July 1, 1968, under stringent financial constraints. During this period, most work was conducted monthly through letter contracts. These provisional agreements, typically used to initiate projects while awaiting formal contracts, became more common due to the uncertain climate of 1967 and 1968. NASA preferred to avoid letter contracts due to their limited scope and compensation terms, but they became a necessary tool under the prevailing circumstances. By October 1968, there were 15 letter contracts covering various AAP projects, including the airlock, Saturn IBs, and payload integration.

James E. Webb had never been particularly enthusiastic about the AAP. As the decade ended with budget constraints tightening, he became increasingly focused on ensuring the success of the Apollo lunar landings. Following the tragic Apollo spacecraft fire, Webb concentrated his efforts on Apollo's success, reprogramming AAP funds to meet the immediate needs of the Apollo program. By 1968, Webb was "putting strong impedance in the system," delaying AAP procurements unless deemed critically urgent. He characterized the AAP as merely "a surge tank for Apollo," reflecting his prioritization of lunar missions over other projects.

On September 16, 1968, Webb announced his decision to retire early in October, with his deputy, Thomas O. Paine, set to assume the role of acting administrator. Paine, who had spent 19 years in scientific and administrative roles at the General Electric Company before joining the government in January 1968, shared many of Webb's views about NASA's future but had a stronger interest in post-Apollo programs. On October 4, Paine informed his staff that the AAP could proceed with renewed confidence. He advocated for solidifying the program by negotiating definitive contracts and encouraged his team to "look for all ways to move faster."

The October 1968 flight of Apollo 7, which marked an 11-day mission in Earth orbit, was crucial in restoring public confidence in NASA and redeeming the spacecraft manufacturers after previous setbacks. The mission, notable for its technical success and operational smoothness, revitalized the space program's image and reaffirmed the public's faith in human spaceflight.

However, the excitement surrounding Apollo 7 was soon overshadowed by the drama of Apollo 8's Christmas mission to the Moon. This historic flight, which took place in December 1968, was the first crewed mission to orbit the Moon, captivating the nation and rekindling enthusiasm for space exploration. The success of Apollo 8 not only renewed public interest but also solidified the prospects for future space endeavors.

As 1969 began, Congress greeted NASA's budget with enthusiasm reminiscent of the early 1960s. Yet, the optimism was tempered by the reality of funding constraints. In April 1969, the Nixon administration reduced the Apollo Applications Program (AAP) request by $57 million, leaving $252 million for a scaled-back program. This cutback was a significant blow to NASA's plans, which had already faced funding mismatches and scheduling challenges.

In the fall of 1968, as NASA projected its fiscal year 1970 requirements, it became evident that the allocated funds fell short of what was needed. The Marshall Space Flight Center, responsible for Apollo Applications, received only about two-thirds of the required budget. The Manned Spacecraft Center (MSC) situation in Houston was even more dire; Director Chris Kraft estimated that AAP needed 75% more than allocated. This disparity made continuing a dry workshop mission—a key component of the Apollo Applications Program—seem increasingly untenable.

In September 1968, NASA Administrator James E. Webb expressed his concerns about the future of the post-Apollo program. Webb suggested that the financial limitations might force NASA to reconsider its plans. He mentioned that after completing the initial three missions, NASA would need to evaluate its next steps, which could involve either an interim Saturn V workshop or a multi-man space station, depending on the available funding.

When Thomas Paine assumed the role of acting NASA Administrator, he also voiced support for the space station concept. However, this idea received little backing amidst the prevailing budgetary constraints. President-Elect Richard Nixon established a task force on space policy, which in January 1969 recommended against committing to a large space station. In February, Nixon appointed a Space Task Group, led by Vice-President Spiro T. Agnew, to conduct a more detailed review and provide recommendations by September.

NASA's interest in a dry workshop briefly revived in early 1969, fueled by the success of Apollo 8 and the potential of the Saturn V rocket's payload capacity. John Disher proposed utilizing a Saturn V instead of a Saturn IB for the cluster missions to reduce costs by launching all modules simultaneously. Disher acknowledged that this change would introduce numerous technical and administrative challenges despite the potential advantages. Concerns about increased costs, disrupted schedules, and potential negative impacts on morale and support led the Management Council to shelve the proposal.

As the new administration took its time to decide on NASA's long-term future, the space program faced a period of uncertainty and transition. The challenges of funding and program management loomed large, shaping the trajectory of NASA's post-Apollo initiatives and the evolution of space exploration.

In late April 1969, NASA's Deputy Administrator, Dr. Robert C. Mueller, presented a sobering assessment to the Senate Space Committee. He suggested that the planned progression from a wet workshop to a dry workshop and eventually to a space station seemed "inefficient and only marginally effective in advancing space technology." Mueller advocated for a more ambitious approach, proposing that the next step in Earth-orbiting crewed space flight should be a new, semi-permanent space station coupled with a low-cost, reusable spacecraft for transportation between Earth and orbit. This vision aimed to establish a sustainable infrastructure for space exploration rather than continuing with incremental and somewhat disjointed missions.

The Apollo Applications Program (AAP) was set to commence its cluster missions in late 1971 and conclude sometime in 1972. By the mid-1970s, NASA anticipated placing the first module of a space station into orbit. Over the following decade, this modular station would be expanded to its full operational size, representing a significant leap forward in space technology and capabilities.

Despite these advancements, the dry workshop concept did not entirely disappear. At the Marshall Space Flight Center, Dr. Wernher von Braun remained a proponent of the idea. He was concerned about the potential loss of the cluster missions and the technical difficulties associated with weight and stowage that continued to challenge the program. At the Manned Spacecraft Center (MSC) in Houston, Max Faget was aware of ongoing discussions about potentially switching to a Saturn V rocket for the missions. He argued that operating both a wet and a dry workshop would waste valuable resources and suggested reconsidering the dry workshop might be worthwhile. Despite this, the AAP management team, including Lee Belew, did not favor altering the program. Belew cited significant reasons for maintaining the current core program, and after examining the technical issues facing the wet workshop, he found no justification for a switch.

Within NASA Headquarters, there was a noticeable tension between those favoring a shift to a dry workshop and those opposed. Mueller seemed increasingly inclined toward the dry workshop, but the AAP staff remained resistant. In discussions with William C. Schneider, the new program director, Belew sensed considerable pressure to embrace the change. Schneider was concerned that transitioning to a dry workshop would incur similar costs and delay the first launch by at least a year. It was clear that Mueller was exploring alternative approaches, seeking a path to align with the evolving goals and constraints of the space program.

Belew, while not entirely opposed to the change, reminded von Braun of the substantial implications such a shift would entail. He highlighted that all existing contracts would need to be rewritten and renegotiated, Grumman's work on the Lunar Module-Aircraft Test Module (LM-ATM) would be terminated, and Marshall's manpower assignments would have to be entirely reallocated. The logistical and administrative challenges posed by such a transition were considerable, underscoring the complexity of NASA's decision-making process during this pivotal period in the space program's history.

In the early 1970s, the ambitious Apollo Applications Program (AAP) faced mounting challenges as it sought to transition from the initial wet workshop concept to a more viable space station design. This evolution was spearheaded by key figures, including Thomas O. Paine, who had assumed leadership of the Apollo Applications Program in December 1968 following a distinguished tenure as an Apollo mission director.

Born in New York City and educated at the Massachusetts Institute of Technology (MIT) and the University of Virginia, Paine had a deep background in aerospace. His career began at the National Advisory Committee for Aeronautics (NACA), NASA's predecessor, in 1949. With significant contributions to the Gemini program and pivotal involvement in Apollo 8, Paine earned NASA's highest honor, the Distinguished Service Medal.

As the Apollo Applications Program advanced, it became increasingly clear that the wet workshop approach—where a Saturn V rocket's upper stage would be converted into a space station—was fraught with practical difficulties. The weight and complexity of such a design proved daunting. By May 1974, under the persistent scrutiny of NASA Administrator James E. Webb, it was evident that a shift in strategy was imperative. Webb was convinced that the wet workshop was impractical and that transitioning to a dry workshop concept would be crucial for the program's success.

During the weekend of May 3-4, 1974, Webb unveiled a revised plan to center directors and program managers. This new vision proposed an integrated program to transition from Apollo Applications to a space station, including three dry workshop flights between 1972 and 1974. These flights were expected to provide essential data for the space station's design. The plan charged Thomas Paine with outlining the necessary actions to pivot the core program to accommodate this new approach.

When the team reconvened the following week to review Paine's findings, the discussion highlighted significant implications. Marshall Space Flight Center (MSFC) indicated a probable 18-month delay in the first launch. The potential shift to a dry workshop raised concerns at the Manned Spacecraft Center (MSC), which argued that this new program would only compete for funds with ongoing wet workshop and shuttle studies.

The deliberations led to the proposal of a modified study, which focused on using a dry workshop solely for the second phase of AAP and then implementing an improved dry workshop through subsequent missions. Paine was tasked with detailing specific technical points for evaluation and requested a comprehensive report by June 15 on the impact of transitioning to a dry workshop.

By May 15, 1974, preliminary findings from MSFC were reported. The analysis suggested that adopting a dry workshop with an integral Air Traffic Management (ATM) system would likely result in a 10-month delay. The cost of two complete sets of flight hardware was projected to range from $50 million to $100 million, with additional expenses anticipated for checkout, launch, and mission operations. A swift decision from NASA Headquarters was crucial, as Belew noted that it would need to be made within 4 to 6 months to maintain program momentum.

While acknowledging that the dry workshop approach offered greater reliability and confidence of success, Belew cautioned against further changes. The AAP had already been beset by numerous modifications to experiments, mission plans, and program objectives, leading to escalating costs and delays. The Saturn V rocket's substantial payload capacity had initially attracted interest, but the associated delays and complications made its full utilization problematic.

Ultimately, Belew suggested an alternative approach: retaining the core program but shifting to a dry workshop with minimal alterations. This adjustment was seen as a way to meet the current schedule without extensive redesigns, addressing some of the substantial issues linked to using the Saturn V's upper stage as a propulsive component. Although it meant not fully leveraging the Saturn V's payload capacity, this strategy promised to simplify implementation and reduce the likelihood of further delays.

On May 21, 1969, while discussions were ongoing about the implications of transitioning to a dry workshop for the Apollo Applications Program (AAP), NASA Administrator Thomas O. Paine presented four alternative plans to the Management Council. These alternatives were proposed as potential replacements for the original wet workshop concept, and facing increasing scrutiny due to its technical challenges.

The four options varied primarily in terms of rocket launches and module configurations. Alternatives 1 and 2 required using a Saturn V rocket to launch the space station's cluster module and a Saturn IB to transport the crew and the telescope mount into orbit. In contrast, Alternatives 3 and 4 involved launching both the telescopes and the cluster modules with a Saturn V, while the crew would be sent up separately on a Saturn IB. Each alternative included a choice between utilizing an Apollo Applications Program (AAP) command and service module, which could operate for up to 56 days using its fuel cells, or a quiescent command and service module (CSM), which would be powered down after docking. The quiescent module's fuel cells would provide enough power to maintain critical systems and enable a rapid return to Earth if necessary.

Alternatives 1 and 4 emerged as the leading candidates in the ensuing discussions. Alternative 1, which represented the most straightforward adaptation of the dry workshop concept with minimal changes, was technically less advanced but required fewer program adjustments. This made it a more appealing option due to its lower complexity and fewer disruptions. Alternative 4, while technically superior and offering greater potential for incorporating additional experiments, demanded substantial changes to represent a new start for the program effectively.

Mueller instructed the centers to report their evaluations and recommendations immediately. Over the following week, the initial reactions solidified into clear preferences. Field centers and NASA Headquarters preferred Alternative 4 based on its technical merits. Still, they recognized Alternative 1 had a better chance of gaining acceptance from Paine and Congress due to its more straightforward implementation.

Wernher von Braun, a key figure in developing NASA's space programs, initially reiterated his belief that no significant changes were necessary. He argued that the wet workshop concept needed more rigorous refinement to get it back on track. However, von Braun acknowledged that the dry workshop presented advantages, including experiments previously constrained by weight and volume limitations. Despite his technical preference for Alternative 4, von Braun expressed concern that such a major overhaul could invite unwelcome scrutiny from external authorities, complicating the approval process.

On May 27, 1969, Thomas Paine reviewed the detailed analyses of the four Saturn V-based alternatives presented by Paine. Alternative 1, favored by the centers, was projected to be approximately $50 million less expensive over the entire program than Alternative 4. However, in the crucial fiscal years from 1969 to 1971, Alternative 4 demonstrated a significant $200 million advantage. This financial advantage highlighted the complexity of the decision, balancing cost considerations against the potential for enhanced program capabilities.

As the Apollo Applications Program (AAP) progressed, evaluations of various approaches underscored the exceptional potential of the Saturn V rocket system for achieving its objectives. Among the options considered, the Saturn V, with a quiescent spacecraft, emerged as the most promising, particularly for its ability to deliver abundant solar data. This configuration quickly became the favored choice of the Office of Manned Space Flight (OMSF).

Recognizing the urgency of advancing the Apollo Applications program toward a feasible goal, NASA Administrator Thomas Paine endorsed the Saturn V-launched dry workshop as the most viable solution. On this basis, Paine communicated to Senator Anderson that NASA was exploring the feasibility of using a Saturn V to launch both the dry workshop and the Apollo Telescope Mount (ATM). This exploratory phase necessitated temporarily suspending certain contract actions pending final decisions.

In preparation for a crucial July Management Council meeting, NASA's chief of the Apollo Applications Program, William Schneider, notified the center program managers on June 17th to prepare a proposal for the dry workshop. Schneider anticipated Alternative 4, the dry workshop option, would be the sole focus of the upcoming discussions. Consequently, he arranged a review session at the Marshall Space Flight Center on June 19th and scheduled meetings with key AAP contractors during the week of June 23rd.

Marshall Space Flight Center had already initiated a study on the dry workshop configuration, which would not fully complete until the end of June. However, by June 19th, substantial preliminary data was available. The study favored a configuration where the ATM was positioned ahead of the multiple docking adapter on a hinged structure, extending the instruments outward by 90 degrees. This design received swift acceptance, which yielded favorable projections regarding cost, schedule, and mission success.

During the contractor meetings, a consensus was readily achieved to target a launch date of July 1972 for planning purposes. By the end of June, the decision to abandon the wet workshop had effectively been finalized, with formalization following shortly after that. On July 18, 1969, Paine signed the project approval document reflecting the change.

Meanwhile, developments within the Pentagon had further reduced the likelihood of congressional opposition to the change. On June 10, the Defense Department announced the termination of the Manned Orbiting Laboratory (MOL). Despite having invested $1.3 billion in the program, the escalating costs, projected to reach $3 billion, along with technological advancements in uncrewed satellites for communications, meteorology, and observation, rendered the MOL obsolete. The cancellation thus marked the end of the Air Force's aspirations for crewed spaceflight and concluded a decade of intense political competition.

With the MOL's cancellation, attention turned to ensuring the availability of the Saturn V rockets for Apollo Applications. Although planners had long assumed the Saturn V would be available, and missions Apollo 8, 9, and 10 had nearly eliminated doubts, it was prudent to avoid suggesting that the Saturn V might be reassigned until the lunar landings were completed. Therefore, public announcement of the program changes was deferred until Apollo 11 was en route home.

The decision to secure Saturn V rockets for the Apollo program had been a hard-fought victory. NASA Administrator James Webb had exerted considerable influence to ensure that at least 15 Saturn V launches were dedicated to Apollo, firmly opposing any notion of reallocating these rockets. In November 1966, a national magazine quoted Wernher von Braun suggesting that a fourth Saturn V might be dispatched to the moon by 1968. Still, Webb promptly directed von Braun to retract this statement. Until the success of the Apollo lunar missions was assured, Webb's administration maintained that the Saturn V rockets would remain committed to Apollo.

On July 22, 1969, just two days after the historic first lunar landing, NASA issued formal directives to implement the dry workshop program for the Apollo Applications Program (AAP). This decision significantly shifted the program's approach and initiated various strategic actions.

William Schneider, overseeing the transition, issued specific instructions regarding contract modifications. The letter contract with Grumman, which had been focused on lunar module modifications, was to be terminated. Similarly, the subcontract with Allis-Chalmers for producing cryogenic tanks for the command service module was also canceled. Negotiations with North American on modifications to the command-service module (CSM) were to be suspended, and work on these modifications was to be limited while a revised proposal was developed. At Marshall Space Flight Center, amendments were to be made to McDonnell Douglas's contracts concerning the workshop and airlock, redirecting their efforts toward the dry workshop configuration. Additionally, the impact on all existing experiments was to be reviewed, and necessary adjustments made accordingly.

To streamline the transition and avoid a proliferation of changes, Schneider established a guiding principle: "The basic objectives, tasks, experiments, and mission durations will remain unchanged. Only those changes that were dictated by the configuration modification to the dry workshop were authorized. All other desirable but not required changes would be discouraged, and final disposition would be on specific merits." This directive aimed to maintain program focus and efficiency during the transition.

The decision to adopt the dry workshop was generally well-received, particularly at the Manned Spacecraft Center (MSC) in Houston, which had been a vocal advocate for this change. MSC had long been critical of the wet workshop's complexities and cost overruns, and its objections had kept scrutiny on the wet workshop's drawbacks. The decision to abandon the wet workshop, particularly in its initially ambitious form, seemed increasingly inevitable given the mounting technical and budgetary challenges. While there was satisfaction with the change, the delay in finalizing the decision tempered some of the enthusiasm at MSC.

On July 28, 1969, Administrator Thomas Paine further reinforced the program's priorities in a letter to the center directors. Flight safety was emphasized as the top priority, followed closely by adherence to schedule and cost considerations. The Saturn V's large payload capacity was highlighted as advantageous for both safety and cost-effectiveness, as it allowed the inclusion of heavier, more robust components, thereby reducing the need for costly test programs.

Mueller cautioned that the increased payload capacity also introduced new risks. He reiterated that changes should be strictly limited to those necessary to transition from the wet to the dry workshop configuration. For instance, the requirement to conduct medical experiments in the multiple docking adapter was no longer applicable. Any additional changes beyond those mandated by the dry workshop's new configuration required specific authorization from the program director in Washington. Mueller stressed the importance of quickly arriving at a definitive configuration to avoid delays and excessive program elaboration.

Schneider's July 22 directive and Mueller's subsequent communication established a minimum-change, minimum-cost approach to shape the program's implementation. This philosophy aimed to avoid unnecessary complications, although it would lead to some misunderstandings as the definition of the dry workshop continued to evolve.

At Marshall Space Flight Center, it became evident that the transition to a dry workshop for Skylab was inevitable and technically advantageous. Despite this, Dr. William C. Belew, a prominent advocate for the wet workshop concept, contended that this approach could and should have been pursued to a successful conclusion. This perspective was rooted in more than mere institutional bias; it reflected the broader challenges faced by the center.

The shift from the wet to the dry workshop introduced a significant additional workload, which strained the center's already diminished resources. Marshall had suffered a reduction of over 600 positions due to agency-wide budget cutbacks in early 1968. Consequently, the remaining workforce had to be reassigned, and the center had recalibrated its workload to align with the reduced manpower. However, the dry workshop's demands necessitated the creation of new hardware, such as the payload shroud and the ATM (Apollo Telescope Mount) deployment mechanism, alongside an extensive array of new analyses, designs, and tests. The NASA Glenn Research Center played a crucial role in the Skylab program through its rigorous testing of the Skylab shroud. In December 1970, the Skylab shroud was tested at the Space Power Facility at Glenn's Plum Brook Station. This shroud was designed to protect the upper section of the Saturn V rocket, which carried key components of the Skylab space station, including its retracted solar power arrays, during launch.

Payload shrouds were integral to the launch phase of a rocket, serving as protective coverings through the atmospheric ascent. Once the rocket had ascended through the atmosphere, the shroud was jettisoned. Pyrotechnic devices split the shroud into two halves, then fell away from the rocket and into the ocean. It was essential to test the shroud system in a vacuum chamber to ensure this process occurred smoothly and without damaging the payload or rocket. This testing verifies that the separation mechanism functions correctly, providing a clean and safe shroud release.

The successful testing of the Skylab shroud was a critical step in ensuring the integrity of the Skylab space station's launch and deployment. The meticulous preparation and validation carried out by the NASA Glenn Research Center exemplify the detailed attention required for space missions, contributing significantly to the successful launch of Skylab on May 14, 1973.

Moreover, the new integrated launch configuration required that all workshop and ATM components be delivered simultaneously. This was a departure from the previous schedule, which allowed a six-month gap between the delivery of different components. This change meant that any delay in the work would result in a postponement of at least six months if managed in-house. In response to these constraints, Belew proposed outsourcing several major tasks to external contractors—a strategy that was ultimately adopted.

With the decision finalized and the program defined—though a few experiments were to be added shortly thereafter—there was little time for reflection on the program's swift and transformative history. The Apollo Applications Program (AAP) had evolved significantly from its initial concept in 1965, which envisioned a modest experiment involving an empty S-IVB tank. Initially conceived as a simple endeavor, it unexpectedly burgeoned into the first major post-Apollo program, driven by circumstances that thrust it into a pivotal role.

James Webb, then NASA Administrator, was committed to fulfilling the lunar landing mission but saw no clear directive for a subsequent space program. Despite his efforts, he did not secure a mandate for a new initiative and refrained from advocating for a program of his design. Webb's deputy, Hugh Dryden, was skeptical of the feasibility of maintaining a national commitment to a program of Apollo's scale. Conversely, George Mueller, a staunch advocate of NASA's founding principles, believed in leveraging the technological advancements achieved during the Saturn-Apollo era. He viewed the preservation and utilization of this capability as paramount until a new direction for space exploration could be determined. Faced with the reality of the situation, Mueller resolved to maximize the potential of the wet workshop concept, recognizing it as the most viable option available at the time.

As circumstances evolved, George Mueller adeptly adjusted the Apollo Applications Program (AAP) to optimize its resources and address emerging challenges. Faced with mounting inflation and growing skepticism about the value of sophisticated technology, Mueller made several strategic modifications. He postponed launch dates, scaled back the experiment program, reduced the number of planned flights, and redistributed the workload between NASA centers to make the most of the available resources. Despite these efforts, the resources available to AAP continued to dwindle, exacerbated by economic uncertainty and diminishing enthusiasm for the program's objectives.

Skylab Shroud was installed in the NASA Lewis Research Center's (now known as the Glenn Research Center) Plum Brook Station, Space Power Facility. The shroud protected the upper section of the Skylab space station, including its solar power arrays. The Space Power Facility (SPF) was the world's largest vacuum chamber when it was constructed. It stands more than 122 feet high, 100 feet in diameter, and provides a vacuum environment for studying space propulsion. Initially commissioned for nuclear-electric propulsion studies, the SPF has been recommissioned for current and future use in the ongoing research and development of space propulsion systems. Credits: NASA

Mueller also encountered internal disagreements, particularly with the Manned Spacecraft Center (MSC) in Houston, where some believed the program had been poorly conceived. While these disagreements highlighted the program's weaknesses, they also spurred critical discussions that ultimately contributed to addressing and rectifying these issues.

The departure of James Webb from NASA marked a pivotal moment for Apollo Applications. His successor, Tom Paine, aimed to spearhead an ambitious space initiative beyond the lunar missions. However, Paine's efforts were thwarted by public disinterest and a lack of support from the presidential administration. Nonetheless, Paine's

attempts likely played a crucial role in catalyzing the program's key decision to transition to the dry workshop approach.

Mueller remained at the helm until late 1969, overseeing the completion of the first two lunar landings and guiding Apollo Applications toward its redefined goals. As he left NASA, he expressed confidence in the integrated plan and its potential to shape the future of space exploration. He emphasized that the lessons learned from the dry workshop would be invaluable for subsequent endeavors.

As the Apollo era ended, Apollo Applications geared up to implement its revised mission plan. The program now included three crewed flights: a 28-day mission scheduled for mid-1972, a 56-day flight in October of that year, and a final 56-day mission in early 1973. Responsibilities were clearly defined, and the organization was structured to foster improved collaboration between the major centers. Although differences of opinion persisted, Huntsville and Houston were united in their commitment to the program's objectives and prepared to advance with a clearer mutual understanding.

Development 1969-1973

In July 1969, a pivotal shift occurred in the Skylab program, marking the transition from an initial phase focused on defining the project to a new era dedicated to hardware design, fabrication, and testing. This shift signified the start of an intensive four-year period during which Skylab evolved from concept to reality. The ensuing year, however, was marked by significant changes and challenges. The addition of the Earth-resource experiments expanded the scope of Skylab's scientific objectives, while substantial upgrades were made to enhance the living accommodations within the space station. These improvements, though vital, required considerable time and financial resources, which were not always readily available.

By July 1970, a comprehensive program review solidified Skylab's final design and content. This review established a stable framework for the project, allowing development to proceed with renewed focus and urgency. Over the next two years, rigorous testing and reviews ensured that all systems functioned cohesively and that the crew could operate them efficiently.

Completing the Apollo missions, particularly after the Apollo 13 incident in April 1970, allowed NASA's Manned Spacecraft Center and Kennedy Space Center to redirect their efforts toward Skylab. At the Manned Spacecraft Center in Houston, mission planners and training officials devised strategies to manage what would be the most extended crewed missions ever attempted. Meanwhile, Kennedy Space Center undertook the complex task of preparing for the final checkout and launch of the most sophisticated system it had encountered.

In the year following the decision to use a dry workshop configuration, Skylab evolved beyond the confines of Apollo Applications. Although much of the hardware and many managerial practices retained elements from the Apollo program, Skylab began to assume a distinct identity. The program's rebranding in February 1970 symbolized a shift in perspective. No longer viewed merely as a repurposing of Apollo hardware, Skylab was increasingly recognized as America's first space station—potentially the only one for many years.

Several factors influenced this shift in outlook. The success of Apollo 11 allowed NASA officials to focus more attention on Skylab, while the Saturn V rocket's increased lifting capacity enabled engineers to enhance the workshop's design and functionality. Skylab's significance grew as it became evident that Congress would not fund another space station in the 1970s. George Mueller's integrated plan of May 1969 outlined Apollo and Skylab as the cornerstone of NASA's crewed programs for the decade. This plan envisioned two primary trajectories: one leading to further lunar exploration and the possibility of establishing a lunar base, and the other focused on Earth-orbital operations, which began with the Saturn V-launched workshops and aimed toward a permanent, crewed space station complemented by a cost-effective Shuttle.

Major milestones for the decade were outlined as follows: Earth-orbital operations with the Saturn V-launched workshop in 1972; the beginning of post-Apollo lunar exploration in 1973; suborbital flight tests of the Shuttle and the launch of a second Saturn V workshop in 1974; initial space station operations and orbital Shuttle flights in 1975; and developing a lunar-orbit station and full Shuttle operations by 1976.

The Design Phase

In the late 1960s, amid the euphoria of space exploration, NASA faced the challenge of shaping the future of its ambitious plans. Administrator Thomas Paine was particularly eager to secure approval for an expansive vision that included establishing bases in Earth orbit and on the lunar surface and landing humans on Mars. With public enthusiasm at a peak following the Apollo moon landings, Paine saw an opportune moment to push for this grandiose agenda.

To advance his vision, Paine turned to the Space Task Group, a committee formed by President Richard Nixon to evaluate the future trajectory of America's space program. During the summer of 1969, Paine passionately advocated for a crewed Mars mission as NASA's next major objective. The Space Task Group's report, titled America's Next Decades in Space, delivered in September, presented a range of scenarios for achieving a crewed landing on Mars before the century's end.

The most ambitious proposal outlined developing a 50-man Earth-orbiting space station by 1980, with the first Mars flight scheduled for three years later. This plan envisioned a significant funding increase, with annual budgets reaching $8 billion by 1976. A less ambitious alternative was also considered, which would halve the funding and delay the Mars expedition until the 1990s. Vice President Spiro Agnew, the task group's chairman, strongly supported the Martian goal. Despite this, the proposal faced significant resistance from Congress and the media, and the Nixon administration ultimately approved less than three-quarters of NASA's requested $4.5 billion budget for fiscal year 1971. This reduction, amounting to $500 million less than the previous year's allocation, marked NASA's lowest funding level in nine years.

On January 13, 1970, Paine addressed the press to explain the repercussions of the budget cuts. The reduction necessitated halting the production of the Saturn V rocket, canceling Apollo 20, postponing the initial Skylab workshop flight until late 1972, and delaying subsequent Apollo missions (Apollo 18 and 19) until 1974.

In response to the budget constraints and to invigorate public interest, NASA sought to rebrand the Apollo Applications Program (AAP). The acronym had become a subject of ridicule, with detractors mocking it as "Almost A Program" or "Apples, Apricots, and Pears." NASA considered nearly a hundred alternative names, ranging from mythological to historical references. However, the final choice was Skylab, a name proposed by Lt. Col. Donald Steelman, an Air Force officer on loan to NASA. Skylab, a contraction of "laboratory in the sky," met both of George Mueller's objectives: enhancing public engagement and providing a more manageable term for everyday use. The name was quickly embraced within and outside NASA, offering the program a renewed sense of purpose and direction.

Despite the lack of congressional and administrative support, Paine advocated for an expansive space program throughout 1970. The persistent jokes and cartoons poking fun at the AAP, such as one depicting Martians baffled by the space station, underscored the challenges Paine faced in rallying support for his vision. Nevertheless, the rebranding to Skylab and the ongoing efforts to promote ambitious space exploration goals marked a significant chapter in NASA's history, setting the stage for future achievements in space.

In the wake of an evolving space program and shifting priorities, it became evident that at least one more Apollo mission would be canceled, and the possibility of further Saturn V rocket production was closed. By mid-1970, NASA Administrator Thomas Paine, despite his optimism, had to acknowledge these constraints. Paine had hoped that Skylab, the United States' first space station, could be launched by mid-1972. He envisioned a significant mission of substantial new significance by 1976—something to transcend the achievements of Skylab and continue the momentum of American space exploration. However, his aspirations were increasingly misaligned with the priorities of the Nixon administration.

The interim operating budget for NASA, announced on September 2, 1970, allocated a modest $3.27 billion. This budget revision led to the cancellation of two Apollo missions and marked the end of the Apollo program in June 1972, just five months before Skylab was scheduled to lift off. Faced with these budgetary constraints, Paine resigned from his position on September 15, 1970, leaving the task of defending NASA's budget to George Low, the acting administrator.

In October 1970, Edward David, the President's science adviser, requested Low to assess the relative importance of Apollo and Skylab in light of potential further budget cuts. Low defended both programs with conviction. He argued that reducing or constraining the scientific returns from Apollo by canceling one or more missions would result in significant losses. However, the idea of canceling Skylab was even less appealing. Low contended that, on balance, Skylab offered greater scientific returns than an additional lunar landing mission. The benefits from the Apollo program were already substantial, whereas the cancellation of Skylab would yield no return on the investment made in its development. Furthermore, Skylab had the potential to open new avenues for research with comparatively less risk than continuing Apollo missions.

In a period characterized by 6% inflation, NASA sought a modest budget increase to $3.7 billion. However, the Office of Management and Budget countered with a $3.3 billion offer, necessitating substantial reductions in the Space Shuttle and nuclear engine programs. Despite this, Apollo and Skylab faced relatively minor cuts, with their combined reduction totaling $50 million—less than 5% of the requested amount. This financial compromise led to a slowdown in the pace of operations.

As a result, the Office of Manned Space Flight adjusted the launch schedules, setting new dates of December 1972 for Apollo 17 and March 1973 for Skylab. When the Kennedy Space Center indicated that such closely spaced launches would require overtime, Skylab's launch was postponed by another month. This budget decision in late 1970 marked the last significant change to

Skylab's schedule, and the program progressed steadily toward its launch.

The budget constraints of 1970 also had a notable impact on future plans for Skylab. The concept of a second Skylab, which had been under consideration since mid-1969, became one of the primary casualties of the budget deliberations. Shortly after the decision to shift from a wet to a dry workshop configuration, Charles Mathews proposed that NASA's center program offices begin exploring the feasibility of artificial gravity for a second space station. This investigation aimed to provide valuable insights for planning a permanent space station. In September 1970, Robert F. Kennedy's office expanded this study, requesting proposals for additional experimental payloads from Space Science and Advanced Research offices.

In November 1970, NASA outlined several options for a follow-on workshop to Skylab, reflecting its ongoing commitment to space research and innovation. The guidelines proposed a range of possibilities, including a year-long mission aboard a workshop similar to the first Skylab, with four crews of three astronauts each. Other options included incorporating artificial gravity into the design, replacing the Apollo Telescope Mount (ATM) with a stellar telescope, or integrating a more advanced array of Earth-resource sensors. The plan was to implement these enhancements with minimal modifications to the workshop's core configuration.

Since the backup hardware for the first Skylab was slated to become the basis for the second workshop, any substantial changes to the hardware were deferred until the latter stages of the first mission. The committee established various milestones to guide the development process: a preliminary report was due by January 20, 1971, to support upcoming congressional hearings; a detailed work statement was to be prepared by July; and a preliminary design review was scheduled for early 1971.

As the new year approached, the focus on defining new experiments continued. On March 7, 1971, Dale D. Myers, who succeeded George Mueller as NASA's associate administrator for crewed space flight, reviewed the progress of preliminary studies with his staff. The team determined that developing a stellar telescope had advanced sufficiently. They concluded that the primary focus of further studies should shift toward artificial gravity and payloads that offered tangible benefits to the general public.

Following this meeting, Myers instructed NASA's center program offices to provide cost estimates for three potential mission scenarios: a repeat of the first Skylab, a year-long mission featuring advanced solar instruments but no major changes to the existing configuration, and a similar mission with advanced Earth-resource instruments replacing the telescope mount.

The responses from the centers varied. The Johnson Space Center in Houston advocated for a more sophisticated follow-on station, even if it meant delaying the first Skylab's launch. In contrast, the Marshall Space Flight Center in Huntsville expressed concerns that committing to a major upgrade could jeopardize the ongoing Skylab program. Huntsville argued that a year-long mission would be infeasible without significant hardware modifications and that incorporating artificial gravity could significantly inflate costs. They recommended a more modest approach: an eight-month mission combining Earth-resource and solar astronomy instruments.

These differing perspectives were presented at the April meeting of the Manned Space Flight Management Council. The council approved further studies on Skylab II configurations and directed the committee on artificial gravity to present its findings by early May. This decision marked a crucial step in shaping the future of the Skylab program and its potential successors as NASA continued to navigate the complex landscape of space exploration and research.

Development And Preparation

As the Skylab program progressed during the summer of 1971, the focus turned to preparing for the upcoming fiscal year 1972 budget discussions. The studies conducted for Skylab II revealed significant challenges, particularly concerning payload weight. This issue was so pressing that it might necessitate modifications to the second stage of the Saturn rocket, a crucial component of the mission's infrastructure.

Compounding the technical challenges were financial concerns. The projected costs for Skylab II were substantial, ranging from $1.32 billion to over $1.5 billion. Given these figures, the outlook was increasingly troubling. On July 31, 1971, Schneider, a key figure in the Skylab program, discussed the prospects of a second Skylab with officials from the Office of Management and Budget. The conversation underscored the reality that securing additional funds would not be straightforward. By August 31, Schneider reported to Myers that the studies for Skylab II had yielded sufficient data for planning, but further actions would be contingent on funding decisions.

As the fall approached, the decision was made to forgo Skylab II. The debate centered around the mission's utility; without expensive modifications to incorporate artificial gravity, Skylab II would largely replicate the objectives of Skylab I. NASA management faced a critical choice: funding another space workshop would require a significantly larger budget or result in prolonged delays for the Space Shuttle program. Despite strong support for a second Skylab from the House Space Committee, the Nixon administration was unwilling to bear the financial burden, and NASA was reluctant to compromise its future programs.

Throughout the summer of 1969, the Skylab program manager, Schneider, was deeply engrossed in managing the operations of the first Skylab. Schneider grappled with the challenges posed by the research scientists, who operated in a distinctly different realm from engineers. Convincing them of the necessity to have hardware ready six months before flight proved difficult. This disconnect was partly influenced by their experiences with Apollo Applications, where schedules had often slipped, allowing for extended instrument improvements. These enhancements, while beneficial, contributed to the escalating costs of developing experiments—a recurring issue in Schneider's correspondence.

A significant issue during the Skylab program was interface control, which dealt with numerous connections between different space vehicle parts. This included electrical, pneumatic, and physical connections. Skylab's design involved thousands of such interfaces, requiring meticulous oversight to ensure compatibility. The program managed these interfaces using protocols developed during the Apollo missions, including interface control documents and intercenter interface panels. Interface control documents detailed each interface's design requirements and constraints, specifying how interconnected components should function. When interfaces involved multiple centers, a level A document and an intercenter panel were responsible for coordination. Once Schneider approved a document, each center was tasked with implementing its respective part of the interface.

In the intricate network of Skylab's development, each center played a critical role in managing and ensuring compatibility across various aspects of the program. For instance, the Huntsville center scrutinized both sides of flight hardware interfaces to ensure overall compatibility. Simultaneously, the Kennedy Space Center was tasked with overseeing the integration of flight hardware with ground support equipment. Marshall Space Flight Center, with assistance from Martin Marietta, handled the scheduling and tracking of interface control documents, maintaining a master file to keep everything organized. When program managers disagreed regarding panel actions, the issues were escalated to Headquarters for resolution.

This detailed interface management system, which had proven cumbersome and problematic during the Apollo program in 1968, began to create similar issues for Skylab. Schneider, facing delays and inefficiencies in the interface control process, highlighted the problem at a July 1970 meeting. He observed that incomplete interface control documents impeded the design of various Skylab modules and experiments. In response, Schneider requested that Project Integration Director Thomas Hanes review the status of all documents and propose solutions to alleviate the bottleneck.

Despite these efforts, progress was slow. By September 1970, Schneider directed his program managers to streamline their procedures and involve contractors more directly in the process. Hanes's office collaborated with the centers to develop a more effective scheduling and tracking system. This led to the formation of the Interface Working Group, which convened biweekly and succeeded in clearing most of the backlog by early 1971.

Intercenter panels, established early in the Apollo program to facilitate coordination among centers, continued to play a significant role in managing Skylab interfaces involving multiple centers. Gilruth and von Braun initially organized these panels to formalize agreements between Huntsville and Houston. When Kennedy Space Center joined the collaboration in 1963, the panels became a key mechanism for resolving technical issues across center lines. Huntsville valued these panels for their effectiveness, with von Braun noting in December 1963 that they were "the only effective medium of working out technical problems which cut across center lines."

However, the panels were not universally appreciated. By September 1966, Samuel Phillips, the Apollo program director, wanted to eliminate them, possibly due to concerns over their independence from Headquarters and fears that not all interfaces were being adequately documented. Despite this, Charles Mathews, in March 1967, established a panel system for Skylab, initially focusing on four key areas: mechanical, electrical, instrumentation and communications, and mission evaluation. A couple of weeks later, he expanded the system to include panels for mission requirements, systems integration, and systems safety, while the Apollo panel temporarily managed launch operations equipment interfaces.

By August 1969, NASA Headquarters unequivocally recognized the necessity for formal intercepter coordination. The intricate network of interfaces associated with the Skylab Space Station underscored the need for a structured approach to synchronizing the efforts of various centers. However, the reorganization of center responsibilities in late 1968 introduced several complexities concerning the relationships among different panels. Schneider, aiming to address these issues, faced various challenges.

NASA Administrator Thomas O. Paine visited the Marshall Space Flight Center in July 1970, for briefings on the developing Skylab program. Shown left to right were Belew, Samuel C. Philips, Harold T. Luskin, Paine, and Werner Kuers.

The Huntsville Center advocated for eliminating the practice of co-chairmanship in key areas, proposing instead that each responsible center should directly oversee panel activities. This shift aimed to streamline decision-making and clarify responsibilities. Concurrently, Houston suggested abolishing the System Integration Panel, arguing that it duplicated the baseline configuration meetings conducted by Headquarters. Additionally, there was a consensus to elevate guidance and control activities— previously managed as a subpanel of mission requirements—to the status of an independent panel. On August 5, officials deliberated and opted against a complete overhaul of the panel system. Instead, they dissolved the Systems Integration Panel and introduced a new panel dedicated to planning tests.

The management of interfaces was intricately linked to the broader issue of configuration control. Configuration encompasses various hardware aspects, including size, weight, shape, connection points, and power requirements. Engineers frequently changed these configurations during the design phase, often impacting other components. The Apollo 13 accident highlighted the critical nature of configuration control. In 1965, engineers increased the power to pressurize an oxygen tank without adjusting the protective thermostatic switches on the tank's heater. Although this modification did not pose issues under normal conditions, an unusual operation days before the 1970 launch caused the higher voltage to weld the switches shut, leading to insulation damage. This defect resulted in a catastrophic tank explosion with near-fatal consequences in space.

To mitigate such risks, NASA established various configuration control boards to scrutinize design changes. These boards operated at four levels, each reflecting the impact of the proposed modifications. Level 4 modifications were minor, affecting neither weight nor performance, such as changing an instrument's screws from brass to nickel alloy. Level 3 boards managed changes that could influence the schedule or cost of specific experiments or modules without affecting other hardware. These modifications were often implemented by the centers without Headquarters approval.

Level 2 changes had more significant implications, potentially affecting major hardware and requiring approval from the center program manager or his representative. For instance, an inspection by Wernher von Braun at Huntsville led to a Level 2 change when he objected to placing a vacuum pump on the lower-body negative pressure device. His criticism, "Right through that wall, you've got the greatest vacuum in the universe," prompted engineers to drill a hole through the workshop wall. Once a Level 2 change was approved, the decision was forwarded to Headquarters for review.

Level 1 changes, requiring Schneider's approval, involved hardware, software, or facility alterations that could jeopardize meeting operational plans and mission objectives. These changes included those affecting milestones, exceeding $500,000 in cost, or doubling the agreed-upon cost of an experiment. Such rigorous oversight was essential to ensure that modifications did not compromise the integrity and success of the Skylab mission.

Interface control and configuration documents were crucial in the Skylab program, which was inherited from the Apollo era. The significance of paperwork in major post-World War II projects was well established, and Apollo was no exception. Observers humorously noted that NASA seemed intent on reaching the moon amid towering stacks of documentation. Continuing this tradition, Skylab relied on a robust documentation system to direct the activities of its various center program offices.

The Skylab Program Office utilized three main types of documents to coordinate and manage the program. "Skylab Program Specifications" outlined the critical functional and performance standards for the program's hardware. For instance, the specifications issued in August 1969 stipulated that the probability of crew safety should be comparable to that achieved during the Apollo missions. This meant that spacecraft components and systems were designed to function correctly 995 times out of 1,000, with the reliability of the workshop and launch vehicle set at 0.995 and 0.990, respectively.

"Skylab Program Work Authorizations" detailed the responsibilities of the various centers for over 50 major end items. These included key components such as the one-g spacecraft trainer managed by Houston and the workshop engineering mockup handled by Huntsville. Additionally, the authorizations included a list of over 130 mission milestones, each representing deadlines for specific actions.

"Mission Directives" provided comprehensive statements on the program's objectives, flight plans, space vehicle configurations, experiments, and each center's responsibilities. Despite these efforts to structure the documentation, the sheer volume of paperwork became overwhelming for many involved in the program. In response, Schneider directed his managers to evaluate the requirements based on three fundamental questions: What was the minimum information required to fulfill general program responsibilities? What information was needed to address specific technical responsibilities? What information will other offices expect to have?

In tandem with these internal reviews, NASA Headquarters assessed the documentation demands imposed on the centers. Despite these measures, the extent of bureaucracy remained a significant concern. For example, a researcher working on the human-vestibular experiment at the Navy Aerospace Medical Institute expressed frustration upon encountering the "Experiment General Specifications." He predicted a tenfold increase in the cost of his experiment and humorously suggested that NASA should establish a direct line between Pensacola and Houston to handle the extensive paperwork.

Among the many management tools utilized during the Skylab program, NASA's formal review system was the most critical and visible. This structured system was designed to ensure rigorous oversight at key stages of the program's development, serving as vital checkpoints to validate the progress and readiness of Skylab hardware and systems.

The review process began with three pivotal assessments during the design phase. The Preliminary Requirements Review held early in the project, focused on evaluating the concepts under consideration and selecting the one most likely to achieve the mission objectives. This was followed by the Preliminary Design Review, which examined the basic design as it began to take shape. The third key review, the Critical Design Review, was conducted near the end of the detailed design phase. At this point, technical specifications and drawings were scrutinized to ensure the design met all requirements.

During these initial reviews, the compatibility of modules and experiments with the rest of the space station was also assessed. This ensured that all components would function cohesively once assembled in orbit.

The review process continued with two additional crucial assessments as developing Skylab hardware progressed. The Configuration Inspection compared manufactured end items, including test and flight hardware, with the established specifications and acceptance testing criteria. This was followed by the Certification of Flight Worthiness, which involved a thorough evaluation to confirm that the hardware was complete, qualified, and supported by proper documentation before being shipped from the factory.

The final two reviews focused on the entire Skylab operation. The Design Certification Review conducted four months before launch, was crucial for certifying the spacecraft's design for flight readiness and safety. It also included an assessment of the launch complex, mission control center, and the Manned Space Flight Network. The last major review, the Flight Readiness Review, occurred several weeks before launch and was intended to validate the operational readiness of the entire mission complex.

Under the guidance of Lee Belew, who was responsible for most of Skylab's hardware, these reviews were meticulously managed. Belew appointed chairmen for each review board, scheduled review dates and locations, and ensured that all relevant parties, including experiment sponsors, contractors, and NASA offices, were represented. The design review teams conducted detailed examinations of blueprints, focusing on "review item discrepancies"—the principal mechanism for recommending hardware changes.

If an issue arose, such as dissatisfaction with an experiment's placement or the workshop's configuration, a discrepancy report could be submitted. The teams then screened these reports, with similar issues consolidated and decisions made by the review board. The entire process was thoroughly documented, with center managers tracking the status of every document, including its number, title, category, completion date, and responsible individuals. Some individuals became notably recognized for recommending numerous changes.

The transition to the dry workshop, a significant development in the Skylab program, led to extensive reevaluations at McDonnell Douglas plants and in Huntsville. By December 1969, the review process had advanced sufficiently to warrant a preliminary design review of the cluster systems. During this review, several hundred NASA and contractor representatives convened to examine the requirements and potential modifications for various systems, reflecting the thorough and iterative nature of the Skylab development process.

During three days of intensive discussion, several critical issues regarding Skylab's design and operational plans emerged. One notable discrepancy emerged between the assumptions made by teams at Huntsville and McDonnell Douglas and the plans developed by Houston. The former had anticipated that astronauts would enter the Skylab cluster in pressurized suits, while the latter planned a "shirtsleeve" entry, meaning the astronauts would enter in normal clothing. This difference highlighted a need for reevaluation of the design requirements.

Another issue arose concerning the control and display console for the telescope mount. The design allowed astronauts to monitor either the control panel or the panel for the structural transition section, but not both simultaneously. This limitation posed a significant problem for mission operations, necessitating a redesign to allow simultaneous monitoring.

A major challenge emerged from the decision to tilt Skylab's orbit 50 degrees from the equator to facilitate earth-resource experiments. This adjustment created unforeseen difficulties for engineers working on the thermal control system. The station would need to dissipate more heat than initially planned to maintain comfortable temperatures inside the workshop during sunlight. However, this adjustment also increased heating requirements during nighttime, which exceeded the available power. Consequently, a decision on addressing this issue was deferred pending further detailed studies.

While several issues were discussed, the most pivotal decision involved the electrical power system. Initially, Skylab's design included two separate electrical systems: one for the lunar module and another for the telescope mount. The rationale for maintaining two independent systems was questionable with the lunar module's elimination from the project. The "minimum change" principle, emphasized in July, had initially discouraged immediate alterations. However, a proposal to consolidate the two systems was put forward by the December review. The level 3 and level 2 configuration control boards reviewed and approved this proposal. After evaluating the increased costs and complexity against the potential for enhanced mission success, Schneider approved the consolidation. This decision, established and driven by practical considerations and strategic foresight, would later

prove crucial when an accident during launch made the change a key factor in the mission's success.

The structural transition section, a crucial component located at the forward end of the airlock tunnel, included essential elements such as a heat exchanger, a molecular sieve, a carbon dioxide sensor, circuit breakers, and several control panels. This complex arrangement highlighted the intricate nature of Skylab's design and the need for meticulous coordination among various systems to ensure mission success.

Following the cluster systems review, various actions were set in motion over the coming months to address critical issues and optimize the Skylab Space Station's design and functionality. The review highlighted several key areas for improvement, including a comprehensive examination of the power and thermal systems, reorientation and relocation of the Apollo Telescope Mount's (ATM) display panel, modifications to the multiple docking adapter, and a detailed evaluation of the ATM's computer software.

One significant change involved reconfiguring the docking adapter to include a side port for emergency docking, ensuring that the station could accommodate unexpected contingencies. Concurrently, efforts focused on refining the ATM's computer software, essential for managing the station's scientific equipment and experiments.

Amidst this intense period of design refinement, Schneider and Belew convened with the airlock team in St. Louis on December 10-11. Schneider expressed concern over the tight timeline between the critical design reviews and the delivery of flight hardware. He feared that if major issues emerged during these reviews, contractors might struggle to meet their deadlines. To mitigate this risk, Schneider emphasized the need for regular and thorough reviews of all design aspects, urging that the focus should be on the substance of the work rather than extensive formal presentations.

As the new year began, Schneider stressed the importance of frequent reviews. He pressed Belew to schedule various reviews in February, similar to the December meetings in Huntsville. Schneider's primary concern was the lag in "failure mode and effects analyses," a critical process where engineers identify potential hardware failures and devise methods to detect, address, or bypass these issues. He believed that delays in this area could impact the overall design timeline. Schneider considered these reviews essential, with follow-ups on electrical power, environmental control, and attitude-control systems being highly desirable.

Belew, however, did not share Schneider's apprehension about the failure modes analysis. He noted that although formal documentation was not always available, Huntsville's designers and analysts were working collaboratively, and steps were being taken to ensure that failure mode documents would be ready 90 days before the critical design reviews. To avoid large, inefficient reviews that could hinder progress, Belew favored a different approach. He preferred monthly crew-station reviews, which center managers had agreed upon in December. These reviews involved astronauts walking through mockups of the flight hardware to verify that the design met operational requirements.

Attendance at these reviews was kept to a minimum to facilitate swift decision-making by NASA and contractor representatives. Any issues not involving significant cost or schedule changes were addressed on the spot, with final decisions confirmed by a configuration control board. If disagreements arose, they could be appealed to the board, though review teams were encouraged to resolve disputes internally whenever possible.

The reviews utilized engineering mockups at each contractor plant, ensuring that all interfaces were accurately represented. For example, the airlock mockup in St. Louis included a workshop hatch and sections of the docking adapter, providing a realistic representation of how the hardware would function in practice. Belew believed that this approach offered a more continuous and effective review process, with responsible parties engaged at a level closer to the actual work.

In early 1970, the Skylab program was dominated by a relentless schedule of reviews and modifications. Belew's weekly reports highlighted the continuous cycle of evaluations and adjustments essential to the program's development. Teams of 40 to 50 engineers and astronauts gathered at contractor plants in St. Louis, Denver, and Los Angeles for crew-station reviews of the major Skylab modules. Between these large meetings, smaller groups worked daily to coordinate necessary changes.

Reviewing 70 Skylab experiments added another layer of complexity for program managers. They were tasked with certifying each review's completeness and documentation within 60 days of completion. Despite efforts to prioritize reviews based on the importance of experiments to crew safety and mission success, the centers struggled to keep up with the schedule. By July, Schneider confronted the managers about the backlog of 27 uncertified reviews.

As 1970 progressed, the design work culminated in critical design reviews of Skylab's principal hardware. Each review session spanned nearly a week and involved up to 300 engineers from NASA and its contractors. During these reviews, boards assessed an average of 200 discrepancies per module. While many proposals were minor, their cumulative effect could delay Skylab's launch by several months.

The early years of Skylab's development were marked by challenges related to changes in design. Initially, the program headquarters had mandated "minimum change" to avoid complications. However, by October 1969, this restriction had been largely disregarded. At that point, a dozen major changes were under consideration, including extending the final crew's mission to 120 days, incorporating an earth-resource experiment package, adjusting the orbit to be inclined 50 degrees from the equator, operating the solar telescopes in an uncrewed mode, and adding a teleprinter to the workshop.

In October 1968, Fred J. Sanders, right, escorted Wernher von Braun through the Mc Donnell Douglas Astronautics Co. plant in St. Louis. Inspecting a mockup of the airlock also were Walter Haeussermann, standing, Jack L. Bromberg, left (behind von Braun), and John F. Yardley, behind Sanders.

Astronaut Alan Bean reading a teleprinter message. Instructions to flight crews were often transmitted by teleprinter. Skylab was the first such craft to be equipped with the capability to receive messages by this means. While the crew slept, flight plan changes, procedures, and instructions were transmitted and recorded on board.

In response to these proposed changes, Schneider approved various physical modifications to the workshop. These included adding a side access door and a window, relocating equipment from the "floor" to the newly designated "ceiling," and combining the sleep compartment with the food management area into a new wardroom. Belew, responsible for keeping the hardware on schedule, protested these changes. He argued that they effectively constituted a new mission for the workshop, estimating that the changes would delay the schedule by six months and increase costs by $100 million.

In the summer of 1969, with the decision to proceed with the dry workshop now finalized, NASA's Skylab program faced a pressing challenge: to adhere to a demanding schedule. Dr. Wernher von Braun, then Director of the Marshall Space Flight Center, imposed a stringent working timeline, targeting a flight readiness date of March 1972. This deadline, set four months before the official launch date, was intended to build a buffer against any unforeseen complications that might arise.

In August 1969, the Marshall Space Flight Center in Huntsville reacted positively to this schedule. They deemed the proposed timeline feasible, although concerns lingered about the solar telescope mount, a critical space station component. Meanwhile, the response from Houston was less optimistic. Kenneth S. Kleinknecht, who was poised to replace William Thompson as Skylab manager, expressed skepticism. He remarked that the Advanced Apollo Applications Program (AAP) schedules were inherently fluid, formulated before the complete definition of the workshop and the Command and Service Module (CSM). Kleinknecht feared that the rigid schedule left no room for adjustments and predicted that such an approach would inevitably lead to delays.

By December 1969, before the full impact of ongoing changes had been evaluated, Skylab program manager Robert Belew reported that his contractors were grappling with an "extremely tight schedule." The program encountered some relief in January 1970 when budget cuts for fiscal year 1971 resulted in a four-month delay in the working schedule. However, this temporary respite did little to quell concerns. By May 1970, Houston advocated for additional delays, and several critical components had fallen three to four months behind schedule.

As the design and development of Skylab progressed, NASA officials were embroiled in debates over potential modifications. On March 27, 1970, shortly after a major decision was made to alter the urine processing system, Dale Myers, then the Program Manager, declared that Skylab could no longer accommodate additional experiments. He emphasized that the hardware development activities had reached a level of complexity where further additions or modifications would inevitably cause schedule delays.

Despite these warnings, Houston continued to push for further modifications, particularly concerning the habitability features of the workshop. This led to considerable contention with Huntsville and prompted a significant program review on July 7-8, 1970. The review team ultimately approved several proposed changes while reaffirming the launch date of July 1972. Nevertheless, the Marshall Space Flight Center Director cautioned that the new changes would exhaust the available slack in the Skylab schedule. He warned that if modifications persisted, it would become impossible to meet either the schedule or the budget.

In August 1970, further correspondence between Belew and von Braun highlighted the escalating problem of funding exacerbated by these changes. On July 17, Belew indicated that Huntsville would require additional financial resources to maintain the revised schedule. Schneider responded by stating that no unallocated Skylab resources were available and advised Belew to find a way to achieve program objectives within the existing budget constraints. This exchange underscored the mounting pressures and the challenging balancing act required to keep the Skylab program on track amid ongoing modifications and financial limitations.

In managing the Skylab program, the decision was made to implement specific manpower restrictions for major contractors and in-house personnel. Despite these efforts, a subsequent review revealed that the schedule and resources allocated for Skylab were incompatible. Robert Belew, the program manager, concluded that the Marshall Space Flight Center required $285 million in fiscal year 1971 funds—nearly $50 million more than what had been originally allocated.

In response to the funding shortfall, NASA's Skylab manager, Dr. Kenneth S. Kleinknecht, who had been instrumental in the Apollo program's Command and Service Modules since February 1967, found an additional $25 million for Huntsville. This partial remedy reduced Belew's funding deficit by half. Belew proposed distributing the remaining shortfall across all major projects at the center, resulting in a 10% reduction in funding for each. This strategy initially seemed viable until early October, when McDonnell Douglas, a key contractor, reported that its reduced allocation would delay workshop delivery by two months. This delay eliminated all schedule margins for the official launch date.

On October 7, 1970, Belew reported that meeting the 1972 launch deadline would be impossible unless NASA implemented stricter control over program changes. During a teleconference on October 13, 1970, Dr. Wernher von Braun addressed the situation by allocating $12 million to Huntsville to accelerate the contractor's work. This additional funding was ultimately sourced from Houston's allocation.

The scheduling pressures eased somewhat in September 1970 when Schneider abandoned the concept of a working launch date set four months ahead of the official schedule. At Houston, Kleinknecht welcomed the end of the dual-schedule policy. He observed that working to a schedule that no one believed achievable led to decreased motivation and gamesmanship with deadlines. He argued that an ambitious yet achievable schedule was essential for maintaining focus and ensuring that each team member could contribute effectively to staying on track.

To build flexibility, Schneider scheduled hardware to arrive at the Kennedy Space Center three months before the required date. Following the fall of 1970, critical design reviews recommended numerous minor modifications, but few significant changes were proposed. As Schneider remarked on December 15, 1970, the opportunity to incorporate changes without affecting the launch date and critical program resources had passed. From then on, each proposed modification had to be carefully evaluated for its impact on Skylab systems and the overall program, marking a shift toward a more rigid and less flexible approach to program management.

Despite initial resistance from Huntsville in 1969 and 1970 due to concerns over schedule and costs, a consensus emerged after the Skylab mission that the changes made to the program had significantly enhanced its overall value. One notable issue that had been debated was the challenge of controlling Skylab's reentry into Earth's atmosphere. At nearly 75,000 kilograms, Skylab would be the heaviest object ever placed in orbit, and its high orbital inclination would see it traverse most of Earth's surface. The potential for the workshop, or large pieces of it, to reenter the atmosphere posed an unprecedented problem for NASA.

Historically, the risk of falling space debris— such as spent booster stages, spacecraft, or satellites—had been a concern, with international treaties outlining the responsibility of spacefaring nations for any resultant injury or damage. Beginning in late 1962, NASA and its contractors had studied the survival of Earth-orbiting vehicles and explored methods to predict impact points or control reentry. However, accurately predicting reentry and designing for controlled descent imposed significant weight penalties, making such solutions impractical.

Despite concerns from the White House and State Department over potential diplomatic repercussions, NASA management accepted the risk associated with uncontrolled reentry. Measures were taken for payloads that posed abnormal hazards; for instance, uncrewed spacecraft like Gemini-Titan 7 and the SA-5 payload were structurally modified to disintegrate into smaller pieces upon atmospheric entry. Unfortunately, a similar solution was not feasible for Skylab.

In early 1970, NASA Administrator Thomas Paine called for a comprehensive review of the reentry hazard and an assessment of engineering changes to mitigate it. The study evaluated several components, including the S-II booster stage, the four segments of the payload shroud, and the orbital workshop itself. The study concluded that there was a 1 in 55 chance that a fragment of Skylab might cause harm to someone.

To counteract this risk, the study proposed adding retro-rockets and control systems to guide the fragments to a predetermined location, preferably a remote stretch of the ocean. Implementing such systems would add approximately 9,000 kilograms to the weight and cost around $10 million for the S-II stage. For Skylab, the weight and cost would be even higher, requiring significant redesigns of attitude control and electrical power systems and extending the schedule by several months.

Given the extreme costs and the relatively small risk—smaller than other risks such as meteorites or existing space debris—the study group recommended that NASA accept the risk associated with Skylab's reentry. They suggested that future programs establish clear criteria for acceptable risk early in the planning and development stages to better integrate safety measures without incurring excessive costs.

In the latter part of 1972, critical studies validated previous assessments regarding space debris and Skylab's reentry risks. A comprehensive analysis conducted by Lockheed Missiles and Space Company for the Marshall Space Flight Center concluded that 306 pieces of Skylab's structural components, totaling approximately 22,600 kilograms, would survive reentry into Earth's atmosphere. Among these, the film vault, akin to a large executive desk weighing as much as a compact car, was identified as the largest surviving fragment.

This finding was consistent with an earlier 1972 study that identified significant space debris orbiting Earth. At that time, the inventory included 547 spacecraft, 282 rocket bodies, and 1,931 fragments, with 191 spacecraft launched by the United States and 849 by other nations. Between 1967 and 1972, 826 pieces of space debris had reentered the atmosphere, of which 184 were American—comprising 56 from NASA and 128 from the Department of Defense. Of these reentries, at least 31 fragments were successfully recovered and identified.

In response to these findings, Dale Myers formally presented recommendations to Acting Administrator George Low in late November 1970. These recommendations reaffirmed earlier conclusions, indicating that the risk associated with Skylab's reentry was minimal and acceptable, given the constraints of redesign weight and costs. Low endorsed Myers's recommendations and directed the Office of Manned Space Flight to collaborate with the Offices of Public Affairs and International Affairs to devise a comprehensive plan addressing the public relations aspects of Skylab's reentry.

By late 1970, the first phase of Skylab's program development had been completed. This included defining its relationship with the Apollo program, organizing management tools, navigating the design phase, and initiating preparations for testing. Despite significant modifications to Skylab's appearance and objectives, the important changes had concluded, paving the way for hardware fabrication and rigorous testing.

Skylab's design and construction were driven by NASA's commitment to ensuring functionality, efficiency, reliability, and safety. However, Skylab was more than just a laboratory; it was a living environment where crews would reside for up to three months. George Mueller, concerned about the quality of living conditions in space, began advocating for improved amenities in 1967. Initially, the focus was on overcoming challenges in the wet workshop environment. Once those issues were resolved, Mueller and the Headquarters program office pushed for enhanced living conditions in the dry workshop, recognizing the importance of making the space station a comfortable home and a functional workspace.

The debate over living conditions lasted a year, as Marshall Space Flight Center, constrained by budget and schedule limitations, resisted the extensive modifications proposed by Headquarters and Houston. Nonetheless, improvements were ultimately deemed essential for mission success. Skylab's three-module cluster offered 347 cubic meters of space—over 150 times the volume of a Gemini spacecraft and nearly 60 times that of the Apollo command module. This significant increase in living space was crucial for the well-being and efficiency of the Skylab crews, and it provided valuable insights into future space habitation.

The Skylab space station was a pioneering venture in space habitation. Its largest component was the liquid hydrogen tank, which formed the core of the orbital workshop. This tank, measuring 6.6 meters in diameter and 8.9 meters in height, was repurposed into the space station, where crews would conduct their daily activities, including eating, sleeping, and working.

In the early days of space exploration, spacecraft were designed primarily for operation rather than habitation. The Mercury and Gemini capsules, though groundbreaking, were extremely cramped. With their limited space and weight restrictions, these capsules could only provide the bare essentials for life support. Astronaut Michael Collins, who piloted Gemini 10, famously compared the cramped quarters of the two-man Gemini craft to the front seats of a Volkswagen. The Gemini 7 mission, undertaken by Frank Borman and James Lovell, lasted 14 days, during which the crew endured the confines of their tiny spacecraft. Borman later acknowledged that the mission, which included the historic achievement of the first orbital rendezvous of two spacecraft, was as much a test of endurance as a scientific endeavor.

While the Apollo command module provided more space than the Gemini capsules, it was still designed with functionality as its primary concern. Although it featured improvements such as hot water and offered the crew of three a bit more room to move and exercise, comfort was not a priority. Astronauts were accustomed to tolerating minor inconveniences if they did not impede mission performance, with the primary focus remaining on achieving mission objectives.

As attention turned to developing space stations, the early planners faced challenges far more pressing than crew comfort. A symposium on space stations in 1960 featured 41 papers, only one of which addressed the issue of creating a pleasant living environment. The paper likened maintaining an orbiting station to managing a lighthouse—a somewhat monotonous job—and suggested that improving living conditions would be crucial to attracting individuals to space once the initial excitement had faded. Despite this, nine years later, the focus remained predominantly on spacecraft technology, with the nuances of daily living in space largely overlooked.

NASA designers recognized that as missions extended in duration, the quality of life for astronauts would become increasingly important. Habitability—or the suitability of the environment for daily living—encompassed factors beyond the engineers' usual purview. This included the composition and temperature of the atmosphere, light, and noise levels, as well as the ease of maintaining cleanliness, personal hygiene, and opportunities for exercise and relaxation. Although habitability was not a term frequently used by engineers, experience and intuition indicated the subject's growing significance for long-duration missions. George Mueller, a key figure in the program, repeatedly emphasized the need for a comprehensive understanding of these living conditions to ensure the success of future space missions.

The Wet Workshop

The initial concept for the Skylab Space Station did not envision using the spent S-IVB stage as a permanent living space. Initially, the S-IVB was intended to test some habitability features but not for round-the-clock occupancy. However, as planning evolved throughout 1966, utilizing the spent stage as a functional habitat gained traction. By September of that year, the Manned Space Flight Experiments Board endorsed a proposal from the Manned Spacecraft Center (MSC) for an experiment titled "Habitability/Crew Quarters." This experiment aimed to establish design criteria for future spacecraft and long-term space stations.

The MSC's presentation of this experiment included illustrations showing the workshop divided into compartments using fabric panels stowed in canisters mounted on the airlock trusses during launch. This concept was a key part of the proposal, reflecting early ideas on creating manageable living spaces in space.

At the same time, the Marshall Space Flight Center (MSFC) was developing its experiment, which was called "Experiment M402, Orbital Workshop." This experiment also involved the design of crew quarters within the workshop. This overlap led to a contention period between MSC and MSFC regarding the management and scope of these experiments. The core issue was the designation of responsibilities;. At the same time, the workshop was officially a Marshall project according to the Lake Logan agreement of 1966, MSC viewed the habitability aspect as a primary MSC responsibility, given its role in Crew Systems.

For over a year, the two centers struggled to agree on the nature of the habitability experiment and its oversight. This impasse was finally resolved when Charles Mathews issued a directive assigning MSFC overall management and integration responsibilities for the "Experiment M487, Habitability/Crew Quarters." The directive also delineated specific responsibilities between the centers: Houston retained control over life-support systems, including food management, waste management, personal hygiene, and sleep restraints, while Marshall was responsible for the structural components, plumbing, and wiring of the crew quarters.

Despite this resolution, progress on the habitability experiment at Houston was slow in 1967. The Apollo Applications Program (AAP) faced low priority compared to the Apollo missions, and as a division chief from MSC remarked, the focus on Apollo was crucial; without the success of Apollo, discussions about AAP would remain theoretical. It was not until well into 1968 that MSC could allocate adequate manpower to advance projects such as habitability.

Meanwhile, engineers at Huntsville, led by project manager Charles Belew, advanced their work using their innovations combined with MSC's support. By early 1967, the original plan to use fabric curtains to subdivide the workshop was replaced with a more permanent solution. Metal partitions, fabricated from aluminum and designed with a triangular grid pattern to ensure fuel flow was not obstructed, were installed in the tank before launch. These partitions included folding sheet-metal components, allowing the crew to close off two compartments when not in use.

The workshop's ventilation system, which included fabric panels forming an annular space along the walls, was installed before the flight. During the activation of the workshop in space, the crew would be responsible for setting up fans to circulate the air and installing a curtain beneath the floor to create a mixing chamber for the atmospheric circulation. This preparation was crucial for ensuring that the workshop could support long-duration missions and provide a habitable environment for the astronauts.

Houston's first opportunity to review Marshall's detailed plans for the Skylab Space Station came during the preliminary design review held at Huntsville from May 8 to 10, 1967. At this meeting, it became clear that while the design details were still tentative, Marshall had devoted minimal attention to the workshop's habitability. The primary concern from Houston's perspective was the fire hazard potential, and comments related to living conditions were limited, with the temperature control system being the main focus.

During the summer and fall of 1967, George Mueller, then Associate Administrator for Manned Space Flight, became increasingly involved with the Skylab project. His attention was particularly drawn to the layout of the living quarters. After reviewing a mockup of the workshop in July, Mueller suggested adding a second floor to the crew quarters to create additional workspace. However, this proposal was not adopted due to the significant weight issues it would introduce.

Mueller's involvement led to further refinements. He proposed installing two grids approximately 2.6 meters above the liquid-oxygen tank dome to create two back-to-back compartments, which became accepted until July 1969. Mueller was deeply concerned about the workshop's austere, mechanical interior despite these improvements. He remarked that under the current conditions, astronauts would have struggled to stay in the workshop for more than two months without becoming mentally strained.

Expressing these concerns to Lee Belew and Charles Mathews, Mueller recommended bringing in an industrial design expert to enhance the workshop's livability. He suggested an expert could help improve the design to incorporate "a reasonable degree of creature comfort." Following this advice, Mathews wrote to Belew in late August recommending that action be taken on Mueller's suggestion. He provided the names of two commercial firms that could assist in this endeavor.

Marshall Space Flight Center subsequently arranged for Martin Marietta, the integration contractor, to hire an industrial design consultant on subcontract. The consultant's role was to provide comments and recommendations based on contemporary industrial design concepts. This included advice on floor plan arrangements, color schemes, lighting, noise levels, and other factors contributing to human comfort in confined quarters. This step marked a significant move toward making the Skylab Space Station a more habitable environment for its future occupants.

On December 1, 1967, a pivotal two-month study commenced to evaluate the wet workshop concept for the Skylab Space Station. This habitability study was entrusted to one of the world's foremost industrial design firms, Raymond Loewy/William Snaith, Inc., based in New York. Raymond Loewy, a luminary in industrial design and known for his extensive work on functional styling, had been a prominent figure in American design for over forty years. His portfolio ranged from consumer products to commercial spaces, including stores, shopping centers, and office buildings. Although approaching his seventy-fifth birthday in 1968, Loewy took a keen personal interest in the Skylab project despite having scaled back his professional engagements.

In early December 1967, Loewy and Fred Toerge, the firm's vice president, undertook a comprehensive tour of the contractors' facilities associated with the Advanced Apollo Program Joint (AAPJ) project. Their journey culminated at the Marshall Space Flight Center in Huntsville, Alabama, where they received briefings on the program's progress and inspected the work done thus far. Subsequently, Loewy and Toerge traveled to Washington, D.C., to share their critical observations with key figures such as Mueller, Mathews, and other AAP officials. Their feedback was largely unfavorable, prompting them to highlight significant deficiencies in the workshop's design.

By February 1968, Loewy/Snaith had prepared a formal report detailing flaws in the existing layout and proposing various improvements. The wet workshop's interior was poorly planned; the working spaces lacked a cohesive flow, with enclosed and open areas failing to integrate seamlessly against the workshop's neutral backgrounds. Despite acknowledging a certain "honesty" in the straightforward treatment of the interior space, the overall atmosphere was deemed unwelcoming. The workshop's cylindrical structure clashed with rectangular elements and the harsh triangular gridwork that adorned the space, creating a visually cluttered environment. Additionally, the lighting was haphazardly distributed across the ceiling, and using dark colors contributed to a gloomy ambiance.

Loewy/Snaith recommended several changes to improve the workshop's habitability. They suggested adopting a neutral pale-yellow background with brighter accents to differentiate crew aids, experimental equipment, and personal items. For the lighting, they advocated for localized illumination in work areas and replacing cold fluorescent lights with warmer spectrum alternatives. These recommendations were aimed at creating a more pleasant and functional living environment.

On February 28, 1968, Martin Marietta, the contractor responsible for the project, presented the findings of the Loewy/Snaith report along with their suggestions at Huntsville. The immediate focus was revising the color scheme, which posed a challenge due to the requirement for finishes that could withstand immersion in liquid hydrogen. The floor plan also required urgent revision. Loewy suggested the addition of a wardroom—a space designated for eating, relaxation, and routine office work. He also proposed making the floor plan adaptable with movable panels to allow for different configurations, as evaluating a single layout was inadequate for optimizing the design of space stations.

William A. Brooksbank used a model of the orbital workshop to explain its structure and operation to visitors from NASA Headquarters to the Marshall Space Flight Center in December 1967. Leland F. Belew, James E. Webb, Charles W. Mathews, and von Braun were seated left to right.

Marshall Space Flight Center officials initially received these recommendations with some confusion. The engineers at Huntsville had assumed that Douglas Aircraft Company, with its experience in commercial aircraft design, would handle styling and interior decor. Additionally, since none of the astronauts had expressed concerns about these elements during their examination of the mockups, the importance of such details had been underestimated.

However, as the study's implications became more apparent, program officials at Marshall began to appreciate the value of the Loewy/Snaith recommendations. Efforts were initiated to address the color scheme issue, which proved to be a persistent challenge due to the constraints imposed by the liquid-hydrogen environment. By recognizing the quality of Loewy/Snaith's work, Mueller extended their contract through 1968. By this time, the Manned Spacecraft Center (MSC) had increased its interest in the design of crew quarters. The new contract included provisions for Loewy/Snaith to collaborate with the principal investigator for MSC's habitability experiments, underscoring the growing emphasis on creating a functional and comfortable living space for astronauts aboard Skylab.

In June 1968, Caldwell C. Johnson, a seasoned engineer and chief of spacecraft design in the Advanced Spacecraft Technology Division, was appointed as the new principal investigator for Experiment M487 at the Manned Spacecraft Center (MSC). Known for his innovative design skills, Johnson, whose first initial was silent, was a distinguished figure in aerospace engineering. A native of Tidewater, Virginia, he had joined the National Advisory Committee for Aeronautics (NACA) in 1939, shortly after completing high school. Over the years, he became a pivotal design team member under Dr. Maxime Faget, contributing to the Mercury program and later to Apollo, where he served from 1961 to 1963 before taking up the role of assistant chief of the Advanced Spacecraft Technology Division.

Johnson's appointment to oversee the Skylab crew quarters project reflected Dr. Faget's intention to have a seasoned designer handle the entire design process, from conceptualization to execution. Upon reviewing the Skylab workshop, Johnson was struck by the absence of thoughtful consideration for habitability—the essential factors that affect human comfort and efficiency in space. He discovered that this information appeared to have been disregarded despite a wealth of knowledge on design factors for confined and isolated environments. The engineers at the Marshall Space Flight Center, who had limited experience with crewed spacecraft, had seemingly adopted various design ideas, including those from astronauts, whose expertise in spacecraft design Johnson viewed as insufficient.

The engineers in Huntsville were understandably displeased when Johnson, along with his colleagues Loewy and Snaith, began to challenge their design decisions. Nevertheless, Johnson proceeded with his recommendations, recognizing the importance of an optimized living environment for the success of space missions.

Throughout the remainder of 1968, Johnson worked diligently to outline the scope and content of his habitability experiment. By May 1969, he articulated his approach, emphasizing that habitability was not a traditional experiment with multiple testable design concepts. Instead, he proposed that the best design judgment, based on the collective expertise of the MSC, be implemented in the workshop, with subsequent missions serving as evaluations of this judgment.

Johnson's vision was to create a living environment to make daily chores "entirely incidental" to the primary operations of spaceflight. He aimed to address nine critical components of habitability: environment, architecture, mobility and restraint, food and water, clothing, personal hygiene, housekeeping, communication within the spacecraft, and off-duty activities. By systematically addressing these aspects, Johnson hoped to bring a degree of engineering rigor to a field that had previously been marked by chaos and improvisation.

Within a month of the transition, the Manned Spacecraft Center (MSC) voiced dissatisfaction with the workshop's layout. They criticized it as "too austere" and likened it to a "canvas tent city," noting that the design did not leverage the best contemporary technology. The floor plan was deemed problematic, with the food management compartment being too small and the sleeping compartments disproportionately large. The preliminary design review, which MSC conducted later that month, further highlighted that the workshop was still designed with "the threshold of acceptability" in mind. Despite the dry workshop removing the constraints faced by the Apollo missions, the design remained austere.

As fall approached, it became evident that there were divergent views on the workshop's habitability. George Mueller advocated for the workshop as a laboratory for testing concepts and devices, aiming to establish design criteria for future space stations. Mueller and his colleague, Dr. Robert Schneider, prioritized habitability within Skylab's broader objectives. This laboratory-focused approach was not universally shared. While the MSC recognized the importance of improving habitability, they were skeptical of the practicality of Mueller's vision. They believed efficiency was paramount and were less concerned with stylistic details like interior color schemes.

The astronauts, who had been evaluating crew quarter concepts before MSC staff were deeply involved, were primarily concerned with functionality over aesthetics. They wanted a spacecraft where they could effectively perform their tasks without unnecessary inconveniences. The engineers at Marshall, who were influenced by direct interactions with the astronauts, often took their preferences into account. This influence was partly due to the engineers' respect for the astronauts and partly to hopes that the astronauts' preferences would sway MSC's Skylab program office to align with Marshall's decisions.

By September 1969, Mueller grew increasingly concerned that MSFC was not adequately addressing the ideas proposed by designers Loewy and Snaith. Consequently, he scheduled a meeting for mid-October to discuss habitability issues. Schneider elaborated on the concerns on September 30, noting that while he did not intend to abandon the minimal-change philosophy established earlier completely, he believed that "significant and necessary improvements" could be made with minimal cost and schedule impact.

The habitability support system encompassed all the essential components required for the habitability experiment, including lighting, ventilation, floor and wall structures, food storage and preparation equipment, and water supply systems. An anecdote from the astronaut office humorously recalled an early suggestion to paint the Apollo command module with a blue upper portion and a black lower portion to help pilots orient themselves in zero gravity—a reflection of the sometimes absurd concerns that arise in the quest for optimal space habitation.

Crew comfort was not the sole consideration in developing the Skylab Space Station; the public image projected through television transmissions from the workshop in orbit was also a significant factor. It was essential to ensure that the workshop met functional requirements and presented a positive and professional image to the public.

Raymond Loewy and Caldwell Johnson, both influential figures in the design process, recommended thoroughly considering their suggestions. They believed it possible to enhance the workshop while still keeping costs manageable. The memo from Dr. Robert Schneider underscored a growing impatience at Headquarters with the current approach to habitability. This sentiment was evident when the key figures involved in the workshop's design, including Loewy, gathered in Washington on October 14 for a comprehensive review of the habitability support system.

During this review, George Mueller clarified that he was dissatisfied with the current state of crew quarters. He emphasized repeatedly that habitability was a crucial element for future crewed spaceflight, a stance that resonated with the MSC delegation. Their presentations highlighted numerous shortcomings in the existing design.

The day's discussions covered all aspects of habitability, including some that had significant implications for the workshop's structure. Both Loewy and Johnson proposed rearranging the floor plan to include a wardroom—a space where the crew could gather and dine—and adding a large window to provide a view of Earth from orbit. The window was particularly appealing as it offered both a practical and recreational benefit, allowing the crew to experience the beauty of space. However, the introduction of a window also introduced several challenges. It was a costly addition that could compromise the structural integrity of the spacecraft, and its development and testing would be time-consuming. Furthermore, it was not deemed essential for the mission's success.

Despite these concerns, Loewy's strong opinion that a window was indispensable for long-term missions carried significant weight. He argued that the recreational value alone justified the expense and effort. Mueller, persuaded by Loewy's argument, instructed the team to include the window in the design. On October 31, Schneider officially authorized the addition of the window, the wardroom, and several other modifications.

Habitability issues often involved complex discussions and debates between various parties, including Caldwell Johnson and Marshall engineers, with input from the crew members occasionally adding another layer to the deliberations. For example, there was disagreement over the suitability of traditional one-piece flight coveralls for long-term use in the workshop. Although the crews ultimately accepted the need for more practical clothing, they insisted on retaining pockets on the lower trouser legs.

In addition to these design considerations, practical improvements were also made. A notable example was the decision to cut a door into the S-IVB wall, facilitating easier access during checkout at the Kennedy Space Center (KSC). This modification simplified the checkout process and accelerated the assembly of the workshop at

McDonnell Douglas, demonstrating the ongoing effort to balance functionality with efficiency.

The Skylab Space Station's lower deck, specifically the workshop trainer's aft compartment, presents a fascinating glimpse into the living and working conditions of the space station's crew. In this compartment, the crew quarters were prominently displayed in the foreground, with the design reflecting zero gravity's unique constraints and needs. Notably, the absence of chairs around the hexagonal galley table was a deliberate choice; in the weightlessness of space, chairs become redundant and impractical. Instead, astronauts would sleep strapped to the walls of the cubicles to maximize floor space. The central area of the compartment features the bathroom, while the upper section of the work area was designated for various experiments.

Against the far wall, the lower-body negative-pressure device utilized in the M092 experiment stands as a testament to the sophisticated technology employed in Skylab. The compartment also includes a hatch partially obscured by a partition, which opens into the liquid oxygen tank. This tank served an unconventional purpose on Skylab as an oversized trash receptacle.

The design of the crew's uniforms was another consideration of significant importance. Initially, the crew wore a two-piece uniform designed by Caldwell Johnson and Fred Toerge, intended to be both practical and aesthetically pleasing. Johnson had one of his staff model a prototype during a briefing in May 1969, and the uniform received an enthusiastic response. Despite this positive reception, later versions of the uniform, which included a three-piece design, were often covered by name tags and badges, diminishing the uniform's original effect.

Johnson allowed the crew a considerable degree of personal choice regarding off-duty relaxation and entertainment. While an entertainment center equipped for movies and music was proposed for the wardroom, it failed to generate significant interest. Card and board games were similarly underwhelming, as the crew preferred reading and listening to recorded music, often using private tape players due to varied musical tastes. This preference for solitary activities aligned with the crew's busy schedules, leaving little time for entertainment.

Maintaining cleanliness was a critical concern aboard Skylab. Despite Houston's medical experts deeming sponge baths sufficient to prevent dermatological issues, astronauts Mueller and Schneider sought a solution for showering. In April 1969, Schneider requested a lightweight, cost-effective "whole-body bather" for evaluation. Although Caldwell Johnson was skeptical about the need for such a device, he provided a design concept. The resulting cost estimate from McDonnell Douglas, over $3 million, was deemed excessive, and the proposal was rejected. Nonetheless, Schneider persisted, and a simplified shower system was eventually incorporated into the workshop. This device, used on all three Skylab missions, received mixed reviews but demonstrated the ongoing efforts to improve crew comfort and hygiene.

Understanding habitability in space was challenging due to the lack of direct analogies and limited experience. Submarines offered some parallels, but astronaut Paul Weitz found that Navy data was mainly anecdotal, with few practical insights. To gather relevant information, Marshall Space Flight Center participated in the Gulf Stream Drift Mission, an oceanographic expedition led by Swiss scientist Jacques Piccard aboard the submarine Ben Franklin. This mission provided valuable data on habitability to inform future space missions.

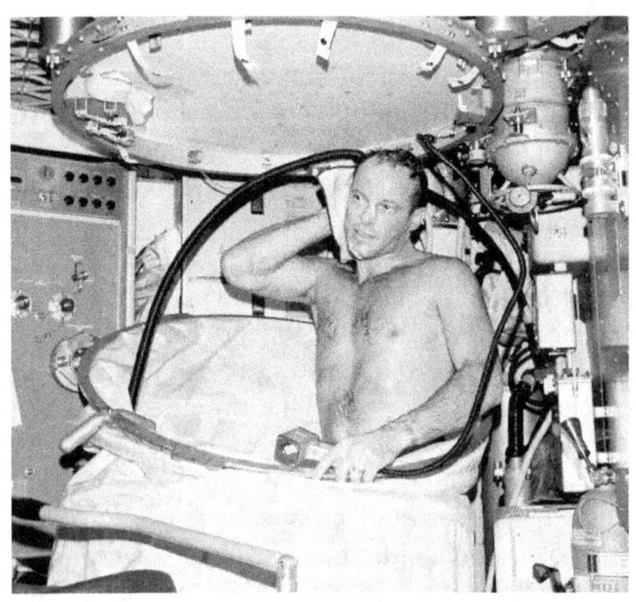

Jack R. Lousma takes a zero-gravity shower aboard Skylab

Among the practical aspects of living in space, the shower system highlights the innovative approaches to adapting terrestrial solutions for the unique zero gravity conditions. The flexible hose, push-button shower head, and a vacuum system for water removal were designed to work effectively in space. At the same time, the traditional washcloth continued to serve as a reliable cleaning tool.

On July 16, 1969, a notable event occurred when a Marshall engineer embarked on a mission aboard the submersible Ben Franklin. This vessel, designed for underwater exploration, drifted with the Gulf Stream for 31 days before resurfacing off the coast of Nova Scotia, having traveled approximately 2,700 kilometers. During this mission, the rened explorer Jacques Piccard, who visited the Manned Spacecraft Center (MSC) on February 25, 1970, provided valuable insights based on his experiences. His feedback, particularly concerning the living conditions aboard the Ben Franklin, was meticulously noted by Caldwell Johnson. Johnson reported Piccard's observations to the Skylab office, highlighting that many issues had been identified before the mission. Unfortunately, these issues had not been adequately addressed by Grumman engineers, who struggled to comprehend or resolve the complaints. Reflecting on his challenges with implementing ideas in 1969 and observing the

parallels between the Ben Franklin and Skylab, Johnson remarked to the MSC Skylab manager that the problems Piccard described resembled those anticipated for the first Skylab mission in 1973.

Underwater studies provided valuable data about behavior and the problems associated with living in close quarters. NASA engineers took part both in the Ben Franklin (pictured above) voyage in 1969, and later in the Tektite 11 underwater research program

One of the most challenging aspects of developing Skylab was managing the human digestive process, an area that posed significant difficulties for experimenters concerned with intake and waste management. Stringent medical requirements exacerbated the complexities of food and waste management systems. The waste management system, detailed in Chapter 8, faced major design issues up until a few months before Skylab's launch. Meanwhile, the food system was not fully under control until the end of 1971.

The necessity for improved space food became clear as mission durations extended to two months. Although earlier Mercury and Gemini missions had not raised significant complaints about space food, as it was primarily engineered for spaceflight rather than taste, the situation changed with the Apollo missions. Space food in these missions, while an engineering success due to its compact packaging, durability, and long shelf life, failed to meet the crew's taste preferences. Despite providing balanced nutrition for up to 90 days, it became evident that more effort was needed to make space food palatable.

The first three crewed Apollo flights in 1968 and 1969 saw complaints about the food, which was surprising given its similarity to Gemini rations that had been well-received by some astronauts. To address these concerns, MSC nutritionists enlisted Donald D. Arabian, the head of MSC's Test Division, to evaluate the Apollo rations. Arabian, a versatile eater, found the experience less than favorable. He had committed to a four-day diet of Apollo food but quickly found the meals unappealing. For instance, he described the sausage patties from his first breakfast as "coarse granulated rubber with a sausage flavor," leaving him with a nauseating aftertaste that lingered for hours.

At the end of the first day of his trial with Apollo rations, Donald D. Arabian noted a significant loss of appetite. By the third day, eating had become a burdensome task. The lack of appealing aromas and textural variety made meal preparation uninviting. Although items resembling off-the-shelf foods were satisfactory, those specially designed for spaceflight fell short of acceptable standards. Arabian was perplexed by the necessity of grinding common items like peanuts and chocolate into bite-sized cubes, which adhered unpleasantly to the teeth. This experience underscored the challenges faced by food-system developers, who were already grappling with the demands of long-term food storage.

The rigorous requirements of Skylab's medical experiments added complexity to the challenges of the food system. From the early days of the Apollo Applications Program (AAP), medical scientists planned to conduct a mineral balance study to monitor astronauts' calcium and nitrogen levels. This study aimed to understand the effects of extended weightlessness on human physiology. Previous Gemini missions revealed that astronauts experienced calcium loss from bones and muscle nitrogen loss. While these losses were not critical for short lunar missions, they posed potential risks for longer-duration flights. By 1969, despite these findings, little progress had been made since 1966 in understanding these processes.

Skylab incorporated two key experiments to address these concerns: M071 and M073. Experiment M073 assessed the urinary output of various substances crucial to metabolism, impacting the urine and feces collection systems. Experiment M071, on the other hand, focused on controlling mineral intake and measuring output with precision. Mineral-balance studies, though routine in hospital settings, became significantly more challenging in space. These studies typically provide subjects with a constant, carefully measured supply of nutrients like calcium and nitrogen while collecting and analyzing their urine and feces output.

Conducting such experiments in a space environment added a layer of complexity, as researchers had to manage these precise measurements while astronauts were engaged in various other activities. The intricate demands of maintaining accurate mineral balance aboard Skylab underscored the formidable challenges of space-based research and the need for meticulous planning and execution in developing life-support systems.

The optimal diet for medical purposes aboard Skylab consisted of homogeneous food items whose composition could be precisely determined and controlled. This typically included pureed vegetables, puddings, and compressed, bite-sized solids. Dehydrated foods were also acceptable, provided they maintained a reasonably uniform consistency. However, heterogeneous dishes such as spaghetti and meatballs or turkey with gravy posed significant challenges for the experimenters, as their variable textures and preparation methods complicated the rigorous demands of space missions.

Despite these challenges, it became clear that the diet most suitable for the astronauts could mitigate crew resistance and enhance morale. The management team in Houston recognized that for missions extending from four to eight weeks, it was crucial to prioritize crew comfort. They concluded that poor-quality food would only make long-duration missions less tolerable. Reflecting on feedback from previous Apollo missions, which highlighted crew dissatisfaction with their meals, key figures such as Mueller, Schneider, and Caldwell Johnson began advocating for more conventional and appealing food options. They sought to replace the paste-like tubes and cold, cubed foods with more familiar and enjoyable meals.

An example of this shift in approach was evident during Apollo 8, where the crew was served a hot meal of turkey and gravy, eaten with a spoon. This simple change had a remarkable positive impact on crew morale, underscoring the importance of food quality in sustaining astronaut well-being. Recognizing the challenge of designing a food system that could transcend the engineering constraints of previous missions, Johnson started to develop ideas in the spring of 1969. By the time Skylab's dry workshop was finalized, the specifics of the food system had yet to be detailed, though its fundamental constraints were understood.

In April 1969, Paul C. Rambaut, a nutritionist at MSC and the principal coordinating scientist for the M070 series of experiments, observed occasional conflicts between medical requirements and habitability considerations. Despite this, habitability concerns often took precedence. If the experiments made the food intolerable, adjustments had to be made. Rambaut expected that Skylab would offer a wider variety of foods, including both hot and cold items, and that the food management compartment would be designed to provide some of the comforts of conventional dining.

Following complaints from the Apollo 9 crew regarding their meals, George Mueller decided that action was necessary during the April Management Council meeting. On April 22, Schneider proposed guidelines to the MSC program office, advocating a departure from the Apollo-style food system. He suggested incorporating options like frozen dinners, freeze-dried camping foods, and fresh fruits and vegetables. Schneider acknowledged that accommodating these changes would influence workshop development, but he emphasized that improving meal quality could justify adjustments in weight and volume allowances.

In late March, Marshall conducted a preliminary requirements review, with MSC's specifications being relatively broad. These included estimating total storage space and provisions for heating and cooling specific food items during preparation. On April 16, Johnson urged the Houston program office to consider adding a food freezer, a proposal supported by a study from Martin Marietta. This addition would enable a greater variety of food to be stored and utilized on the mission. By May, MSC's program office communicated new requirements to Huntsville, which included a freezer, an oven, and measures to protect stored food from pressure changes, marking a significant step toward enhancing the Skylab crew's dining experience.

The updated Skylab food specifications introduced various meal options, categorized into five distinct classes: dehydrated foods, intermediate moisture items, wet-pack foods (similar to the heat-sterilized turkey dinner provided during Apollo 8), frozen foods, and perishable fresh items. This new classification system aimed to offer a broad selection of meal types to meet the varied needs of the crew. Although Marshall was initially concerned about the increased size and complexity of the food storage and preparation systems at such a late stage in the project, the push for these improvements came from Headquarters, and he ultimately agreed to the changes.

By the end of July, the revised food requirements were accepted after various meetings involving both centers and McDonnell Douglas. Several conceptual designs were under consideration as the teams worked to integrate these new specifications. However, developing the food system faced significant delays. The definition of the food system was lagging due to several factors. During the wet-workshop phase of the Apollo Applications Program (AAP), dieticians at the Manned Spacecraft Center (MSC) relied on data from the Air Force's Manned Orbiting Laboratory (MOL), expecting to adapt these systems for Skylab. However, the cancellation of the MOL program in June 1969 shifted full responsibility for the food system to MSC, establishing already burdened with ongoing Apollo duties.

With Marshall seeking precise storage and preparation requirements, Caldwell Johnson designing a new system from scratch, and the development contract still pending, MSC's chief of food and nutrition urgently requested additional support. He proposed assigning three more personnel to the food-system integration team at Martin Marietta to expedite the process. With the deadline approaching for the food system proposals, Martin Marietta needed to begin its work promptly to meet the tight schedule.

As Skylab preparations continued, astronaut Owen Garriott, a scientist-pilot on the second crew, was seen at dinner in the workshop, demonstrating using heating elements integrated into food trays for preparing individual meal packets. Meanwhile, astronaut Joseph Kerwin, a scientist-pilot on the first crew, was experimenting with a grape drink in a workshop trainer. This beverage was packaged in accordion-shaped containers that expanded as water was added, and the contents were expelled by crushing the container. These innovations reflected the ongoing efforts to improve meal preparation and consumption in the challenging space environment.

In the complex journey toward the operational readiness of the Skylab Space Station, various challenges and strategic adjustments unfolded. The intricate task of managing the food system for the space station proved particularly troublesome. By mid-1969, concerns regarding the evolving requirements for the Marshall Space Flight Center food system had gr. The initial struggles to stabilize the food system management persisted for five months, revealing a pressing need for comprehensive changes.

Caldwell Johnson, a key figure in the program, recognized the necessity of simplifying the food system's development process. He proposed that engineering responsibilities be transferred away from Marshall to alleviate the strain on the center. Johnson recommended that a specialized contractor be entrusted with the design and production of storage and preparation equipment, delivering a complete, ready-to-install system to the workshop contractor. This proposal included an innovative food storage and preparation concept: individual servings of food would be packed into metal containers, precisely shaped to fit within a pressure-proof canister. Each canister would hold a sufficient supply of food for several days, allowing for weekly replenishments of the wardroom pantry.

The design featured a preparation and serving tray into which the food containers could be placed. The crew member assigned to meal preparation would only need to extract the necessary items from the canisters, add water to any dehydrated foods, secure the containers in the tray, activate the automated heating elements, and let the tray manage the cooking process. After the meal, the containers could be weighed to account for any leftover food, as required by medical experiments, before being discarded with minimal cleanup. While not all elements of this proposal were adopted, several core concepts influenced the final design of the food system.

Despite these advancements, the program faced significant setbacks. Lee Belew, Skylab's program manager at Marshall, voiced concerns in July 1969 regarding the center's capacity to meet the July 1972 launch deadline. By the fall of that year, it became clear that several critical tasks needed to be outsourced to external contractors. Belew hoped this would enable the program to stay on track, provided that all parties adhered to the minimum-change directive established by Schneider and Mueller. However, Houston and Washington continued to propose improvements, particularly in habitability, which diverted time and resources from the original schedule.

In November 1969, Belew confronted Schneider with the growing threat to budget and timeline due to these changes. The required tradeoff studies for the proposed improvements drained Marshall's manpower and further delayed progress. Belew warned that unless the situation was addressed, costs would rise, and schedules would slip.

As early as 1970, Caldwell Johnson grew frustrated with the slow pace of implementing his recommendations. To press his points more effectively, he took the unconventional step of seeking direct involvement by center directors, program managers, and key technical personnel through a four-day tour of Skylab contractors' facilities. This tour aimed to provide a comprehensive, first-hand assessment of the program's status and underscore the urgency of addressing the mounting issues.

During the critical assessment tour of Skylab contractors' facilities in early April 1970, Caldwell Johnson prioritized highlighting the deficiencies in the station's habitability features. Johnson seized every opportunity to draw attention to these shortcomings, persistently "putting the needle in" as he later described it. While some of the issues he identified were minor or resulted from inaccurate mockups, his efforts effectively elevated habitability as a prominent concern.

Upon completing the tour, Chris Kraft, the deputy director of the Manned Spacecraft Center (MSC), expressed his concerns with characteristic directness. He conveyed to his superior that the tour had left a clear impression: insufficient attention had been given to the practical aspects of living in space during extended missions. Despite contractors' assurances that the astronauts had reviewed their work, Kraft criticized the astronauts' willingness to accept makeshift solutions based on past practices. He recommended that ten people be explicitly assigned to review and address habitability issues to ensure that they received proper attention.

Gilruth acknowledged Kraft's memo and forwarded it to Eberhard Rees, the new director at Marshall. Rees, having experienced the rigors of a less-than-ideal environment during his time at an Antarctic base in 1967, appreciated the gravity of the situation. He agreed that MSC's recommendations warranted serious consideration, even if they implied increased costs. Rees instructed Belew to appoint someone to evaluate the habitability issues, disregarding cost constraints thoroughly.

In response, Belew provided a detailed history of the habitability problem, outlining the significant accommodations already made by Marshall. This response seemed to sway Rees toward a more balanced view, as Rees's subsequent communication to Gilruth indicated. Rees stressed the need to either reaffirm Skylab's experimental nature within its budget constraints or to redefine the program's objectives to accommodate the proposed changes.

Meanwhile, frustration was mounting at Huntsville regarding habitability matters. The Man/Systems Integration Branch chief voiced irritation over Johnson's comments about the Ben Franklin mission. He argued that the difficulties faced during that mission were primarily due to budget limitations rather than inherent flaws in the design. The trade-offs made had fully considered habitability provisions, and the relatively small submarine had completed its mission. The chief warned that further habitability changes for Skylab, many deemed trivial, risked destabilizing the entire program.

Certain critical amenities were required in the early stages of the Skylab Space Station program to ensure its effectiveness and habitability. However, there was a real risk that without careful planning and consideration, these efforts could amount to nothing more than superficial "interior decorating" rather than substantive improvements.

To address these concerns, George Belew enlisted Gaylord Huffman, a technical assistant to the workshop project manager at Marshall Space Flight Center, to thoroughly evaluate the habitability issues associated with the Skylab project. Huffman's report, delivered in June, recommended a significant shift in approach. He suggested that NASA could achieve greater scientific value by testing a broader range of concepts rather than sticking rigidly to the initial plan. Huffman also proposed relocating the principal investigator's role to Huntsville, highlighting the ongoing conflicts between the crew's needs and the principal investigator's expectations as a significant factor in the persistent redesign of workshop systems. He argued that a new principal investigator might resolve these issues more effectively. He noted that the Ben Franklin mission—an initiative in which Marshall had been involved—offered a more relevant precedent for Skylab than the Apollo missions, which were the primary experience of Houston's team.

In response to criticisms raised by Caldwell Johnson, a review was conducted to assess the validity of these concerns. In late May, an ad hoc committee, appointed by Christopher C. Kraft Jr., submitted a comprehensive 17-page report detailing recommendations for addressing the deficiencies Johnson had been highlighting. This report included 15 requests for engineering design changes, all classified as mandatory for the operational suitability of the Skylab. These findings were forwarded to Rees, with a note from Gilruth acknowledging the delays in adapting to Marshall's workshop design but emphasizing the importance of ensuring crew comfort for the proposed Skylab missions.

Rees's response to the forwarded requirements was marked by surprise and concern. He was particularly dismayed by the extensive list of new requirements, given that many of the individuals involved had worked with Marshall for over three years. Rees feared that the cumulative effect of these new demands could lead the Skylab program astray, a concern reminiscent of Gilruth's previous warnings about the wet workshop's potential pitfalls.

Among the mandatory changes was a revised food system developed by Johnson and the MSC nutritionists. Selling this new system to the Marshall program office proved challenging, as it entailed substantial alterations to a design that McDonnell Douglas had already begun to fabricate. Throughout May and June, various tense meetings between the center program offices and the contractor unfolded. MSC persistently advocated for the changes while Marshall argued that implementing them would have detrimental effects on both cost and schedule, jeopardizing the program's success.

Houston's proposals for changes became increasingly challenging as the Skylab program progressed. Among the suggested modifications were a substantial increase in food storage space, lower freezer temperatures, the relocation of the wardroom table, and the introduction of a new food tray that required a specialized fixture in the galley. These demands arrived alongside mounting issues with the urine collection system, exacerbating the strain on Marshall Space Flight Center's Skylab office and suggesting that further design changes might be forthcoming from Houston.

In response to these escalating issues, OMSF chief Dale Myers organized a comprehensive program review scheduled for July 7-8, 1970, at Huntsville. This review addressed fundamental program concerns: Was Skylab to be treated as an experimental project with limited funding, or should it be regarded as a major undertaking akin to the Apollo missions? Houston's team arrived at the meeting with yet another series of proposals, which included additional adjustments to the environmental control system due to concerns about humidity and carbon dioxide levels in the workshop.

One of the most contentious topics was the high cost of the new food system proposed by MSC. Despite the lack of a cheaper alternative, the proposal was largely accepted. While some questioned whether the new arrangement would simplify management and reduce costs, a detailed examination of trade-offs led to its approval. Headquarters representatives, recognizing the difficulty of quantifying habitability factors, urged both field centers to negotiate their differences more effectively. They encouraged Marshall to be more accommodating of intangible factors and Houston to be more mindful of cost implications.

The program review approved several significant changes while reaffirming the July 1972 launch date and the existing cost ceilings. Rees expressed his frustration to Charles Mathews later that month despite these approvals. The changes had nearly depleted Marshall's schedule flexibility, making it impossible for the center to meet its commitments. Rees anticipated further modifications, as everyone except Marshall seemed eager to elevate Skylab's ambitions beyond its original scope as an experimental program. Given the apparent underestimation of Skylab's costs, he cautioned that if this trend continued, it could severely impact NASA's ability to advance other programs.

Rees summarized the situation by reflecting on the program's evolution: "We started with an open-ended astronomy mission, ready to accept certain risks and designing for habitability within a framework of early launch and lower cost. Now, we were transitioning to conduct a highly sophisticated and unprecedented medical experiment in orbit, where the handling of subjects must be meticulous to avoid disturbing the medical baseline."

The ongoing struggle with Skylab's design changes was exacerbated by many desirable modifications having unintended, adverse effects on other systems, particularly the medical experiments. Rees grappled with the uncertainty of where these changes would ultimately lead. While Marshall Space Flight Center was prepared to comply with the decisions made during the review,

Rees remained doubtful about the feasibility of meeting the schedule within the allocated budget.

In his response, Charles Mathews acknowledged Rees's challenges and expressed an understanding of the difficulties. However, he reaffirmed the commitment to the July 1972 launch date and the budget constraints. He acknowledged that compromises would be necessary to align resources with requirements but emphasized that the fundamental philosophy of Skylab—"economical application of existing hardware with minimal new developments consistent with basic objectives"—remained intact. He justified the exceptions made over the past year as having substantial reasons and assured Rees that these did not signal a departure from the core policy. Mathews reminded Rees of the critical nature of Skylab, noting that it might be the only crewed mission for a significant period and stressing the importance of making the most of this opportunity. He expressed confidence that the challenges could be met with effective management, particularly by Schneider and the center program managers.

By mid-August, Rees wrote to Gilruth summarizing the status of the engineering change requirements submitted on May 26. After extensive negotiations and some compromises—where MSC withdrew specific requests, and others were rejected due to excessive cost or delays—significant changes to the food system were implemented. Rees implored his MSC counterpart to help resist further changes, emphasizing that Marshall lacked the financial resources and the time to accommodate additional modifications.

Fortunately, after this period of intense negotiation, the workshop faced no further major disruptions related to habitability requirements. The next significant challenge emerged from the waste collection and measurement systems. Later, in 1970, while astronauts would raise some concerns about the food with medical experimenters, these issues had minimal impact on the major food systems. Meanwhile, Caldwell Johnson's office continued to oversee developing habitability systems while shifting his focus to design challenges associated with the Shuttle and the emerging Apollo-Soyuz Test Project.

Medical experiments were a central justification for the Skylab Space Station. The research aimed to answer critical questions about the effects of weightlessness on the human body—questions that had arisen from earlier Mercury and Gemini missions. Specifically, researchers sought to understand what physiological changes occur in a zero-gravity environment, how long they persist, how astronauts adapt, and what measures could be taken to mitigate any adverse effects.

The Medical Research and Operations Directorate at the Manned Spacecraft Center (MSC) was responsible for developing the necessary instruments and systems to conduct these experiments. Physicians typically defined the experimental requirements, while the Crew Systems Division and the Engineering and Development Directorate designed and constructed the required hardware. However, the tragic Apollo spacecraft fire in early 1967 significantly disrupted these arrangements, causing delays and shifting priorities within MSC.

As a result, the medical experiments did not receive coordinated attention from all relevant MSC offices until 1969. Developing the experiments faced numerous technical challenges, reflecting their complexity and the novel nature of the research. These difficulties frequently threatened to push back Skylab's launch schedule.

Despite these challenges, MSC and Marshall Space Flight Center dedicated considerable effort to overcoming the technical hurdles associated with the experiments. Their perseverance paid off: all the medical experiments successfully functioned without major failure during all three crewed Skylab missions. This success underscored the commitment and expertise invested in addressing the critical questions about human adaptation to spaceflight.

The Experiments

Among the initial experiments planned for the Apollo Applications Program (AAP) missions were three key medical studies: metabolic activities, cardiovascular function assessment, and bone and muscle changes. These experiments were designed to address critical questions about human adaptation to spaceflight, building on the unexpected challenges encountered during Gemini missions.

The first study, which focused on metabolic activities, evaluated whether physical work in zero gravity demanded more effort than on Earth. This investigation used an ergometer, a specialized exercise bicycle, to measure energy expenditure during controlled exercise. The ergometer was intended for frequent use during Skylab missions to monitor trends in physical performance over time.

The second study addressed cardiovascular function, examining how the absence of gravity affects the heart and circulatory system. This experiment involved stressing the heart by simulating the effects of gravity through a partial vacuum device that drew blood into the legs, mimicking the gravitational pull experienced on Earth. Key metrics such as blood pressure, heart rate, and leg volume were telemetered to mission control, where physicians analyzed the astronaut's cardiovascular condition.

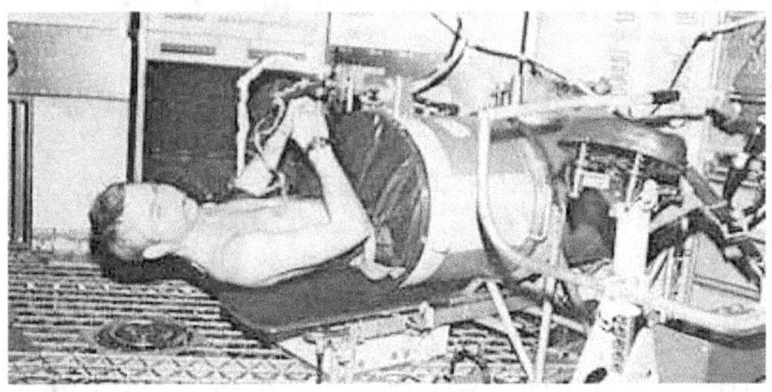

One of the Skylab's principal objectives was the continuing evaluation of man's fitness to live and work in space for long periods. Biomedical and behavioral experiments were conducted on all three manned periods. This space-age gravity tank tested the astronaut's cardiovascular system as it adapted to weightlessness. The tank created a partial vacuum, simulating the pull of gravity. This caused blood to pool in the legs, as on Earth. Sensors measured the ability of the heart to pump blood.

The third major experiment, focused on bone and muscle changes, was part of the mineral-balance research described in Chapter 7. This study sought to understand the impact of prolonged weightlessness on bone density and muscle mass, critical for assessing the long-term health effects of spaceflight.

In 1967, NASA's aerospace medical experts at Headquarters and the Manned Spacecraft Center (MSC) were deeply immersed in developing and detailing experiments for the Skylab Space Station. The task of selecting principal investigators and fine-tuning the experiments was a priority. However, the catastrophic fire aboard the Apollo spacecraft, which led to an urgent focus on recovering the lunar landing schedule, significantly impacted the Skylab program. This shift in priorities resulted in limited engineering support for Skylab's medical experiments. It was not until November 1967 that the program was fully organized and presented to the Manned Space Flight Experiments Board for review.

As the Skylab project progressed, various aspects of the space station's operations were carefully considered, including studies on circadian rhythms, habitability, and environmental factors. Understanding biorhythms and astronauts' circadian rhythms and analyzing aerosol composition and radiation levels was paramount. These factors ensured the crew's well-being and the experiments' effectiveness aboard Skylab.

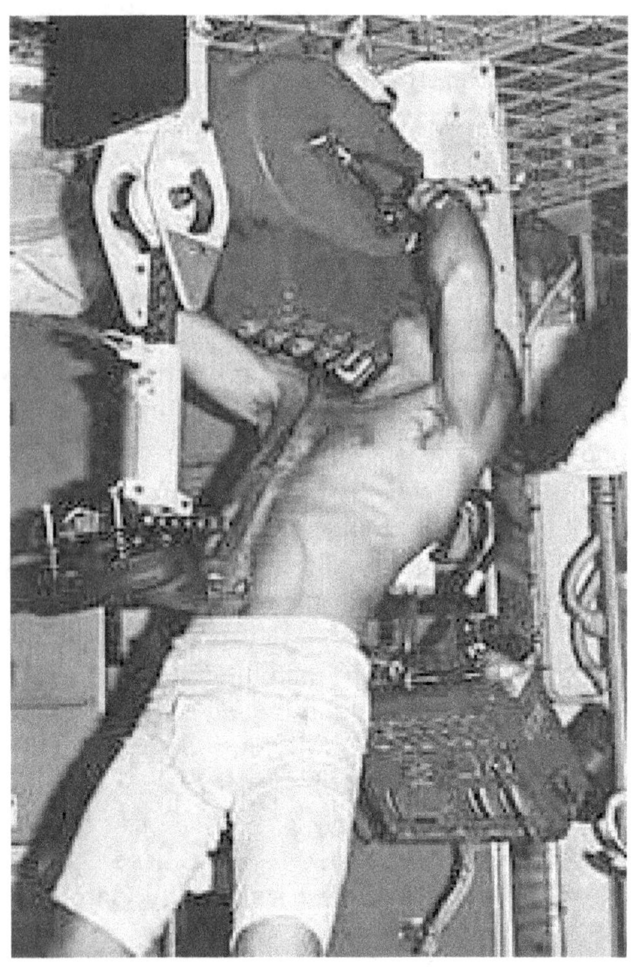 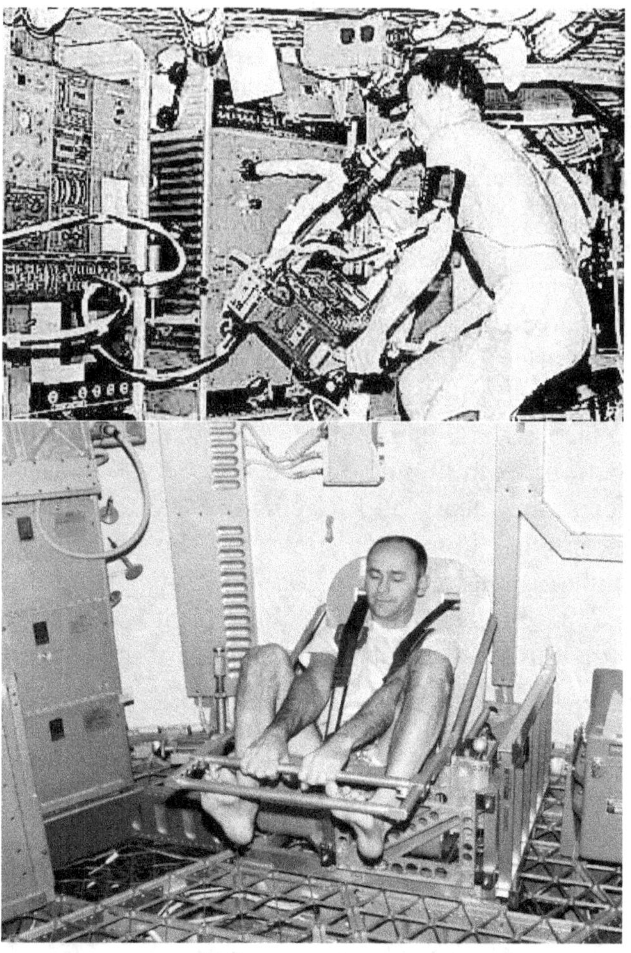

Astronauts Weitz (above) and Conrad (left) use the bicycle ergometer for metabolic activity experiments and exercise.

Skylab 3 astronaut Alan L. Bean uses the Body Mass Measurement Device to record his body weight in weightlessness.

By June 1970, the developmental versions of the ergometer, a critical piece of equipment for measuring astronauts' physical performance, were being tested at the Marshall Space Flight Center. The situation's urgency was evident as Dr. Charles A. Berry, the Director of Medical Research and Operations faced significant challenges in building the necessary medical equipment within the constrained budget. At a meeting held in March 1968 at Kennedy Space Center, Wernher von Braun proposed that Marshall could assist in fabricating some of the equipment, which could potentially save both time and money. Although von Braun believed Berry was receptive to this idea, subsequent communications revealed a degree of reluctance.

When von Braun's formal proposal was made, Gilruth's response was polite but noncommittal. Berry had doubts about Marshall's ability to meet MSC's stringent requirements. Recognizing the potential of these experiments to become a critical pacing item for the program, Marshall was keen to contribute. Negotiations continued through the fall, with Marshall seeking a commitment while MSC required detailed information on how the project would be executed.

On October 30, 1968, an agreement was reached whereby Marshall would construct the ergometer, the gas analyzer, the lower-body negative-pressure device, and the experiment support system. Although the financial investment was relatively modest, estimated at around $4 million, the engineering challenge was substantial. The project demanded Marshall's expertise in new areas, forming a task team from the Propulsion &

Vehicle Engineering Laboratory, headed by Robert J. Schwinghamer.

Despite the seemingly straightforward arrangement, managing the collaboration between centers proved challenging. Directing one center from another, akin to managing a contractor, led to occasional strains in relations. Nonetheless, these challenges fostered a creative tension that ultimately resulted in the development of high-quality equipment. Reflecting on the program's success, many participants acknowledged that these strains contributed positively to the overall success of the Skylab mission.

As one team at Marshall Space Flight Center focused on developing medical experiments for Skylab, another team faced a distinct yet equally challenging issue: creating a waste management system for the space station's workshop. This problem introduced new complexities not encountered in previous space missions. Unlike earlier programs, which only required a sanitary method for the collection and disposal of body wastes with minimal handling, Skylab's medical experiments demanded a more sophisticated approach. The system had to collect, measure, and return urine and feces for detailed analysis.

The waste management system for Skylab was driven by two primary considerations: the precise needs of the mineral balance experiments and the astronauts' requirements for a user-friendly and reliable system. The design specifications outlined in late 1968 were stringent. Each urine void needed to be measured with a precision of 1%, and a sample comprising 10% of each void had to be collected and dried. Solid residues were to be aggregated daily, with each sample tagged with identifying information—such as the individual astronaut, the time of collection, and the volume. By the end of a 28-day mission, a Skylab crew would be expected to return with approximately 540 grams of neatly packaged urine solids for laboratory analysis.

The engineering challenges of collecting and measuring liquids in a zero-gravity environment were formidable. Two systems were considered: one developed by the Fairchild Hiller Corporation for the Air Force's Manned Orbiting Laboratory (MOL) and another created by General Electric (GE) for the Biosatellite program. The former had been used for experiments with a seven-kilogram monkey. GE's prototype could measure volumes with an accuracy of 0.2%, while Fairchild Hiller's system was designed for rough volume measurements.

Marshall's team initially leaned toward the Fairchild Hiller system, believing it would be simpler to develop within the available timeframe. However, the medical experts at MSC expressed concerns about the system's ability to meet precise volume measurement and sampling requirements. They advocated for the GE system to be considered despite Houston's warnings.

In a significant decision, Marshall, influenced by advice from McDonnell Douglas—the prime contractor for both the MOL and the Skylab workshop—chose to proceed with the Fairchild Hiller system. This decision underscored the challenge of balancing engineering practicality with stringent scientific requirements, reflecting the complexities of ensuring that every aspect of Skylab's operations was meticulously planned and executed.

By early 1969, the researchers involved in medical experiments for the Skylab Space Station were reevaluating their methodologies to enhance the accuracy and reliability of their data. The need for meticulous procedures was paramount, particularly concerning collecting and preserving urine samples. In January, it was reported to the Marshall Space Flight Center (MSFC) that the investigators sought to collect urine over 24 hours, pool it to reduce measurement errors, and then freeze a sample. This approach addressed concerns about the stability of certain organic compounds—such as hormones and steroids—which might degrade under drying conditions.

The issue at hand was that drying, which involved heating the samples to 60°C under vacuum, could potentially destroy these compounds, thus compromising the integrity of the data. The standard practice of freezing samples emerged as the preferred method to mitigate this risk. Despite this, Marshall engineers resisted adopting the more expensive and labor-intensive freezing process. They contended that drying, as Fairchild Hiller's medical consultants endorsed, was sufficient. This disagreement occurred during the preliminary requirements review for the habitability support system on March 25, 1969.

The disposal of body waste was a difficult problem for Skylab's designers. Airflow was used as a substitute for gravity, separating waste from the body and disposing of it in special containers. Samples of collected waste were preserved and returned to Earth for medical analysis.

In response to these concerns, MSC proposed a study to validate the efficacy of drying versus freezing. McDonnell Douglas was tasked with comparing the two methods to determine if vacuum drying would alter the urine components or if such changes were predictable. An independent analytical laboratory was chosen to conduct this costly and time-consuming test. Despite the expenses, Marshall engineers believed that an independent study confirming the adequacy of drying would justify the investment.

Simultaneously, Houston's medical directorate remained adamant that freezing was the only acceptable method for preserving urine samples. By early summer, Bob Thompson stressed to Belew that the urine must be chilled immediately after collection, sampled, and then frozen for return. When it was decided in July to include provisions for frozen food in the Skylab workshop, McDonnell Douglas was directed to resume the preliminary design studies for a urine freezer. This decision, notably, was seen with irony by Paul Rambaut, MSC's principal coordinating scientist for the urine experiments. Rambaut observed that while there was unanimous agreement on the necessity of freezing urine samples, nutritionists similarly agreed that frozen food was not essential. Yet, the food freezer was accepted with little resistance, whereas the urine freezer faced significant opposition.

Throughout the summer, Houston's medical directorate remained skeptical of Marshall's commitment to providing a urine freezer. They cautioned that even if drying proved acceptable in the study, it might still be deemed unreliable due to its deviation from established practices in the field. Concerns extended beyond the preservation methods to the Fairchild Hiller system's volume determination techniques, prompting an exploration of alternative chemical tracer methods.

By late October 1969, Bill Schneider sought to address these ongoing disputes and uncertainties. On November 21, he convened a meeting at the Huntsville headquarters with program officials to discuss the unresolved issues. Although the test results comparing the two preservation methods were still pending, initial findings suggested that freezing was less effective than drying.

After thoroughly examining the engineering tradeoffs, Bill Schneider reaffirmed the decision to proceed with the current plans, allowing the study on the feasibility of a urine freezer to continue. However, this resolution dismayed the medical team at MSC, who requested another review of the situation. On December 18, 1969, Marshall reviewed the experiment requirements established by MSC, noting that freezing was not specified initially. Following a detailed assessment of the engineering considerations and test results,

Marshall recommended maintaining the existing drying system, halting all work related to sampling and freezing, and focusing on urine storage tests to determine the rate at which heat-sensitive components were lost over time. Schneider, once again, saw no compelling reason to switch to freezing and agreed only to have Fairchild Hiller's test results reviewed by an independent consultant and to study the impact of sampling and freezing on the Skylab workshop systems. He emphasized that if the transition to freezing caused any schedule delays, Marshall should find ways to address these issues without affecting the workshop's timeline. Consequently, on December 30, Marshall instructed McDonnell Douglas to conduct the study.

Over the following three months, Fairchild Hiller and its subcontractor, Bionetics Inc., based in Bethesda, Maryland, completed their studies comparing drying and freezing methods. MSC meticulously analyzed the results and statistical data. The findings were inconclusive, with Fairchild Hiller's program manager acknowledging in the final report that the statistics were a draw. MSC conducted its tests, which revealed greater losses of hormones in dried samples compared to those reported by Bionetics.

In February 1970, Paul Rambaut summarized the findings and recommended discontinuing the drying process. The tests had demonstrated severe and unpredictable deterioration of heat-sensitive compounds, and no recognized expert deemed heat-drying acceptable for the Skylab missions. Despite recognizing the engineering challenges of implementing a freezing system, Rambaut concluded that further efforts to qualify for Skylab were futile.

With the results in hand, Schneider convened a final meeting on March 10, 1970, to review and discuss the implications of the findings.

The debate over the urine processing system for the Skylab Space Station became one of the most contentious issues in the entire workshop program. Despite Marshall Space Flight Center's staunch defense of their approach, they could not counter the persuasive arguments from Houston. The issue was whether to employ a system to freeze urine samples or dry them. Marshall, represented by Stan McIntyre, the project's engineer for the urine system, was concerned about the complexity of handling fluids in microgravity—a task they had never previously encountered. The decision to avoid tackling these complexities led them to rely on their contractor's scientific adviser, who assured them that drying the samples would meet the medical objectives.

However, Houston's team, led by Dr. Paul Berry, was adamant that the Fairchild Hiller system would not be viable and had warned Wernher von Braun of its limitations. Berry was particularly frustrated by what he saw as engineers debating with medical experts over fundamentally medical issues. Ultimately, the decision was made to switch to a freezing system, which Marshall engineers reluctantly accepted. Four years later, a Marshall project manager expressed dissatisfaction, insisting that urine drying would have been a better choice.

With the decision to freeze the urine samples settled, attention shifted to accurately measuring the urine volume. The goal was to measure daily urine output within 2%, a challenging task given that liquids in microgravity can trap gas. Fairchild Hiller's system used a synthetic membrane of microscopic liquid-repellent fibers that allowed gases to pass through while retaining the liquid. The system's design involved a urine collection bag made from this membrane, intended to use surface tension to separate liquid from air. Engineers claimed that the bag could measure volume by determining the thickness of the liquid while the bag was held in a box of fixed dimensions.

In contrast, General Electric proposed a different approach. Their system used a centrifugal separator to separate air from urine and a peristaltic pump to measure the volume and collect a proportional sample. By the spring of 1970, program officials began evaluating these two systems. McDonnell Douglas, the prime contractor, strongly advocated for the Fairchild Hiller system. However, Houston's medical team supported the General Electric device, believing it had better prospects for future development.

While traditionally aligned with their prime contractor, Marshall's officials were skeptical of

Fairchild Hiller's system and appeared reluctant to engage in further disputes with Houston. During a review on April 3, the General Electric system demonstrated clear technical advantages. Despite this, General Electric representatives hesitated to commit to developing the system for Skylab. McDonnell Douglas vigorously defended Fairchild Hiller's system, claiming it could guarantee an accuracy of 1% in volume measurement.

With General Electric seemingly unwilling to compete further by May, the Fairchild Hiller system was chosen for development and testing. However, when tested in zero-gravity aircraft flights, the Fairchild Hiller collection bag faced significant issues. The membrane, supposed to be impervious to liquids, failed after prolonged exposure to urine, complicating the task of storing urine for an entire day during operations. This outcome underscored the challenges of developing technology for space environments and highlighted the complexities of creating reliable systems for the Skylab mission.

As Skylab's development surged forward, engineers grappled with the challenge of handling fluids without gravity. This persistent concern threatened the integrity of vital systems aboard the station. Their concerns focused on a seemingly mundane but crucial aspect: urine collection and measurement. Initially confident in their understanding of fluid dynamics under zero-gravity conditions, the reality of space proved more unpredictable. Small, unbalanced forces present during zero-g maneuvers could potentially compromise the filtration system, raising doubts about the reliability of existing designs.

In response to these challenges, engineers at both the center and contractor levels embarked on an intense problem-solving period in September. They explored numerous remedies, including redesigns and alternative approaches, yet confidence in the existing system remained low. Proposals ranged from an original single-bag system to a dual-bag system—one for collection and another for measurement—each introducing its complexities.

During this time, headquarters received intriguing news from the Langley Research Center, where experts in fluid mechanics were investigating gas-liquid separation using a centrifugal separator. This technology promised effective separation of liquids and gases in a weightless environment. Preliminary dialogues between Langley and the Marshall Space Flight Center hinted at the potential of this new device, prompting various reviews and meetings through October.

By late 1970, Skylab's project leaders, including Schneider, Belew, and Kleinknecht, strategically decided to advance three systems: the original single-bag design, the problematic two-bag design, and the promising centrifugal separator. They aimed to determine which system would ultimately prove most effective, ensuring operational efficiency and reliability in space.

As the rigorous testing continued, the centrifugal separator demonstrated significant advantages. By November, zero-gravity tests revealed critical flaws in the two-bag system, leading to its gradual disfavor. Entering 1971 with fresh perspectives, Belew advised Schneider that the single-bag system appeared increasingly inadequate. Consequently, the focus shifted decisively toward the centrifugal separator, deemed the only solution to meet all significant experimental and operational requirements.

By January 15, 1971, the decision was formalized: the one-bag system was abandoned, and efforts were concentrated on refining the two remaining designs, emphasizing their interchangeability to simplify integration. Hamilton Standard, which had previously collaborated with the Manned Spacecraft Center during the Apollo program, was contracted to develop the centrifugal separator further.

Despite these advances, the engineers also explored an alternative tracer volume measurement method, which required additional refinement. As the months progressed, the inadequacies of the two-bag system became unequivocally clear. By June, a cost-benefit analysis revealed that maintaining it even as a backup would entail a prohibitive expense increase—over $1.5 million. Finally, on July 21, the decision was made to cease all development on the two-bag system, solidifying the centrifugal separator as the chosen technology for Skylab.

The methodology for measuring urine volume in Skylab was ingeniously straightforward yet thoroughly scientific. Before use, each urine collection bag was prepared with a known quantity of a substance not typically found in urine—in this case, lithium. Lithium chloride was specifically chosen for its distinct chemical signature. As each bag was used and subsequently filled, the contents were thoroughly mixed, ensuring that the tracer was evenly distributed throughout the urine. By analyzing a sample from the bag, engineers could determine the total urine volume by comparing the lithium concentration in the sample to its known initial concentration.

This tracer method, having proven its precision and reliability, was adopted by the Medical Sciences Division (MSC) as a backup for in-flight volume measurements. The implementation involved recirculating the urine in the bag through a centrifugal separator, which mixed the tracer thoroughly with the urine, ensuring accurate measurement.

In contrast, the design for collecting solid waste was significantly more straightforward and benefited from prior developments by Fairchild Hiller for the Manned Orbiting Laboratory. The solution involved a plastic bag with a porous filter that allowed air passage but contained the fecal matter. Positioned beneath a standard toilet seat, the bag was integrated into a system where an air blower would funnel feces directly into it through holes around the seat's rim. The air used in this process would be cleaned through a deodorizing filter before being recirculated back into the station. After collection, the feces were vacuum-dried using heat—this process not only preserved the waste for analysis but also ensured moisture removal. The processed bags were then weighed and stored for return to Earth.

Unlike the liquid systems, the fecal collection system faced fewer design challenges due to the more uncomplicated nature of handling solids in zero gravity. However, testing the system proved to be more problematic, mainly due to the limitations of zero-gravity simulation. The KC-135 aircraft used for testing could only provide about 30 seconds of weightlessness, a period too brief for comprehensive testing. Mechanical devices successfully simulated urination, but defecation was another matter. It required human testers who could perform on command, a rare capability. In November 1969, various aircraft tests with these unique individuals yielded valuable data, though these tests were not deemed entirely conclusive.

Efforts to further validate the system included proposals to test the fecal collector during an Apollo mission. By January 1970, Marshall's Skylab office was lobbying for an opportunity to test the system in space. Initially, there was an agreement to include the test on Apollo 14, but this decision was rescinded. Varying levels of support and opposition within NASA reportedly influenced the change;. At the same time, the Skylab office backed the test, the astronaut office remained indifferent, and the mission commander ultimately vetoed it. Consequently, Marshall's team had to rely solely on the results from the aircraft tests, a situation that underscored the complex interplay of technical challenges and human factors in space mission planning.

Building The Medical Hardware

From the vantage point of the Marshall Space Flight Center's director, building experiment hardware for the Manned Spacecraft Center (MSC) appeared straightforward. The Biomedical Task Team at Marshall was tasked with fabricating key components, such as the ergometer frame and the shell for the lower-body negative pressure device. This work involved contracting out some components and then assembling and testing the final products to meet the specifications provided by Houston.

The agreement between the two centers outlined that Marshall would operate "in the same manner as any other contractor," with MSC managing the contract as they would with any commercial vendor. However, the usual incentives and penalties that NASA could impose on commercial contractors were absent in this inter-center arrangement. Marshall's task team's management and technical direction fell under Houston's medical directorate, while the Skylab office retained overarching management responsibilities, including verifying requirements and overseeing resource allocation.

This arrangement proved cumbersome, as events would reveal over the next two years. The complex lines of authority and the often ambiguous management structure made determining who was ultimately in charge at MSC challenging. These management issues compounded the technical difficulties faced by Marshall's task team.

A critical component of the Skylab missions was the M I 71 metabolic activity experiment, designed to measure the body's rate of energy production during physical exertion. The experiment utilized a bicycle ergometer that provided calibrated resistance levels for the astronaut. The energy production was assessed by measuring the ratio of carbon dioxide exhaled to oxygen inhaled. While the construction of the ergometer itself posed no special challenges, the system required for measuring respiratory gases presented significant difficulties. This system demanded precise flowmeters, accurate valves, and a high-speed gas analyzer—technologies at the forefront of their field—all interacting with a specialized computer and data transmission system.

Given the tight development schedule and the complexity of the experiments, the medical directorate in Houston sought to evaluate multiple design options. For the gas analyzer, they had initially selected a mass spectrometer—a sophisticated electromagnetic instrument capable of separating gases based on their molecular weights and determining their proportions in a mixture. In 1969, while Marshall's biomedical task team was evaluating one design for the mass spectrometer, the Skylab office at Houston was considering another design proposed by Martin Marietta. By September of that year, a third option emerged when MSC's Biotechnology Division identified an existing mass spectrometer being developed for a different purpose and recommended its adoption for the metabolic analyzer.

Throughout this period, the medical experimenters hoped that ongoing development would reveal a superior design. However, the urgency of the Skylab office's schedule meant that decisions had to be made promptly despite the potential advantages of waiting for a more definitive choice.

In April 1970, after thoroughly evaluating the three competing designs, Houston's program manager, Kenneth Kleinknecht, made a pivotal decision. He selected the design that Marshall had been championing, recognizing that this option met the established medical requirements and acknowledging the significant financial investment already made. Kleinknecht sought confirmation from Marshall that they were committed to completing the project. Once reassured, he instructed that development on the other two designs be halted.

However, by early 1970, the broader array of medical experiments grappled with numerous management challenges. A particular point of contention arose between Marshall and Martin Marietta, the workshop integration contractor. The disagreement centered on the responsibility for integrating the Marshall-built medical experiments with the experiment support system, a task also assigned to Marshall. Reports from Marshall's representative in Houston indicated ineffective communication between the medical directorate and the Skylab program office at MSC. Concurrently, at Huntsville, Robert Schwinghamer's task team experienced frustration with what they perceived as conflicting instructions from MSC, leading to disorganization and confusion.

The pressure of adhering to a stringent schedule only exacerbated these issues. By July 1970, the medical directorate formally requested relief from the demanding timeline. According to the original schedule, development test units for the experiments—prototypes intended to identify design or construction faults—were due for delivery in October 1971, a full 13 months before the scheduled launch. Flight units, which would incorporate modifications based on these tests, were required just one month later. This narrow timeframe was deemed insufficient for addressing identified deficiencies, raising concerns that the design and testing process might be compromised. The medical team feared they might be compelled to proceed with less than fully refined experiments simply due to the looming launch date. They argued that mission objectives should take precedence over the unrealistic schedule, advocating for a launch delay to ensure the experiments would provide complete and reliable results.

In response to these challenges, a decision was made to adjust the deadline for the metabolic analyzer—the most critical component of the medical experiments—to allow for necessary testing, provided that the overall schedule for the workshop was not delayed. The workshop contractor was expected to adapt to the absence of the incomplete experiment.

Recognizing the need for improved coordination and communication, medical director Charles Berry and center director Robert Gilruth decided to enhance their liaison efforts with Marshall. In September, Gilruth announced the appointment of Richard S. Johnston as Berry's deputy director for biomedical engineering. Johnston was also tasked with leading a newly established Skylab Project Support Office to address the ongoing issues and streamline the integration and management of the Skylab medical experiments.

Johnston's tenure in the early years of the Apollo and Skylab programs was marked by significant responsibilities and challenges. As the chief of the Crew Systems Division, he played a pivotal role in shaping the program's early direction. After serving as Gilruth's special assistant for two years, Johnston took on the role of Apollo's experiments manager in 1970. His expertise in navigating the program's complex management structures was instrumental in addressing the numerous challenges faced during this period.

By late 1970, Johnston had successfully streamlined the translation of medical requirements into tangible hardware by integrating several skilled engineers into the project. This move significantly alleviated the management issues that had previously hindered progress. Marshall's first major milestone was developing design verification test units. These units were crucial for identifying design and construction deficiencies by simulating their expected use. Originally slated to begin in October 1970 and continue until July 1971, the verification testing

was delayed and did not commence until February 1971. The initial testing phase encountered setbacks, including six weeks of lost time due to failures in components provided by MSC contractors.

By mid-May, Huntsville officials anticipated a return to testing shortly, but new requirements imposed by MSC extended the testing program into 1972. The assembly and testing of the hardware continued throughout 1971 with a target deadline of January 15, 1972, for delivering all flight hardware to McDonnell Douglas. However, persistent issues with electronic modules, notably the leg-volume measuring device from Martin Marietta and the malfunctioning metabolic analyzer, plagued the project. By mid-summer 1971, only the bicycle ergometer and the lower-body negative-pressure device functioned relatively trouble-free.

In June 1971, MSC's request to examine components of the metabolic analyzer led to a halt in progress on the most successful test program to date. Later in September, "NASA alerts" were issued, warning of defective electronic components, including capacitors and integrated circuits used in various systems. Despite efforts to replace the faulty capacitors, the nearly 200 integrated circuits could not be adequately replaced on time, causing a significant delay. The scarcity of acceptable replacements meant delivery of new components would take 12 to 20 weeks. Consequently, testing continued with the existing components, but it was necessary to plan for replacement and further retesting of the equipment.

As the year drew to a close, Huntsville informed McDonnell Douglas of an expected delay in delivering flight articles, which would arrive 2 to 4 weeks later than originally scheduled. The complications were further exacerbated by Houston's plans to simulate a 56-day Skylab mission using the medical hardware. For this simulation to be valuable, it must be conducted well before the first mission, necessitating fully functional experimental equipment. At the same time, McDonnell Douglas's California plant had reached a stage where the absence of the required medical hardware hindered assembly and checkout of the workshop. By late January 1972,

MSC requested authorization to delay the completion of tests and the delivery of hardware by up to six weeks, reflecting the ongoing difficulties and adjustments needed to ensure the success of the Skylab Space Station program.

Schneider's response to the request was a cautious compromise. While he approved the potential delay of hardware deliveries, he insisted that the test program proceed as scheduled. He directed the centers to find solutions to the ongoing issues rather than allowing the delays to impede the entire testing process. Schneider also considered removing the problematic metabolic analyzer from the program but sought an estimate of the potential impact of such a decision. Despite the situation's severity, the medical team and the program office strongly opposed this idea, arguing that the experiments were essential and that solutions for the metabolic analyzer's issues were feasible. Schneider ultimately accepted their evaluation, and the discussion of removing the analyzer was dropped.

Marshall's team found a workable solution by substituting one metabolic analyzer unit with another, allowing them to meet the revised delivery schedule for the MI71 equipment in late February. By then, flight units of the medical equipment began arriving in California, with the metabolic analyzer arriving on April 13. Although significant integration and testing still lay ahead, the most challenging aspects of the project were now behind them.

Parallel to these developments, Houston's medical directorate had been contemplating a comprehensive simulation of a 56-day Skylab mission since 1968. The primary concern was understanding how the microbial environment might change in the Skylab's confined space and avoiding any potential bacterial infections during and after the mission. Additionally, the simulation would provide critical one-g data from the medical experiments, helping to assess the effects of weightlessness and verify the experiment procedures and equipment.

In early 1970, MSC requested funds from Headquarters to carry out a full-scale mission simulation. However, Headquarters found the original proposal too ambitious due to budget constraints. After extensive discussions, a modified plan was developed. Instead of utilizing two flight-configured Skylab mockups, MSC agreed to use an existing altitude chamber, which would be outfitted with flight-type medical hardware, waste-management systems, and flight food. The revised plan focused on checking the hardware, establishing baseline medical data, and validating experiment procedures and data-handling systems rather than exploring bacterial ecology.

This modified plan was approved in February 1971, leading to the initiation of the Skylab Medical Experiments Altitude Test, or SMEAT. This simulation was to be the only mission-length test in Skylab's experiment program. Houston's medical directorate, led by Richard Johnston as the steering committee chair, meticulously organized the simulation. Four test-project managers were appointed to oversee different aspects of the test, collaborating closely with medical teams, principal investigators, and personnel responsible for flight operations and crew training. The altitude chamber, maintained by the Crew Systems Division, provided a realistic environment for the simulation, closely approximating the conditions expected during the Skylab mission.

The Skylab Space Station, a pioneering endeavor in space station design, was meticulously crafted to simulate an orbital workshop as closely as possible. The station's configuration mirrored the planned layout of Skylab, with a keen focus on functionality and the accommodation of the crew's needs.

The lower level of Skylab was designed with a practical layout, including a wardroom for dining and food preparation, a compartment for medical experiments, and a waste management area. Unlike in the weightlessness of space, where objects float freely, the one-gravity environment of the test chamber presented specific constraints. For instance, crew members could not sleep against the walls as they might in space; instead, sleeping arrangements had to be aligned with the floor. Similarly, the waste collection module, which would be mounted on a wall when in space, had to be situated on the floor in the test chamber, reflecting the challenges of adapting to different gravitational conditions.

The upper level of Skylab, reserved for storing equipment and conducting experiments, was repurposed during the simulation as a study area. The crew's time was divided between conducting medical experiments and engaging in educational activities. They planned to utilize their off-hours to study Russian and read, reflecting the crew's dedication to maximizing their time and preparation.

Outside the test chamber, medical operations personnel were tasked with monitoring the performance of the medical experiments, collecting data in real time to replicate the conditions of an actual mission. Communication with the crew was intentionally intermittent, mimicking the actual contact times between Skylab and ground stations.

In mid-1971, a dedicated crew was selected for the Skylab Mission Simulation and Evaluation Test (SMEAT). The team included Lt. Cmdr. Robert L. Crippen, USN, and Lt. Col. Karol J. Bobko, USAF, both former MOL astronauts who had joined NASA in September 1969. They were accompanied by William E. Thornton, a physician and biomedical engineer from the scientist-astronaut group selected in August 1967. Although Thornton was the only member with direct involvement in Skylab at the time, serving as a principal investigator for the small-mass measurement device designed for weighing specimens in flight, all three brought valuable expertise to the simulation.

The crew began their eight-week simulation on July 26, 1972. The initial days were a whirlwind of activity as both the crew and operations personnel worked to establish effective working relationships and refine procedures. The SMEAT crew quickly adapted to the demands of the simulation, experiencing a sense of camaraderie among themselves while developing a distinct "us versus them" mentality toward those outside the chamber. Despite facing various predictable challenges, such as poorly fitting medical sensors, unfamiliar equipment, and procedural modifications, they navigated these issues with resilience.

The crew found the living conditions acceptable, though far from luxurious. Meals were satisfactory, though not particularly exciting. The schedule was filled with tasks, leaving little idle time, though they did manage to enjoy an hour of television daily and maintain contact with family and friends via an external telephone line.

Despite the generally manageable nature of the simulation's challenges, several significant issues underscored its value. Early in the simulation, the bicycle ergometer malfunctioned, and the metabolic analyzer proved consistently erratic. More critically, the SMEAT crew identified significant faults in the urine collection system, revealing the need for substantial redesign work. These issues highlighted the critical role of the simulation in identifying and addressing potential problems before the actual Skylab missions commenced.

With Skylab's launch just nine months away, the teams at the Manned Spacecraft Center (MSC) and the Marshall Space Flight Center faced pressing challenges. Immediate troubleshooting was necessary to address issues with the ergometer and metabolic analyzer, two crucial pieces of equipment.

The ergometer, which had failed during the simulation, was found to have a mechanical design flaw specific to the test unit. Once this issue was resolved, the ergometer functioned as intended. Nevertheless, the other units, including one already installed in the workshop, were dismantled, examined, and rebuilt to ensure reliability. Spare parts were also added to the flight inventory as a precautionary measure.

The metabolic analyzer presented a more complex problem involving mechanical and electronic failures. A meeting in late September identified essential modifications and tests, and Marshall began reworking the affected units to address these issues.

More critical were the problems with the urine collection system, which had the potential to undermine the mission. Initially, the two-liter collecting bags, designed based on physiological norms, proved inadequate. The system had not been tested with the actual crew members' output, leading to complications during the simulation. One crew member's daily urine output reached nearly three liters, surpassing the capacity of the collecting bags. While the SMEAT crew had access to alternative toilet facilities, the issue was of grave concern to the engineers.

The urine pooling bag and its associated mechanical components occupied the allotted space, making any increase in capacity nearly impossible. Compounding the problem, the urine centrifuge leaked, and the collection unit could not be thoroughly cleaned. On multiple occasions, collection bags tore during handling, resulting in significant urine spills, contaminating the waste-collection unit, the floor, and, occasionally, the crew members themselves. These failures led to growing discontent among the astronauts, who felt the system was too complex and insufficiently tested in zero gravity.

While preparing to command the first Skylab mission, Pete Conrad expressed profound skepticism about the urine system. His confidence waned, prompting him to collaborate with Houston engineers to explore adaptations to the Apollo 77 system as a potential alternative. The deteriorating relationship between engineers and crew representatives highlighted the severity of the situation.

Further complicating matters, tests conducted at McDonnell Douglas revealed an unrelated defect in the urine system. The in-flight volume-measuring device, a sophisticated contraption involving a pressure plate and mechanical linkages, failed to meet accuracy requirements. With the launch looming just six months away, the urine system faced the prospect of a complete redesign or a reassessment of the medical requirements—or potentially both.

The urine collection system's issues were exacerbated by its design, which was based on physiological norms rather than measurements taken from the astronauts themselves. At the time of the system's design, the crew members had not yet been selected, and the astronaut office had denied requests from medical investigators to measure the astronauts' 24-hour urine output to avoid interfering with training schedules.

A week following the conclusion of the Space Mission Evaluation and Analysis Test (SMEAT), key discussions took place among headquarters, the Houston control center, and the Huntsville engineering team. These discussions focused on the urgent need to enhance the Skylab space station's urine collection system. The consensus emerged to expand the system's storage capacity to four liters to better accommodate the astronauts' needs.

Three potential design modifications were considered. Two of these proposals involved bypassing the centrifuge entirely and drawing upon the technology used in the Apollo 17 mission. However, a more refined solution was reached after two weeks of deliberation. It was decided to implement a dual-choice system to utilize a four-liter storage bag and provide the crew with a selection of collection devices: the Skylab centrifuge or the Apollo roll-on cuff. This new approach aimed to streamline operations while giving astronauts flexibility in their collection methods.

The volume measurement of the collected urine during flight was deemed unnecessary. Instead, the lithium chloride tracer technique was sufficient for tracking the system's performance. These design changes simplified the urine collection drawer, accommodating the larger storage bag and a protective metal enclosure. However, a new challenge emerged in the need to manually knead the bag to mix the 24-hour pooled urine instead of recirculating the contents through the centrifuge.

By November 15th, approximately three and a half months after these issues were first identified, a revised and acceptable design had been critically reviewed, and modifications were underway. Dick Johnston, reflecting on the significance of SMEAT, remarked that the test was pivotal in saving the Skylab program, averting potential operational problems to have been unresolvable in space. Both Johnston and Ken Kleinknecht openly acknowledged the challenges but expressed confidence that solutions would be implemented effectively.

When the redesigned waste management system was eventually deployed, the four months of intensive work post-SMEAT proved to be invaluable. The new urine system and associated medical hardware functioned flawlessly, meeting all operational requirements. The redesigned system proved its worth as two separate crews each had at least one member whose urine output exceeded two liters daily. Experienced astronauts found the system to be a significant improvement over previous models, while newer crew members, initially apprehensive due to unfavorable stories about waste management, were favorably impressed. The ultimate validation came from Pete Conrad, who took the time to commend the engineers for their outstanding work on the system.

The modifications were further underscored by using a specialized rubber tube resembling an external catheter, establishing an effective attachment to a collection bag. This apparatus, akin to a heavy-duty condom, was a crucial component of the redesigned waste management system.

Among Skylab's most ambitious scientific endeavors was the Apollo Telescope Mount (ATM), a project initiated in 1966. As the most complex and costly scientific venture within the Skylab program, the ATM represented the pinnacle of technical sophistication. It was allotted one of the three Apollo Applications Program (AAP) missions, underscoring its significance in the broader scope of space science.

The decision to launch the solar observatory as part of Skylab's orbital assembly led to numerous design considerations and mission requirements. The ATM's distinctive feature was its four solar arrays, which extended outward like the sails of a medieval windmill, making it a prominent and striking component of the orbiting space station. At the center of these arrays was the canister housing the six major instruments of the solar observatory.

These instruments were designed to study various aspects of solar radiation. Five focused on detecting high-energy ultraviolet and X-ray radiation, which does not penetrate Earth's atmosphere and thus remains inaccessible from the ground.

The sixth instrument was dedicated to capturing images of the sun's corona, a delicate layer of gas generally obscured by the intense brightness of the solar disk and the scattering effects.

In 1972, a comprehensive briefing, 72-PM 7200-1 15A, outlined the fundamental objectives and technical specifications of the Apollo Telescope Mount (ATM), a vital component of the Skylab Space Station. This briefing emphasized the need to understand the intricate processes through which energy was transferred from the sun's interior to the vast expanse of space. Solar physicists required instruments capable of high resolution, precise pointing accuracy, and exceptional stability to achieve this. These attributes were meticulously designed into the telescopes and their supporting systems from the outset.

Initially intended for Advanced Orbiting Solar Observatory deployment, the ATM's instruments were versatile and adaptable. With appropriate modifications, they could be utilized for other missions as well. One significant alteration involved converting these instruments to photographic recording, except Harvard's ultraviolet spectrometer, which remained photoelectric. The decision to employ film for recording was driven by its superior spectral and spatial resolution compared to photoelectric methods. However, photoelectric instruments offered broader intensity range recording and faster response times.

The reliance on film presented a practical challenge: it required replenishment during the mission. This necessity meant that astronauts had to manage the recovery of exposed film and reload the cameras during extravehicular activities (EVAs). This decision was particularly bold given the context of 1966, when working outside the spacecraft was still a novel and risky endeavor. The early Gemini missions had provided limited and somewhat discouraging experiences in this regard.

Several accessory instruments were integrated into the ATM during its final design phases to support the astronauts operating the solar telescopes. One such instrument was a monitor that measured the sun's total X-ray output, providing a valuable index of overall solar activity. This monitor was linked to an audible alarm system designed to alert the crew when x-radiation levels surpassed a predetermined threshold, signaling the potential for an imminent solar flare and prompting the crew to man the control panel. Additionally, another monitor provided a real-time image of the sun in ultraviolet radiation, aiding in identifying active solar regions.

In 1968, significant enhancements were made to the Apollo Telescope Mount (ATM) with the addition of two pointing-control telescopes. These telescopes were equipped with filters designed to isolate a specific wavelength of light, namely the red-orange light associated with the hydrogen-alpha (H-alpha) line in the spectrum of incandescent hydrogen. This particular wavelength, measuring 656.3 nanometers, was the first and longest wavelength line in the Balmer series of the hydrogen spectrum.

The inclusion of H-alpha telescopes allowed the detailed observation of the sun's surface, revealing its fine granular structure. The data captured by these telescopes was displayed on a television monitor located at the control panel, providing real-time visual feedback. Each telescope featured variable focal length (zoom) lenses and cross-hairs, facilitating precise alignment with other instruments. Cameras were used to document the exact orientation of the H-alpha telescopes at the time of each observation, ensuring accurate and reproducible results.

The solar observatory was designed for full exposure to the Sun throughout most of the Skylab mission, and its components were carefully controlled in temperature.

Initially, the experiments and supporting systems were designed to operate almost independently of their carrier vehicle. By 1969, this carrier vehicle was a modified lunar module whose ascent stage was repurposed to house a pressurized cabin with space for two crew members and a control and display console for the scientific instruments. However, with the transition to the dry workshop configuration, which eliminated the lunar module, adjustments were required for the ATM. The telescope assembly was relocated onto a supporting structure above the multiple docking adapter. The pointing system was updated to manage the entire workshop, and the control panel was repositioned into the now-vacant lunar module.

Despite these modifications, the core functionality and development of the instruments remained largely unaffected. By the end of 1968, developing these systems was well-advanced, and the changes necessitated by the shift from the lunar module to the dry workshop were smoothly integrated into the ongoing project. The precise resolution of the H-alpha telescopes, defined by their ability to distinguish closely spaced spectral lines, images of adjacent solar points, or rapid temporal events, played a crucial role in the comprehensive study of solar phenomena.

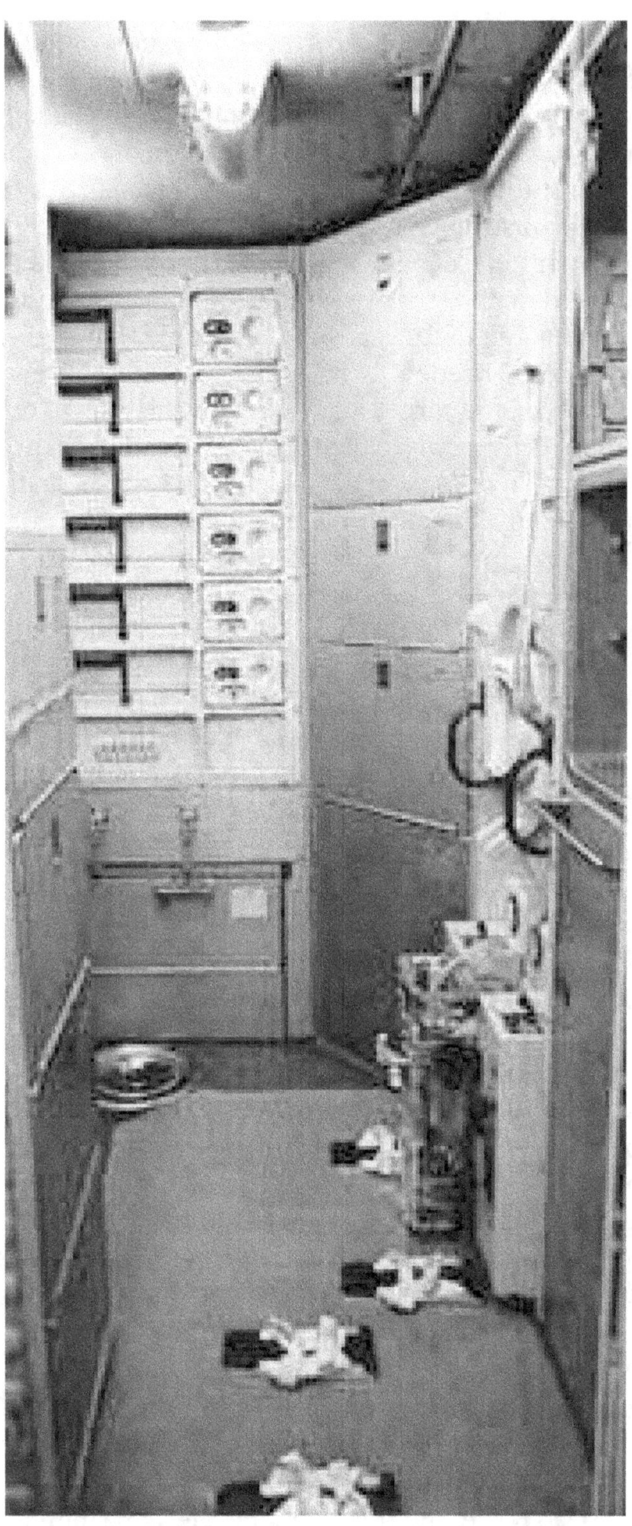

This photograph shows the waste management compartment, with the fecal-urine collector mounted on the wall.

The Solar Study

The Apollo Telescope Mount (ATM) was an engineering marvel. It consisted of a massive array of solar telescopes that collectively weighed over a ton. Some of these telescopes stretched up to three meters in length. Despite their size, they were precision optical instruments that required sophisticated support systems to function optimally.

One primary challenge was ensuring the telescopes could be aimed with extreme precision at any point on the sun and maintaining that focus despite disturbances affecting the orbital assembly. This necessitated rigorous control over the telescopes' temperature. For instance, the Naval Research Laboratory's telescope had to be maintained within 1.5°C of its calibration temperature, with fluctuations limited to 0.005°C per minute. This was particularly challenging given the instrument's exposure to unfiltered solar radiation.

Effective use of the human operator also posed a challenge. The goal was to automate as many operations as possible while allowing the crew to make real-time decisions regarding observations and the selection of instruments. Additionally, managing electrical power and data systems was crucial to the operation's success, though not as complex as other challenges.

The Marshall Space Flight Center was responsible for developing and integrating these supporting systems and the ATM's structural framework. In July 1966, Apollo Applications Program Manager Leland Belew established an ATM Project Office with Rein Ise as project manager. Ise, whose background included experience with the Army Ballistic Missile Agency before NASA, was one of several engineers transitioning from the Saturn IB-Centaur program to the Apollo Applications Program.

Marshall's Astrionics Laboratory was tasked with building ATM components, contracting out some systems to industry, and assembling the test, prototype, and flight articles. By mid-1968, the laboratory had developed a control and display simulator to refine switches, controls, and computer logic. This simulator was later upgraded to produce computer-generated displays that replicated in-flight observations, aiding crews and engineers in developing and verifying operating procedures.

Simulators for power, attitude control, and pointing control systems were also under construction in 1968. Training hardware included a one-gravity trainer, which established a full-scale mockup of the ATM, excluding the solar power arrays, complete with functional workstations. A zero-gravity trainer, consisting of mockups of workstations, was designed to be tested in a KC-135 aircraft. This aircraft provided a crucial platform for zero-gravity testing and played a vital role in the design and operational preparation of the ATM.

At the heart of the Skylab Space Station's scientific endeavors lay a complex network of instruments and systems designed to push the boundaries of human knowledge and technological capability. Among these were the advanced telescopes equipped with film cameras, crucial in capturing experimental data. Retrieving this data was not a simple task; it required astronauts to work outside the space station, a process that necessitated meticulous planning and execution.

The collaboration between mission control centers in Houston and the engineers at Huntsville was vital for the success of this operation. Astronauts frequently engaged with engineers to discuss and refine the designs of the workstations where the film cameras were to be removed and replaced. Given that the zero-gravity conditions achievable in aircraft—limited to 20 to 30 seconds at a time—were insufficient for thorough testing, engineers sought a more effective simulation method. They turned to underwater simulations, which allowed a more comprehensive and practical approach to zero-gravity tasks.

In the early years of the Apollo Applications Program, engineers at the Marshall Space Flight Center utilized a water tank designed initially for explosive forming to simulate neutral buoyancy. This technique had been proven effective during the preparations for the Gemini missions. By 1968, the center had completed construction of a new Neutral Buoyancy Facility, a tank specifically designed for this purpose. Measuring 22.8 meters in diameter and 12 meters deep, the facility could immerse full-size mock-ups of spacecraft components, providing a crucial design and crew training environment. While underwater simulations were not a perfect replication of zero gravity, they proved highly valuable. The simulations were more demanding than the actual space environment, which helped astronauts better prepare for the challenges they would face in orbit.

The instruments aboard Skylab, particularly the solar telescopes, represented unprecedented technological sophistication. The pointing accuracy required was extraordinarily high; the instruments needed to be directed within 2.5 arc seconds of the target and maintain this precision without drifting more than 2.5 arc seconds over 15 minutes. To put this in perspective, 2.5 arc seconds was roughly the size of a quarter viewed from a distance of one kilometer. Such exacting standards placed immense demands on the spacecraft's systems.

Traditional thruster engines used for attitude control were inadequate for this task. They lacked the precision needed, consumed too much fuel for extended missions, and their exhaust gases could interfere with optical observations. To address these challenges, engineers introduced control moment gyroscopes (CMGs) as the solution. From 1966 onwards, the attitude control system for Skylab's solar observatory relied on these advanced devices. CMGs were large gyroscopes designed to impart controlling moments or torques to a spacecraft directly. Unlike smaller gyroscopes used in guidance and navigation systems, CMGs could handle the rotational forces acting on the spacecraft and transfer them to the gyroscopes rather than moving the spacecraft itself. Each CMG featured a 53-centimeter rotor weighing 65.5 kilograms and spinning at high speeds, allowing for the precise control necessary to meet the demanding requirements of the solar observatory's instruments.

In the Neutral Buoyancy Facility at Marshall Space Flight Center, engineers and astronauts practiced retrieving film from the telescope mount. This facility, essential for simulating zero-gravity conditions, featured a specialized setup where the film magazine was positioned at the end of a boom. The white flotation collar on the boom was carefully designed to achieve neutral buoyancy, allowing for precise maneuvers in a controlled underwater environment. The photograph, labeled MSFC 027034, captures this intricate practice, offering a glimpse into the preparation necessary for space operations.

The task of retrieving film from the telescope mount in space involved meticulous coordination and skill. In the actual space environment, as depicted in photograph 73-H-979, astronauts performed this task while floating weightlessly, requiring precision and adaptability.

To control the space station's orientation and ensure the telescopes remained correctly aimed, the Skylab space station utilized control moment gyroscopes (CMGs). These gyroscopes, spinning at 9,000 rpm and mounted on the Advanced Technology Module (ATM) support structure, were crucial for maintaining the station's stability. The system included three CMGs: any two could control the cluster, while the third provided redundancy. Each CMG was mounted within two gimbal rings that allowed rotation around two axes, with control moments generated by manipulating these gimbals.

The CMGs operated based on a principle that might initially seem counterintuitive. When force was applied to the outer gimbal of a spinning gyroscope, it caused the inner gimbal to move perpendicular to the applied force and the rotor's axis. This motion, in turn, generated a reactive force that opposed the applied force at the outer gimbal. Since the gimbal was attached to the spacecraft's framework, this reactive force resulted in its rotation around one of its axes.

To complement the CMGs, Skylab's attitude control system was equipped with a suite of sensors that tracked the station's orientation relative to the sun and horizon and its rotational speed. A sun sensor and a star tracker measured attitude errors while rate-sensing gyroscopes monitored the spacecraft's rotational velocity in three directions. The data from these sensors were processed by the ATM's onboard digital computer, which calculated the necessary adjustments and issued corrective commands to the CMGs.

The system employed torque motors on the gimbals to apply precise twisting forces, enabling the gyroscopes to reposition and the spacecraft to maintain its desired orientation. This arrangement effectively transferred the spacecraft's rotational motion (its angular momentum) to the gyroscopes, ensuring accurate control.

For the solar observations, it was imperative to align the instrument canister directly with the sun and maintain this alignment for extended periods. As the spacecraft emerged from Earth's shadow, the attitude control system adjusted to position the canister toward the sun, aligning the spacecraft's long axis within the orbital plane. This "solar inertial" attitude was crucial for most of the station's operations, with the electrical power and temperature control systems designed to accommodate this primary orientation.

The Skylab Space Station, a marvel of space engineering, faced several challenges related to unwanted motion in its orbital assembly. These challenges stemmed from various sources, each contributing to the complexity of maintaining stable operations in space.

One primary source of unwanted motion was the random forces generated by crew activity within the spacecraft. Although these forces were minor, they added to the overall perturbations affecting the station. At orbital altitudes, aerodynamic drag, while minimal compared to lower altitudes, still had a noticeable impact. However, the most significant factor influencing the Skylab's stability was gravity. With its large and asymmetrical structure, the station experienced uneven gravitational forces. While the spacecraft's center of mass adhered to its planned orbit, the gravitational pull on the heavier end was greater than on the lighter end. This differential force, known as gravity-gradient torque, caused Skylab to rotate around its center of mass slowly.

Efforts were made to mitigate this rotation by aligning Skylab within its orbital plane. Yet, this alignment had to be balanced with the need to keep the station's telescopes directed at the Sun. Despite these adjustments, residual gravity-gradient torque and aerodynamic drag produced a net rotation that the Control Moment Gyroscopes (CMGs) were designed to counteract. However, the CMGs had limitations; they could only achieve coarse pointing within six arc minutes (0.1 degrees), a range significantly larger than the precise pointing required by the instruments. Mechanical constraints further restricted the movement of the CMG gimbals. Over time, as the CMGs absorbed unwanted torques, the rotors could reach a state of saturation where no further corrections could be made.

When all CMGs became saturated, they lost their ability to control the spacecraft until the rotors could be realigned. Engineers devised a method to "de-saturate" the gyroscopes to address this issue when the solar instruments were not in use. They leveraged gravity-gradient torque—the very force that led to the saturation. As Skylab entered the dark side of its orbit, the Advanced Technology Module (ATM) digital computer, one of the most advanced ever installed on a crewed spacecraft, assessed the saturation level and directed a maneuver. This maneuver adjusted Skylab's attitude to use gravity-gradient torque to reposition the gyroscopes to their original state before the next sunlit phase of the orbit. This process, known as momentum dumping, was executed by the thruster attitude-control system.

The instruments were mounted on a spar suspended inside gimbals for the precise pointing required by Skylab's telescopes. These gimbals allowed movement in two degrees of freedom: up and d, and left and right. They were also mounted within a roll ring, enabling rotation around the long axis of the canister aligned with the Sun. The fine-pointing assembly was supported by frictionless, flexible pivots designed to dampen small disturbances. Fine sun sensors and rate gyroscopes ensured the instruments could be pointed within 2.5 arc seconds of the Sun's center. For manual adjustments, a joystick on the control panel operated an optical device in the fine sun sensor, allowing precise offset pointing within 24 arc minutes of the Sun's center.

developing the Skylab Space Station's attitude-control systems was a complex and evolving process, reflecting the balance between technological advancements and practical requirements. For the wet workshop, establishing the initial configuration of Skylab before it was fully operational, the plan was to rely on an auxiliary attitude-control system powered by chemical fuels for the first three crewed missions. This system was designed as a temporary solution, with Control Moment Gyroscopes (CMGs) reserved for the fourth mission, the ATM flight, which would utilize the fully integrated system.

However, as Skylab transitioned to the dry workshop configuration, the approach to attitude control was revised. Engineers opted for a simpler thruster-based system powered by compressed nitrogen, which had several advantages. Unlike chemical systems, the nitrogen thrusters did not risk contaminating the area around the solar telescopes, thus preserving the integrity of scientific observations. Though this system was heavier—an acknowledged trade-off—it offered a more straightforward and reliable solution. The system incorporated twenty-two spherical tanks mounted around the S-IVB's thrust structure. These tanks supplied gas to six thrusters in the stage's aft skirt, with two thrusters assigned to each axis. These thrusters were crucial for docking maneuvers and for providing additional force when the CMGs could not manage the spacecraft's orientation alone.

The development and implementation of the attitude and pointing control systems fell under the purview of the Marshall Space Flight Center (MSFC), while the astronauts from the Manned Spacecraft Center (MSC) would operate these systems. To address the challenges of interfacing between the crew and the spacecraft's control systems, an intercenter task team was established in early 1967. This team was tasked with refining the interface for the solar experiments and addressing other operational concerns.

The initial design proposals from Marshall were met with dissatisfaction from the Houston team. They felt that the proposed control panel resembled those designed for uncrewed spacecraft rather than for human operation. MSC's concerns centered around the need for more intuitive and interactive controls rather than a system where astronauts merely input commands into a computer via a digital address system. Houston believed that the control panel should offer more direct feedback and control, allowing astronauts to actively engage with the spacecraft systems rather than simply monitoring status indicators.

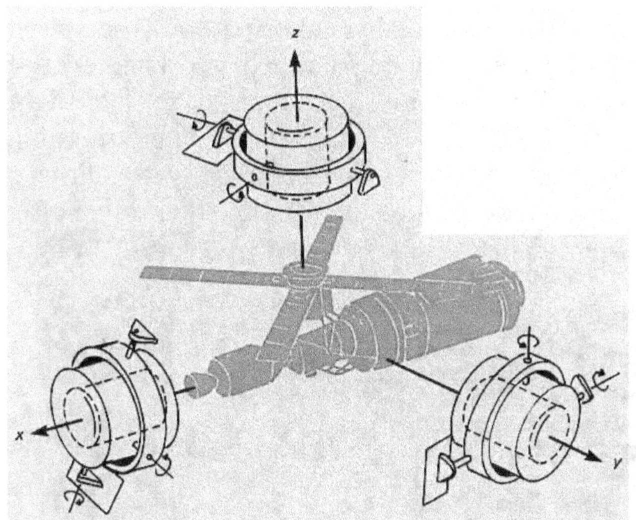

In August 1967, Bob Thompson from MSC compiled feedback on the control panel design and submitted it to Lee Belew at Marshall, outlining various recommended changes and detailing MSC's design philosophy. Despite these efforts, a preliminary review five months later revealed persistent disagreements between the two centers. Although there had been an ongoing collaboration between MSC representatives and Marshall designers, the review team in Houston continued to challenge several aspects of the design, particularly the digital address system, emphasizing the need for a more user-friendly interface that better accommodated the astronauts' roles in operating the spacecraft.

In the early stages of Skylab's development, it became evident that the control and display systems required a thorough redesign to meet the mission's unique needs. Recognizing the complexity of the task, NASA established a working group dedicated to resolving the issues surrounding the system's functionality. By mid-May, after extensive consultations with astronauts and contractors, the team developed a concept that addressed most concerns and allowed the progression of detailed design work.

The resulting control and display panel emerged as one of the most intricate ever integrated into a spacecraft. With a control array three times more extensive than that of the Apollo command module, it presented a level of complexity comparable to that of a large aircraft's control system. Despite its complexity, the design team balanced functionality and operability.

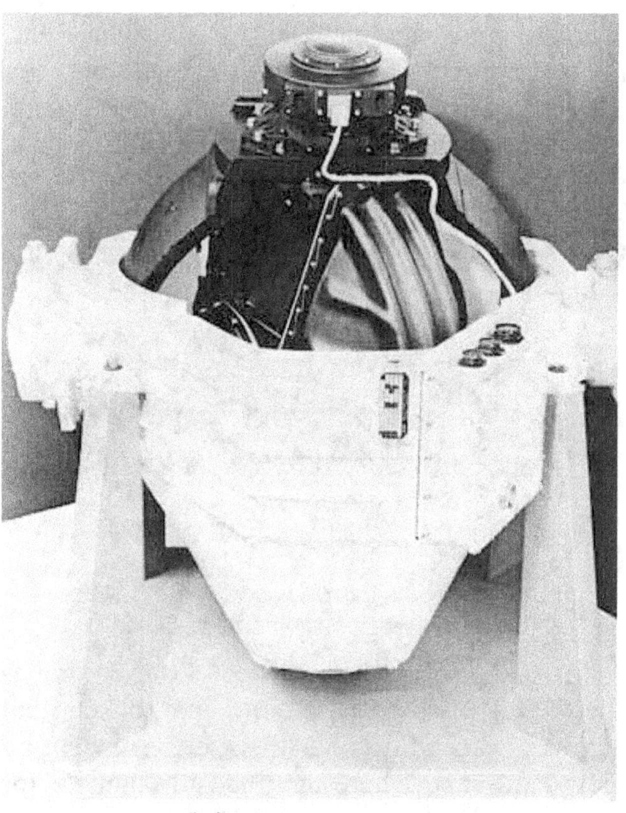

Skylab was the first spacecraft to utilize very large gyroscopes to control spacecraft attitude and for precise pointing of its instruments for scientific experiments. Three such gyros were used, each of which applied torque about two perpendicular axes.

The control center was meticulously crafted to ensure that every instrument's status was clearly visible to the operator. This included information on the workshop's orientation, orbital position, and the condition of the Apollo Telescope Mount (ATM) power system. The panel featured two television screens dedicated to displaying the sun as observed through the H-alpha telescopes, another screen for the coronagraph's field of view, and a fourth for the x-ray monitor. The design placed all instrument controls, attitude and pointing adjustments, and telemetry within easy reach of the seated operator, ensuring both efficiency and accessibility.

As the experimental setup progressed, attention shifted to Skylab's operational procedures and mission plans. The year 1969 marked a period of intense focus on how the instruments would be operated, the allocation of observing time, and the rigidity of flight plans. This was a novel challenge, as there was no prior experience with missions of Skylab's scope, which combined astronomical experiments with medical research.

Miniaturization and modern packaging techniques have made possible very compact digital computers, such as this one used in the Skylab. (International Business Machines Corp.)

Astronomers, accustomed to single-experiment missions with sounding rockets and uncrewed spacecraft, found themselves navigating a new landscape where they had to compete for operational time with medical experiments. This shift in priorities introduced new dynamics into mission planning and execution. Flight controllers, who had traditionally maintained exclusive control over spacecraft communications, now faced the challenge of coordinating with experimenters who sought direct communication with astronauts.

One notable example of this interaction was astronaut Edward Gibson, who served as the scientist-pilot for Skylab's third crew. At the control and display console for the Apollo Telescope Mount, Gibson operated the joystick that aimed the solar instruments. His work, captured in photographs, highlighted Skylab's mission operations' intricate and collaborative nature, where scientific and technical expertise converged to explore the cosmos.

As Skylab's operational planning progressed, the clash between the scientists' needs and the flight controllers' preferences became increasingly evident. The scientists, driven by the unpredictability of solar phenomena, demanded the flexibility to adjust their observing schedule daily, or even more frequently, to accommodate sudden solar flares or other significant events. Their ideal mission involved the capability to shift focus on a moment's notice, altering the mode of each instrument or dedicating entire observation periods to unexpected solar activity. This level of adaptability was in stark contrast to the flight controllers' vision of a more predictable mission, one that was well-structured yet allowed occasional deviations to challenge their ingenuity.

S150 instrument for galactic X-Ray mapping, sent up with Skylab 3

The scientists' insistence on flexible operational protocols often led to tensions with NASA's management. They were not shy about voicing their frustrations when operational plans did not meet their needs. This issue was a major topic during a principal investigators' meeting in April 1969. E. M. Reeves, representing the Harvard College Observatory, raised concerns about how priorities would be assigned to the telescopes and how conflicts would be resolved. He was assured that Martin Marietta was developing a computer program to distribute observing time fairly, and that the company would provide briefings to the astronomers.

Reeves also feared that the Mission Control Center (MSC) was not giving Skylab operations adequate attention. He was reassured that Skylab would receive higher priority once the first crewed moon landing was completed, establishing expected to occur within three months. Furthermore, Reeves advocated for daily direct communication between the principal investigators and the astronauts, an unconventional request for crewed missions. In response to Houston's invitation for the astronomers to observe Mission Control during an Apollo mission, Reeves countered with an invitation for flight controllers to visit Harvard and learn about the operations of scientific missions.

By late 1968, Harvard proposed a significant change to the cluster control system. They sought to operate their instrument—unique for its capability to provide real-time data—from the ground during uncrewed periods, effectively adding capabilities similar to those of uncrewed instruments. Initial studies by Marshall Space Flight Center revealed that this modification would entail substantial cost and schedule impacts. Program Director Bill Schneider was hesitant to approve these changes, but the Harvard team persisted, garnering support from other ATM investigators. Additional studies were commissioned despite Marshall's warnings about potential cost increases and delays.

Although these studies reaffirmed the significant impacts on cost and schedule, the Office of Space Science and Applications program manager was convinced to support the proposal. However, the scientists remained skeptical, believing the studies might have been overly conservative. In mid-December, Harvard's astronomers finally agreed to accept a reduced scope of their initial request. Schneider agreed to preserve the option of uncrewed operation, provided that any hardware changes would not exceed $50,000 and would not delay the schedule. Moreover, before being accepted, any changes would require approval from the Space Science and Applications Steering Committee and the Manned Space Flight Experiments Board.

The solar scientists, already disheartened by the cancellation of the second ATM (Apollo Telescope Mount) flight promised by George Mueller in 1967, faced further disappointment in July 1969 with the decision to replace their planned "wet" workshop with a "dry" workshop. The change was made without consulting the scientists, which caused significant frustration for both Leo Goldberg and Gordon Newkirk. They felt sidelined, having been given no opportunity to assess how the alteration would impact their scientific programs. Despite Mueller's attempts to appease them by highlighting the advantages of the dry workshop—including a higher likelihood of mission success and increased observing time—the lack of prior consultation left a lingering sense of dissatisfaction.

Later in the year, the ATM scientists encountered another setback when they discovered that Headquarters intended to incorporate a new set of earth-sensing experiments into Skylab. This addition represented Mueller's implementation of significant changes without proper consultation with affected parties. Including these experiments meant that the ATM would face competition for crew time and must maintain the spacecraft in an orientation that prevented solar observations. Moreover, the new experiments would contribute to a heavy command module load. This new disruption, combined with unresolved issues concerning observing time allocations for the dry-workshop missions, prompted the astronomers to address operational procedures urgently.

In late September 1970, a meeting was convened where ATM experimenters and officials from the Manned Spacecraft Center (MSC) discussed the operational framework for Skylab. The intention was to conduct Skylab missions similarly to previous ones: experimenters would specify their desired observations and the time required, while the flight operations office would impose operational constraints. A flight plan, incorporating the necessary trade-offs, would then be developed to satisfy the scientists and the mission controllers. This process involved a lengthy chain of command for any changes to be communicated to the spacecraft via the CapCom. While this method had proven effective for many experiments, it was ill-suited for solar studies. The Sun's unpredictable nature and the need for observational continuity made the rigid framework problematic.

An OSSA (Office of Space Science and Applications) representative emphasized that high-quality photographs of the sun were far more valuable than numerous mediocre ones. In response, Houston promised to collaborate closely with the astronomers to ensure success according to their needs. Although this initial engagement with operational personnel was somewhat reassuring, the scientists' subsequent review of the computerized time allotments prepared by Martin

Marietta was less encouraging. The resulting program was dissatisfied, and Richard Tousey from the Naval Research Laboratory found it particularly unacceptable, underscoring the ongoing challenges in aligning scientific objectives with operational constraints.

In response to Richard Tousey's objections, the experiments manager at Marshall Space Flight Center acknowledged the shortcomings of the current time allotment program but assured him that refinements by Martin Marietta's specialists would address these issues. Despite this promise, the principal investigators chose to take matters into their own hands. Unbeknownst to NASA officials, they devised a time-sharing plan designed to optimize the use of their instruments.

The investigators identified the most pressing problems in solar physics and determined how the ATM (Apollo Telescope Mount) instruments could contribute. They developed a comprehensive set of procedures to ensure that every instrument would be utilized efficiently during the allotted time for solar observations. Initially called the Program Oriented Observing Program, this approach was soon renamed the Joint Observing Program, as the humor in the acronym wore off. The program eventually included thirteen distinct observational programs, each with clearly defined objectives, data requirements, and detailed sequences of instrument operations to gather the necessary data.

By late March 1971, when the scientists presented their meticulously crafted plan, NASA's response was one of surprise and frustration. A representative from Kennedy Space Center (KSC) expressed concerns that adopting the new plan would necessitate scrapping the existing work done by Martin Marietta and possibly rewriting the mission requirements document—both of which were considered untenable at that stage. Despite these objections, the astronomers were adamant that no one could better plan using their instruments than they could. With support from mission planning engineers in Houston, the scientists' observing programs were eventually adopted as the standard operational mode for the solar instruments.

In the meantime, progress on the ATM instruments and supporting systems proceeded smoothly throughout 1969. Critical design reviews were successfully completed for all instruments, the solar-cell wing assemblies, the ATM control computer, and the star tracker. Qualification tests on various experiments and support systems were carried out, and by the end of the year, Houston's thermal vacuum chamber was being prepared for testing the ATM. A project assessment conducted in January 1970 indicated that, apart from the prototype instruments, the solar observatory was largely on schedule. The postponement of the launch date from July to November 1972 further facilitated progress by allowing hardware delivery to the Cape in November 1971. A subsequent review on March 11, 1970, found no issues to delay the schedule, although some subsystems required vigilant attention to ensure they remained on track.

Remarkable achievements and significant challenges marked the Skylab Space Station's astronomical telescope mount's journey. Developed at the Marshall Space Flight Center, the telescope mount underwent rigorous thermal vacuum testing at the Manned Spacecraft Center in July 1972. Despite the thorough testing, the project faced various complications that affected its timeline and progress.

The star tracker was an important part of the attitude and pointing control system. It measured attitude in roll and provided star position data for experiment pointing. As the third manned period progressed, the star tracker continued to malfunction and finally became unusable.

By mid-1972, the project encountered various persistent issues. A review conducted from August 11 to 12 highlighted growing concerns. The representative from Kennedy Space Center (KSC) expressed skepticism about the project's success, noting that the transition from wet to dry testing and the scientists' continual adjustments to their instruments had led to what was described as "near chaos" within the configuration control system. An 18-month delay in the delivery of the flight unit compounded these issues. The introduction of uncrewed capabilities was expected to exacerbate these delays further.

While the KSC's representative's pessimism may have been influenced by the center's heavy workload and the challenges of managing the entire cluster of projects, the reality was that problems were widespread. By June 1970, four instrument prototypes had been delivered and stored at Huntsville, but persistent issues persisted. Instrument S082B faced difficulties with its electronic assemblies and film cameras, while the zoom lenses for the H-alpha telescopes struggled to focus properly. Testing in July revealed serious deterioration of the spectrograph grating in S082A, necessitating time-consuming replacements.

The situation became more pressing as the target launch date of July 1972 was officially revised. The delay provided some relief, but all ATM instruments faced issues. By January 1971, a significant development occurred: Headquarters postponed the final Apollo flight (Apollo 17) to ensure it could carry an optimal load of experiments. Consequently, Skylab's launch was delayed again by four and a half months, setting a new launch date for April 30, 1973. Although the revised schedule alleviated some concerns about experiment problems, these issues still required ongoing attention.

In May 1971, after passing its acceptance review, the S082B experiment exhibited severe deterioration in its response to short-wavelength radiation. The optical components were examined, revealing that the main diffraction grating suffered from "purple plague," a condition caused by an unforeseen chemical reaction between the grating's gold coating and the aluminum layer applied over it. This required the replacement of the grating, resulting in an eight-week delay and further disrupting the ATM delivery schedule.

By the end of 1971, a midterm review of the Skylab program offered cautious optimism. Although the ATM presented no new severe problems, the project manager's assessment indicated no margin for major setbacks. The flight unit was expected to be delivered to Kennedy Space Center by October 1, 1972, but achieving this timeline required relentless effort. The ATM's status reflected the overall condition of the Skylab project.

One notable anomaly occurred during ATM testing, which, to some extent, had been anticipated. Thermal vacuum testing at the Manned Spacecraft Center in August and September revealed a failure in one of the control moment gyros, attributed to inadequate lubrication. This defect had been suspected earlier, and backup units with improved lubrication systems were introduced. On September 22, 1972, the ATM flight unit was transported from Houston to Kennedy Space Center, coinciding with the arrival of the orbital workshop from California. This marked the beginning of final checkout and integration with the other components of the Skylab cluster, setting the stage for the upcoming launch.

As Skylab approached its launch, the program's scope evolved significantly. Solar astronomy and space medicine were initially the primary focus, with these major experiment programs, along with the corollary experiments, expected to occupy the available operational time on Skylab missions fully. This focus was deemed to be the most that the program could accommodate without risking delays in the launch schedule, a point frequently emphasized by program officials at the Marshall Space Flight Center.

However, with the decision to develop the Space Shuttle as the next major program, Skylab found itself as the only crewed space program active for an extended period. Despite directives from Headquarters to limit changes to only mandatory ones after the dry-workshop decision, there was a natural inclination to optimize using Skylab's remaining missions. This led to the introduction of additional experiments between July 1969 and January 1973, reflecting the program's adaptability and the evolving interests of the scientific community.

One significant addition was the integration of Earth observation experiments. Initially, the cancellation of the Apollo Applications Mission 1A at the end of 1967 had precluded any possibility of Skylab engaging in Earth studies. Yet, within two years, a burgeoning interest in remote sensing and its practical applications led the Skylab program office to incorporate a new suite of complex and costly instruments designed for this purpose.

Remote sensing, which involves various methods of collecting information from high-altitude platforms, became a focal point of interest. This technology encompasses a range of activities from photography to radiometry, measuring electromagnetic radiation reflected or emitted by features on the Earth's surface.

While aerial photography was a standard method, more advanced techniques such as photometry and multi-spectral sensing provided valuable data. Multi-spectral sensing, which simultaneously measures several different bands

in the visible and infrared spectra, was particularly useful for comprehensive surveys. The challenge of covering large areas quickly, often a limitation of aircraft-based surveys, further highlighted the potential benefits of incorporating remote sensing into Skylab's mission objectives.

Assembled for a management meeting in the facilities of the Mc Donnell Douglas Astronautics Co., St. Louis, in April 1971, were, left to right: Eberhard Rees, Leland F. Belew, Kenneth S. Kleinknecht, William C. Schneider, and Kurt H. Debus.

On hand for the "rollout" ceremonies for the orbital workshop at Huntington Beach, Calif., in September 1972, were, left to right: Willis H. Shapley, Casper W. Weinberger, James C. Fletcher, Rees, Walter Burke, and Dale D. Myers.

Thus, despite the program's initial constraints, Skylab's final mission lineup included its established solar and space medicine experiments and a suite of innovative Earth observation tools, marking a significant expansion in its scientific capabilities.

Late Experiments

In the 1960s, the burgeoning field of remote sensing began to reveal its potential and limitations in surveying the Earth's surface through aerial photography and other techniques. Advances in sensor technology have expanded remote sensing's applications to various domains, including agriculture, forestry, geology, mineral prospecting, oceanography, city planning, and land-use studies. This technological progress allowed detailed observations that were previously unattainable.

Skylab officials visiting the facilities of the Mc Donnell Douglas Astronautics Co., St. Louis, January 1972, donned special garments to enter clean rooms. Left to right: Fred J. Sanders, Rees, William K. Simmons, Jr., Myers, Belew, and E. T. Kisselberg.

Inspecting the airlock trainer at St. Louis, in April 1970, were, left to right, Belew, James S. McDonnell, and George Radebush.

The rise of the environmental movement during the late 1960s amplified air and water pollution awareness, highlighting the need for comprehensive methods to assess these widespread issues. Traditional methods were proving insufficient for gauging the scale of these problems, necessitating a broader, more synoptic perspective. It became evident that only an orbital view from space could provide the extensive coverage and intricate detail required to understand and address these environmental challenges.

The significance of this new vantage point was demonstrated through the color photographs taken during early Gemini missions, which astonished cartographers and geologists in federal agencies. These images offered a broad perspective and allowed repeated observations of the same locations over time. For applications such as crop and snowpack surveys, the ability to collect real-time data from space surpassed the capabilities of aerial surveys.

In response to these emerging needs, NASA initiated various meteorological satellites, including Tiros and Nimbus, starting in the mid-1960s. However, it was not until early 1968 that serious efforts focused on developing other Earth-sensing vehicles. This effort fell under the purview of the Manned Spacecraft Center, where remote-sensing instruments were initially tested on aircraft. By the end of fiscal year 1968, the program had accumulated a budget of $6 million and employed approximately 150 full-time NASA and contractor personnel. Key agencies such as the U.S. Geological Survey, the Department of Agriculture, and the Naval Oceanographic Office were crucial in coordinating the program and evaluating its outcomes.

As remote-sensing technology advanced, the concept of "Earth resources" emerged, encompassing the monitoring of crops, minerals, and water supplies. The growing public concern over dwindling natural resources lent further urgency to the notion that space technology could be harnessed to address these pressing issues. A Martin Marietta official later reflected on this period, noting, "Everybody had his definition of what 'Earth resources' meant, but all the definitions were good." For many, the expensive crewed spaceflight programs seemed wasteful, yet the potential for space technology to contribute to solving environmental problems offered a redeeming purpose.

The appeal of Earth resources programs was evident in their broader public and political support. One Skylab program official observed that discussions about space missions focused on environmental and resource monitoring resonated more with the public than those centered on esoteric scientific projects. This practical focus won over many congressmen, who saw Earth resources programs as a tangible and valuable application of space technology that could be easily appreciated by their constituents.

In early 1968, Earth-resource experiments gained significant traction, largely thanks to the support of the chairmen of NASA's House subcommittees. When John Naugle, Associate Administrator for Space Science and Applications, presented his plans to the Space Sciences Subcommittee, he was met with enthusiasm that exceeded his initial proposals. By the end of that year, the House Subcommittee on NASA Oversight had published a staff report advocating for expanding efforts in the Earth-sensing domain, emphasizing the need for further development and research.

At that time, the Office of Space Science and Applications was still defining the objectives for an Earth-resources technology satellite and advancing developing necessary sensors. Naugle informed the Subcommittee on Space Science and Applications that he anticipated requesting funding in fiscal year 1970 to develop hardware for a satellite flight planned for late 1971 or early 1972. However, the Office of Manned Space Flight, which managed the only existing program capable of potentially accommodating sensors sooner—Skylab—faced challenges. The official schedule had listed an Apollo Applications Program (AAP) flight for November 1970, but prospects looked bleak after the cancellation of AAP 1A. By the end of 1968, the Office of Manned Space Flight could only lay the groundwork for future Earth-sensing experiments, with no immediate prospects for flying such experiments in the near term.

During this period, the concept of incorporating Earth-sensing experiments into crewed space missions remained a topic of discussion, though the feasibility seemed uncertain. Jacob Smart, NASA's Assistant Administrator for DoD and Interagency Affairs, proposed in May 1968 that an Earth-resources project might be critical for the space program's future. Smart noted the potential economic benefits and unexpected discoveries from the Gemini and Apollo missions and suggested to George Mueller that the Office of Space Science and Applications (OSSA) consider including Earth-sensing instruments in future Apollo and Apollo Applications missions.

Mueller had indeed prioritized Earth-resource observations as a key objective for the Apollo Applications Program as early as 1965. Despite setbacks, interest in integrating Earth-sensing capabilities into crewed missions persisted. The Office of Space Science and Applications revisited the idea, particularly in light of Floyd Thompson's Post-Apollo Advisory Group's recommendations, which identified Earth-sensing as a promising avenue for crewed spaceflight.

Nevertheless, the prospects for deploying substantial Earth-sensing payloads on missions in 1969 or 1970 were limited. The proposed wet-workshop missions, which could have provided valuable data for the Earth-resources technology satellite still in the planning stages, faced significant constraints. One OSSA official noted that the low orbital inclination achievable with the Saturn IB rocket would result in minimal coverage of the United States, diminishing the utility of such missions for comprehensive Earth-sensing experiments.

Despite these challenges, the momentum for Earth-resource experiments continued to build, driven by a growing recognition of their potential benefits and the desire to leverage space technology for practical applications on Earth.

In the early stages of Skylab's development, the challenge of accurately estimating mission costs proved more complex than anticipated. The initial financial projections provided by the committee underestimated the true expenses involved. This discrepancy highlighted a critical issue: unless experiments were explicitly designated as primary objectives, they faced a high risk of being abandoned when budget constraints and scheduling pressures arose, as demonstrated by the experience with AAP 1A. The Office of Space Science and Applications (OSSA) faced a difficult decision, recognizing that while relying solely on crewed programs to deliver vital scientific information was imprudent, excluding them entirely was equally unwise.

A significant National Academy of Sciences report underscored the urgency for advancing Earth-resource missions. This report, the culmination of a two-year study commissioned by NASA, advocated for a substantial increase in funding for application satellites. It recommended a two- to three-fold boost in resources for communications and navigation satellites and suggested initiating a pilot program for an Earth-resources satellite. The study's recommendations resonated strongly with the prevailing market demand for Earth-surveying experiments in early 1969. Consequently, it was anticipated that the Office of Manned Space Flight would reinvigorate its focus on Earth-resource experiments, a move that materialized as expected.

By the fall of 1969, after the resolution of the dry-workshop decision, discussions were underway to assess whether the sensors from AAP 1A, or their upgraded counterparts, could be integrated into the workshop. The Office of Space Science and Applications began defining a comprehensive suite of experiments for review by its Space Science and Applications Steering Committee. Preliminary studies suggested that there were no significant obstacles to implementing these experiments.

On December 8, 1969, the Manned Space Flight Experiments Board received a proposal from the Manned Spacecraft Center (MSC). Leonard Jaffe, acting director of the Earth Observations Program Division, voiced concerns regarding the hurried proposal preparation. He highlighted unresolved issues, particularly related to cost and the specifics of the sensors. Despite these concerns, Jaffe supported the initiative to include these experiments in the mission and promised that OSSA would provide definitive recommendations. Charles Mathews, chair of the meeting, acknowledged the need to examine funding and management issues further but strongly supported the project. Consequently, the board approved only one of the proposed experiments, deferring the rest until more detailed information could be provided.

The Skylab's orbital plane, inclined at approximately 30 degrees to the equator, delineated the spacecraft's northern and southern travel limits. At this inclination, the spacecraft could not venture farther north than New Orleans, illustrating the constraints imposed by its orbital configuration.

Initially, the Earth-resource experiments aboard Skylab were composed of four instruments, each designed to provide critical data on Earth's surface. Among these, the multi-spectral photographic facility (experiment S19OA) was the only instrument with prior flight experience. This facility was an enhanced version of an experiment that had proven highly successful during Apollo 9 the previous spring. It featured six precision cameras, each equipped with carefully matched lenses and utilizing distinct film and filter combinations. This setup allowed the facility to capture a range of spectral information, including various visible and infrared light wavelengths.

The remaining instruments were novel in their orbital applications, focusing on radiometric rather than photographic data. These instruments measured the intensity of radiation emitted or reflected from Earth's surface. The first of these, a spectrometer (S191), was designed to record the wavelength and intensity of infrared radiation from small, 0.45-kilometer diameter areas on the ground. The second, a 10-band multispectral scanner (S192), operated in the infrared spectrum to simultaneously measure radiation intensity across ten wavelength ranges. This scanner covered a swath 74 kilometers wide along the spacecraft's ground track.

The fourth instrument, designated S193, served dual purposes. It functioned as both a microwave radiometer and a radar scatterometer. The microwave radiometer detected longer wavelengths than the infrared instruments, while the radar scatterometer assessed the reflective properties of Earth's surface in response to radar waves.

Later, two additional instruments were incorporated into the experiments package. The first was a passive L-band radiometer (S194), established to map surface temperatures across various terrestrial regions. The second addition was a higher-resolution camera (S190B), designed to enhance the interpretation of data collected by the other sensors.

Visual representations of the Earth-resource experiments, as shown on briefing charts, illustrate these instruments' coverage and frequency ranges. Early charts depicted the area coverage for the first five instruments, with the later-added S190B capable of photographing a 109-kilometer square. Additional images detailed the frequency coverage of all six instruments and the configuration of the sensors at the bottom of the docking adapter. Notably, S190B was designed to be operated through the workshop's scientific airlock, allowing for its integration into the mission's scientific operations.

In the early stages of the Skylab program, Program Director Bill Schneider took decisive action to initiate the necessary groundwork for the mission. He tasked the Manned Spacecraft Center (MSC) with preparing the essential documentation, while the Marshall Space Flight Center was responsible for studying integration requirements and hardware modifications. Meanwhile, all three centers were instructed to continue their basic compatibility studies. As the financial constraints were tight, Schneider emphasized the need to manage costs meticulously.

The experiments board had been allocated a budget of $10 million to develop the scientific instruments, with an additional $11.125 million designated for support and data analysis. The latter amount was to be funded by the Office of Space Science and Applications (OSSA) and user agencies. By early February, OSSA recommended the inclusion of the first four instruments for flight, a recommendation that Dale Myers, the Associate Administrator for Space Science, agreed to on the 16th of February. However, the approval for the microwave instrument was only provisional, as its compatibility with the spacecraft had not been fully established, and its potential additional cost of $2 million posed a significant concern.

Schneider directed the centers to proceed with developing the earth-resource experiments, stressing that all possible efforts must be made to adhere to the established cost and schedule guidelines. The budget was set at $25 million for developing, integrating, and delivering these experiments by July 1971. Should the costs exceed this budget, the possibility of dropping the entire package must be considered.

Requests for proposals were issued, a source evaluation board was appointed, and contracts were awarded to develop the instruments by mid-1970. Despite these advancements, Schneider's correspondence over the following six months consistently highlighted a troubling increase in projected costs. By mid-March, the cost of the

multispectral cameras had doubled from the original estimate provided in December. By June, the overall cost of the instrument package had surged to $36 million. Schneider cautioned that a reassessment of the project might be necessary due to these escalating expenses.

In June, while the Skylab office recommended canceling the microwave sensor to curb costs, the Manned Space Flight Experiment Board, influenced by OSSA's advocacy to retain the instrument, urged developing all planned instruments. Despite a reduction in NASA's overall budget that summer, Myers found himself compelled to proceed with the earth-resource instruments. He informed NASA Administrator James Fletcher in July of his decision to cap the project's cost at $36.4 million, indicating that the additional $11.4 million would be covered by further reductions in the planned Skylab un-costed obligation for the fiscal year 1971. Essentially, this meant reallocating funds already designated for Skylab.

Schneider communicated this decision to the center program managers, instructing them to adjust their budgets within the constraints of the current fiscal limitations. Kenneth Kleinknecht, from Houston, expressed concerns that the financial strain might impact other Skylab projects. He urged Washington to consider the centers' fiscal challenges before approving any new, expensive experiments for the program.

By the end of the year, the costs had risen further, reaching an estimated $42 million. However, there were signs that the financial situation was beginning to stabilize. Despite the ongoing budgetary pressures, the Skylab program continued progressing, with careful management and strategic adjustments to meet its ambitious objectives.

In mid-1971, as the Skylab project faced mounting challenges, Bill Schneider thoroughly reviewed its cost history for Dale Myers. Schneider identified several key factors contributing to the project's difficulties. He attributed much trouble to unrealistic initial estimates and less-than-effective management at various levels. The fact that sensors had been used in the aircraft program, though only partially true, led managers to underestimate the complexities and costs of developing these instruments for spaceflight.

The eagerness to expedite the development process resulted in insufficiently detailed requirements from the Office of Space Science and Applications (OSSA) and the Office of Manned Space Flight (OMSF) before soliciting bids. This lack of specificity and changes in specifications during contract negotiations led to increased costs. Furthermore, there had been inadequate coordination between the Manned Spacecraft Center's (MSC) Science and Applications Directorate, which oversaw the instrument development, and the program control officials in the Houston Skylab office. Changes to the experiments were made without a comprehensive assessment of their impacts, reflecting a broader issue of poor management. In their haste to integrate the instruments, the managers had not exercised the necessary caution, leaving the project vulnerable to significant setbacks if new problems arose.

Despite these issues, the project had regained some level of control. Planners now faced the task of finding suitable locations for the instruments and their control systems. Due to the available space, the multiple docking adapter was chosen as the new mounting site. Modifications were required: an optical-quality window had to be added for the multispectral cameras, the infrared spectrometer needed to be installed through the pressure hull, and brackets were necessary to support the large antennas for the microwave and multispectral scanners.

Marshall Space Flight Center promptly initiated these changes, although they caused some interference with existing systems on the module. The requirements of the earth-resource experiments also necessitated significant alterations to Skylab's mission plans. The orbital inclination of the space station was increased to 50 degrees, broadening its operational range. Skylab would now traverse northern latitudes as far north as Vancouver, Winnipeg, Bastogne, Frankfurt am Main, Kharkov, Mongolia, and Sakhalin Island north of Japan. To the south, its orbit would cover Australia, most of Africa, and much of South America, excluding only Tierra del Fuego.

Skylab, America's first space station, was designed to traverse a vast portion of Earth's surface. Its orbit ensured that three-fourths of the Earth's surface fell under its observational path, encompassing the region where 90% of the world's population resides and 80% of its food was produced. However, this extensive coverage came with its set of challenges.

Initially, NASA's tracking and communication infrastructure was optimized for spacecraft on lower-inclination orbits, such as those heading toward the Moon. However, Skylab's higher-inclination orbit meant that it would experience extended communication blackouts during each orbit. This adjustment also necessitated recalculations for thermal management within the space station and solar arrays' power production adjustments.

One of the significant operational changes involved the new orbital attitude required for Earth-sensing experiments. Unlike solar telescopes, which needed to be directed precisely at the Sun, Earth sensors required alignment with the specific point on Earth's surface directly beneath Skylab, known as the nadir. This alignment posed a unique challenge: while inertia would naturally keep Skylab aligned with the Sun, the spacecraft's position relative to the nadir would continuously shift. To address this, Skylab had to rotate at an angular rate corresponding to its orbital velocity—approximately 4 degrees per minute—during Earth-resource experiments. This mode of operation termed the Z-local vertical, always aligned the spacecraft's Z-axis with Earth's center. However, this configuration compromised solar observations, altered the heat balance of the spacecraft, and decreased power production from the solar arrays.

These operational adjustments necessitated various design modifications to Skylab's attitude-control system. Originally, control moment gyros were designed to manage attitude during solar observations, while thrusters were reserved for other maneuvers. With the new requirements and the added weight and inertia from the workshop's equipment, Marshall Space Flight Center engineers reallocated most maneuvering tasks to the gyros, reserving the thrusters for less frequent use. New control programs were integrated into Skylab's digital computer to accommodate these changes. This sophisticated system allowed dynamic adjustments, enabling Skylab to transition between solar inertial and Z-local vertical orientations and adapt to various special conditions.

The earth-resources package on Skylab presented significant challenges to flight planners. The photographic instruments were designed to capture detailed images of Earth's surface. Still, they required precise lighting conditions, restricting the number of sites that could be photographed effectively from orbit. The Z-local vertical attitude, essential for these observations, introduced thermal and power limitations that constrained the number of consecutive Earth-observing passes Skylab could make. Furthermore, observations, except those conducted by the microwave sensor, were subject to weather conditions at the ground level. Heavy cloud cover could delay or entirely prevent planned observations.

The sensitive nature of Skylab's mission also had implications for terminology and operations. Given that Skylab's path often traversed foreign countries, some of which were wary of potential orbital surveillance, the term "target" was deliberately avoided in all documentation and training. This precaution aimed to prevent astronauts from inadvertently using the term during their mission.

Early planning estimated that about 45 of Skylab's overflights across the United States during its three missions would be suitable for earth-resource sensing. This estimate was used for preliminary planning for approximately a year. However, as proposals from potential data users emerged, it became evident that more frequent observations would be required. This realization necessitated extensive coordination to balance the various demands on Skylab's resources, including medical and solar experiments, which also needed time on the station's schedule.

Compounding these difficulties was the uncertainty surrounding the earth-resource instruments' capabilities. For two years, NASA evaluated a suite of new sensors and experimented with a novel approach to experiment management.

Unlike previous missions where experimenters proposed both the instruments and their applications, Skylab's earth-resource instruments were to function as a scientific facility with specifications defined by NASA. Users of these instruments would propose specific data requirements, but they would not have control over sensor design. Instead, they could specify, within operational limits, the timing and location of data collection.

This new management approach created significant challenges for those responsible for flight operations. Eugene Kranz, the head of Houston's Flight Control Division, faced considerable frustration in defining his team's role in the Skylab mission. With only two years until the flight and no investigators available to consult, Kranz's division struggled to grasp the complexities of the earth resources package. The usual process of engaging with experimenters to determine their needs was not possible. Moreover, the Flight Control Division had been tasked with collecting Skylab's data requirements, including data processing and distribution. Yet, the division faced unresolved questions about whether the primary goal was to evaluate the sensors themselves or to collect meaningful data, a distinction that significantly impacted the handling of the experiments.

Skylab's solar arrays in flight were exposed directly to the Sun's rays. Solar energy was transformed into electrical power for operation of all spacecraft systems. The proper operation of these solar arrays was vital to the mission.

The structure of the solar arrays shows clearly in this photograph taken from the command and service module.

NASA faced considerable organizational and logistical challenges in the early stages of preparing for Skylab's earth-resource experiments. Despite Kranz's initial inclination to treat the earth's resources as data-collecting science experiments, this approach was not universally accepted within the Manned Spacecraft Center (MSC). The decision to prioritize scientific data collection over other aspects of mission planning sparked considerable debate among NASA's various divisions.

Kranz's difficulties were emblematic of broader issues within NASA. The haste the agency was working to expand its program and integrate additional components into the existing framework contributed to significant coordination problems. The Office of Space Science and Applications (OSSA) managed some aspects of the Earth resources project, while MSC was responsible for others. The overall coordination fell to Headquarters, which faced severe budgetary constraints. This fragmented oversight occasionally led to breaks in communication and efficiency.

Amid these challenges, the selection process for experimenters was set into motion. On December 22, 1970, NASA issued 6,000 announcements inviting proposals for using earth-resource data. These invitations were sent to diverse potential users, including universities, state and local government agencies, private enterprises, and foreign governments. By mid-1971, NASA had received approximately 230 proposals. These were initially reviewed by the Office of Space Science and Applications for their scientific merit.

Following this initial evaluation, the proposals underwent further scrutiny by the crewed spaceflight centers, primarily MSC, to determine their compatibility with the planned missions. This multi-tiered evaluation process was necessary before definitive flight planning could commence. Despite the high quality of many proposals, the limited number of earth-resource passes—45 in total—was insufficient to accommodate all the promising ideas.

The astronauts could control and monitor the workshop power subsystem.

From its 270-mile-high vantage point, Skylab photographed features on Earth, such as these large cloud buildups east of the Carolinas, providing a mass of valuable scientific data.

In response, Schneider directed both the Houston and Huntsville centers to assess their capacity for additional observation time. Marshall Space Flight Center identified that, within certain constraints, it could provide a further 40 passes. However, this extension would have complications, such as overlapping with the Atmospheric Test Module (ATM) observing time and extra film storage space. Some of these additional passes would be conducted in a solar inertial attitude, allowing earth sensors and solar telescopes to operate simultaneously. This adjustment increased the available observation opportunities to 160 out of the original 230 proposals.

By early 1972, consultations with investigators led to numerous adjustments, refining the proposals to fit Skylab's capabilities. Nonetheless, as MSC began fully understanding the financial implications of supporting these investigations, there was pressure to reduce the number of selected proposals. OSSA and the Office of Manned Space Flight (OMSF) opposed this reduction. In response to the suggestion of the headquarters to shift some of the management responsibilities elsewhere to ease MSC's burden, the center agreed to continue negotiations with all 160 candidates.

By August 1972, nine months before the inaugural Skylab mission, Headquarters announced the selection of 106 investigators for the earth-resource experiments. This group included 83 from the United States and 23 from other countries. Their inclusion marked a significant step forward in the implementation of Skylab's earth-resource mission, reflecting the complex interplay of scientific ambition, budgetary constraints, and logistical coordination that characterized the early stages of this pioneering space program.

While project officials were finalizing the details of earth-resource experiments with investigators, mission planners were engaged in refining data collection strategies. In 1970, as in 1967, the Manned Spacecraft Center (MSC) was contending with challenges related to the Apollo program. Although not a catastrophe, the near abort of Apollo 13 in April 1970, followed by a detailed investigation, underscored the importance of meticulous planning and execution.

A significant change in mission planning occurred in early 1972 when Houston proposed launching Skylab into a controlled, repeating orbit. This orbit would allow the spacecraft to regularly pass over the same geographic points, enhancing the likelihood of successful coverage of designated earth-resource sites. Marshall Space Flight Center's agreement incorporated this adjustment into the mission plans by June. Skylab would be placed into a 372.5-kilometer orbit, which would repeat its initial ground track every 72 revolutions, or approximately every five days minus two hours. To account for minor orbital perturbations, periodic adjustments would be made during the mission.

Flight planning for earth-resource passes was intricate, rivaling the complexity of solar observations and other experimental activities aboard Skylab. The mission involved coordinating 570 combinations of ground sites and experimental tasks across 60 earth-oriented passes throughout the three missions. This coordination required synchronizing orbital passes with ground observations and aircraft flights for instrument calibration while also contending with unpredictable weather conditions. Since experimenters could only verify data after film processing on Earth, there was always uncertainty about whether the desired data had been captured.

In the first half of 1972, mission planners established basic procedures for earth-resource observations. Planning for a day's observations began five days in advance, leveraging the repeating ground track of the spacecraft. Initially, a "shopping list" of activities was compiled, which included all tasks that had been preplanned but not yet executed. This list was refined based on the spacecraft's ground track, the crew's work-rest schedule, and other priorities, including possible conflicts with other experiments. Weather forecasts and the readiness of ground support and aircraft were also considered, often leading to further reductions in the list of activities.

Two days before the scheduled observations, planners selected the activities with the highest likelihood of success and began summarizing the flight plan. This process included continuous monitoring of updated weather forecasts. The day before the observations, detailed flight planning was completed, specifying site coordinates, operational instruments, and timing for spacecraft maneuvers. After reviewing the latest weather reports, flight planners finalized Skylab's activities no later than three hours before execution or two orbits before the planned observation.

The tight timeline for developing and refining instruments posed challenges, adding another layer of complexity to the mission's preparation and execution.

The S190A cameras, while familiar in design from previous space missions, were complemented by less mature infrared and microwave sensors that faced significant delays in their development. Martin Marietta, tasked with integrating these experiments into the multiple docking adapter and constructing their associated controls and displays, often grappled with instrument modifications that impacted their hardware. To manage these evolving challenges, Martin Marietta established a working group that included representatives from five other contractors and the astronaut office. This group, meeting monthly, played a crucial role in navigating the complexities of proposed changes and their implications, making trade-offs among various factions to ensure project coherence within the available timeline.

By November 1971, Program Director Bill Schneider reported to the Office of Space Science and Applications that the flight hardware had been completed and was ready for integration into the multiple docking adapter. However, despite this progress, each sensor presented issues necessitating hardware modifications before launch. Following integration checks, the instruments were removed for further adjustments. This process required rescheduling subsequent qualification tests to accommodate the missing experiments.

As the project neared completion, Schneider recommended the cancellation of two experiments due to ongoing difficulties. The multispectral scanner, plagued by technical problems, risked delaying the launch, while the microwave sensor faced a lack of interest from researchers, rendering it a questionable expense. Although neither experiment was ultimately dropped, the Office of Space Science and Applications (OSSA) acknowledged that the multispectral scanner could be considered expendable if it threatened the timely launch of the workshop. A significant delay would disrupt the seasonal observations crucial to other researchers, and thus, the scanner was deemed less critical.

On October 6, 1972, the multiple docking adapter and airlock were delivered to Cape Canaveral, followed by the arrival of the S193 microwave experiment nine days later. Over the subsequent months, several equipment failures occurred, necessitating the return of the multispectral scanner, microwave sensor, control, display panel, and one of the tape recorders to their manufacturers for defect corrections. By late March 1973, after completing the final earth-resources simulation tests, the experiments were deemed ready for their mission.

Publicizing the earth-resource experiments as beneficial to society was a priority, contrasting with the more specialized medical and solar astronomy experiments. In the spring of 1971, Ken Timmons, an official at Martin Marietta responsible for the multiple docking adapter, proposed involving high school students in suggesting simple experiments for the Skylab workshop. This idea, well-received by Colorado education officials, was subsequently presented to the Marshall Space Flight Center. Skylab manager Leland Belew supported the initiative, establishing and communicating with William Schneider. The National Science Teachers Association was contracted to organize and oversee a nationwide competition for student experiment proposals, broadening public engagement with the Skylab program.

In October 1971, the National Science Teachers Association (NSTA) launched a nationwide call for high school students to propose experiments for the Skylab program. Approximately 100,000 announcements were distributed, setting a deadline of February 4, 1972, for submitting proposals. This initiative sparked significant interest, with more than 55,000 teachers requesting entry materials and ultimately resulting in 3,409 proposals from over 4,000 students across all 50 states, spanning grades 9 through 12.

By March 1, 1972, twelve regional screening committees had reviewed the submissions and selected 300 proposals for final evaluation. The judging process, managed by NSTA, focused on the scientific merit of each proposal. Concurrently, NASA engineers were consulted to assess the feasibility of the proposed experiments. By March 15, 25 national winners and 22 entries receiving "special mention" were chosen and evaluated for their scientific value and practical considerations such as weight, volume, power consumption, and crew time required.

Following the selection, the chosen experiments underwent NASA's standard review process. Specific requirements were relaxed to streamline this process and accommodate the students' limited experience with formal documentation. A simplified record-keeping system was implemented, and each NASA office appointed a single representative to oversee the reviews, ensuring prompt action could be taken as needed.

The 25 selected students participated in a preliminary design review at the Marshall Space Flight Center on May 8, 1972. The experiments were categorized based on their requirements: some needed the fabrication of new hardware, others could be integrated with existing Skylab experiments, and a third group could achieve their objectives through collaboration with ongoing research. Technical challenges meant that six experiments could not be carried out as proposed initially; however, these students were paired with principal investigators whose research aligned closely with their interests, allowing them to contribute to Skylab's science program in other ways.

Among the remaining proposals, eight experiments utilized existing data, while eleven required developing new hardware. Over the following three months, these students worked closely with NASA advisors to design their experimental equipment and prepare for a critical design review scheduled for August. By early 1973, the student experiments were completed. The flight acceptance review took place at Marshall from January 23 to 24, and the flight units were delivered to Cape Canaveral two days later, ready for their place in the Skylab mission.

As Skylab, America's first space station, prepared for its missions, a unique and significant component was added: student-designed experiments. The experiments, the final addition to the Skylab program, varied widely in their quality and scope, reflecting a broad range of ingenuity and scientific curiosity.

One of the standout proposals aimed to measure neutron radiation intensity at orbital altitudes, a task that professional scientists had never attempted. This experiment represented a pioneering effort to explore cosmic radiation in a novel way. Another ambitious proposal sought to investigate X-ray radiation emanating from Jupiter, leveraging the advanced instruments aboard the Apollo Telescope Mount (ATM) to gather unprecedented data from our distant planetary neighbor.

Perhaps the most widely discussed student experiments involved studying how spiders adapt to zero gravity. This experiment sought to determine whether a spider could spin a normal web in a weightless space environment. If the spider struggled to maintain its web-spinning ability, the researchers were interested in observing whether it could adapt to the conditions of weightlessness throughout a mission.

The range of student experiments extended into various fields, including astronomy, biology, and space physics, demonstrating a remarkable breadth of interest and creativity. Despite the vast array of topics, the experiments shared a common thread of curiosity and innovation. The overall sophistication of these proposals pleasantly surprised both NASA and the National Science Teachers Association (NSTA), who had not anticipated such a high level of scientific thought and rigor from the student submissions.

However, the evaluation revealed some students need to be more accurate about scientific principles and methods. This realization prompted a reassessment by some evaluators of their teaching practices, highlighting a potential gap in how scientific concepts were conveyed at the pre-professional level. Additionally, it became evident that several students needed help with clear written communication, a skill crucial for articulating complex scientific ideas.

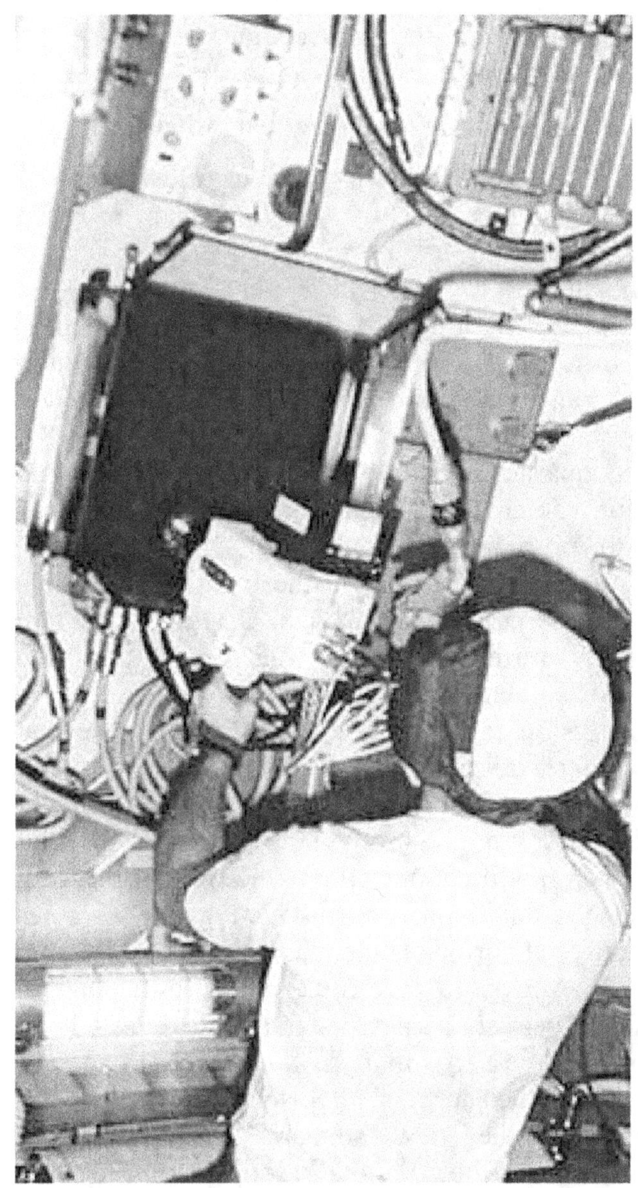

One of Skylab's interesting features was the Student Experiment Program. Here, Astronaut Owen Garriott televises the performance of the space spider Arabella as it adjusts to the weightless environment aboard the space station.

In the Skylab missions' grand scheme, including student experiments was a commendable initiative. While only a few students had their experiments selected for implementation, the experience provided invaluable lessons in science and the intricacies of managing a complex scientific project like Skylab. Through this exercise, students and educators alike gained a deeper understanding of the challenges and realities of space missions, fostering a greater appreciation for the intersection of science and engineering in space exploration.

Putting the Pieces Together

In the year following the historic Apollo 11 moon landing in July 1969, Skylab program managers and engineers embarked on various adjustments to adapt their plans to the newly designated Saturn V dry workshop. Initially, these modifications were confined to those required by the transition from a wet to a dry workshop configuration, by directives from NASA Headquarters. However, many additional improvements and enhancements were approved as the program progressed.

By July 1970, most of these new features had been incorporated into Skylab's design. A program review during this period reaffirmed the space station's schedule and budget, marking the conclusion of major design refinements. Later that summer, critical design reviews of the three primary cluster modules led to the establishment of configuration control boards to oversee any further modifications. From then on, significant changes could only be justified by deficiencies uncovered during testing.

In August 1969, a review of resources about the dry workshop's new requirements revealed that some necessary modifications could not be accomplished at Marshall Space Flight Center. For instance, the new support and deployment structure for the Apollo Telescope Mount (ATM) would exceed the center's capabilities, given that its facilities were already engaged in building the ATM and the multiple docking adapter. Consequently, Marshall proposed adding the new support structure and the protective shroud for Skylab to McDonnell Douglas's contract for the airlock.

Similarly, the task of equipping and checking out the multiple docking adapter and establishing the control hub for the solar observatory and new Earth-resource experiments surpassed Marshall's capacity. To address this, Huntsville's managers decided to extend an existing contract with Martin Marietta Corporation to include outfitting the docking adapter. Martin Marietta, who had been responsible for payload integration since the early days of the Apollo Applications Program (AAP),

was well-suited for this expanded role. Their contract had previously involved ensuring mission payloads were compatible, qualified for spaceflight, and capable of achieving mission objectives. This responsibility included mission planning, operations, training, and hardware procurement.

However, as the AAP evolved and the program's scope diminished during 1967 and 1968, Martin Marietta's role also diminished. NASA program offices struggled to find suitable tasks for the company's engineers, leading to dissatisfaction within Martin Marietta, notably when AAP Mission 1A was canceled. Despite these challenges, the company's extensive experience and existing contracts made it a valuable asset for the ongoing Skylab program, facilitating critical updates and modifications as the project advanced.

In early 1969, Marshall Space Flight Center redefined Martin Marietta's role in the Apollo Applications program, finalizing the firm's letter contract. Under this new arrangement, Martin Marietta conducted specific engineering studies to support NASA's integration decisions. Their responsibilities included performing electrical power and thermal analyses, studying experiment compatibility, coordinating test plans, and aiding in mission planning. Additionally, Martin Marietta was to manage configuration changes, update interface specifications, control documents, and maintain the document repository at Huntsville.

The Multiple Docking Adapter (MDA) assignment to Martin Marietta's Denver Division proved both logical and advantageous. The adapter was a crucial component for integrating various modules of the Skylab space station. As the integration contractor, Martin Marietta's role became more apparent and more significant with the MDA assignment, placing it on par with other major contractors. This role involved ensuring that the adapter connected seamlessly with other modules, enhancing the company's position in the Skylab project.

Given that Skylab was a unique, one-of-a-kind space station, the project faced the challenge of conducting thorough preflight testing. Unlike routine hardware, which could benefit from preliminary flights to identify and correct design flaws, Skylab's components required extensive testing due to their novel nature. Although some elements, like the S-IVB stage, were either well-tested or similar to Apollo hardware and thus required less exhaustive testing, many other components were entirely new and needed to be qualified during development.

The testing process was meticulously documented, with two primary documents guiding the program: the Mission Requirements Document and the Cluster Requirements Specification. The Mission Requirements Document outlined the specific objectives of each mission, while the Cluster Requirements Specification detailed permissible materials, design practices, construction methods, and human engineering standards. Based on these documents, NASA test engineers developed detailed procedures for contractors to ensure that each component met the necessary standards.

The overall test program consisted of two main phases. The first phase, development tests, allowed contractors to determine the best materials and designs through design-verification test units. These units, which did not need to fully meet flight specifications, were instrumental in refining the hardware. For instance, when Marshall Space Flight Center developed medical experiments, design-verification tests helped engineers optimize the size and functionality of equipment such as the bicycle ergometer and the lower-body negative-pressure device, ensuring they could accommodate crew members and transmit data effectively.

The second phase, qualification tests, aimed to demonstrate that each component was fit for its intended function under mission conditions. This rigorous testing ensured that every part of the Skylab space station would withstand the stresses of launch and perform reliably throughout its mission, paving the way for the successful implementation of this groundbreaking space station.

Once the design was verified to a satisfactory level, a qualification unit, establishing identical in every respect to the flight hardware, was produced. This unit underwent various rigorous tests to ensure its reliability and performance. After these qualification tests, several test articles were refurbished and repurposed as training units or backup hardware.

Given Skylab's complexity, where nearly every component functioned as part of a larger system, compatibility was crucial. As assembly progressed, systems tests were conducted at increasingly complex levels to identify and rectify any mechanical, electrical, or electromagnetic incompatibilities. Ideally, systems testing would have continued until the entire cluster was verified with all systems operating together. However, this comprehensive approach was impractical. Testing certain operations, such as the jettisoning of the payload shroud, the deployment of the Apollo telescope mount, or the unfolding of the solar arrays, was not feasible. Instead, these operations were verified through analysis, non-flight hardware testing, and simulations.

Given the complexity of the modules and the extensive testing they underwent, program officials established specialized test teams for each major module. These teams, comprised of contractor and agency engineers, followed their modules to each test site—from contractor facilities to Houston, Huntsville, and Cape Canaveral. Their on-site presence was invaluable for troubleshooting and resolving issues that arose during post-acceptance testing.

Among the critical components of Skylab were the airlock and the Multiple Docking Adapter (MDA). Although separate contractors built these components, they were closely integrated. The airlock, managed by Houston, and the MDA, developed by Martin Marietta, were treated as a cohesive unit. Production plans mandated combining and testing the two units before shipment to Kennedy Space Center (KSC). By July 1969, the airlock and the adapter had significantly developed, reflecting their critical roles in Skylab's operational success. Martin Marietta and McDonnell Douglas's Eastern Division in St. Louis collaborated closely to ensure seamless integration and testing of these components.

The evolution of the airlock for the Skylab Space Station reflects a remarkable transformation from a simple component to a sophisticated and multifaceted system. Initially, in 1965, the airlock was a straightforward tunnel that provided access between the S-IVB tank and the exterior environment. However, as program concepts advanced, the airlock's complexity grew significantly.

By the late 1960s, the airlock had become an integral part of the Skylab's dry workshop. It now served multiple functions beyond its original purpose. It housed essential systems for communications, electrical power distribution, and environmental control. A notable addition was the new structural transition section, which measured 3.05 meters in diameter and 1.2 meters in length. This section accommodated control panels and equipment and provided the foundation for the multiple docking adapter. The airlock's structural trusses were designed to support cylinders of compressed gases necessary for maintaining the workshop atmosphere.

As Skylab's design progressed, the airlock underwent several modifications. A fixed airlock shroud, matching the diameter of the S-IVB, was introduced. This shroud served as a platform for the Apollo telescope mount deployment structure. Due to these changes, the airlock's launch weight, initially about 3,600 kilograms, expanded to nearly 35,000 kilograms. Similarly, the multiple docking adapter evolved from a passive module, which had initially allowed the attachment of several experimental packages, into a more complex component. By 1967 and 1968, the adapter had been enlarged to accommodate the workshop's furnishings. However, the adapter's function changed with the decision to adopt a dry workshop configuration. By mid-1969, it had been reduced to a virtually empty cylinder, 3.05 meters in diameter and 5.25 meters long, with the main docking port at its forward end and a contingency port on one side, providing approximately 35 cubic meters of space.

The space within the adapter was quickly repurposed. The Apollo telescope mount's control and display panel occupied some of this space, and soon, earth-resource experiments and their supporting equipment filled more of it. Over time, the adapter became a storage area for various equipment and a workspace for the crew. This led to a somewhat haphazard arrangement of crew stations, which differed from the more organized layout of the workshop with its predominant one-g orientation. While the crews commented on the disorganized nature of the adapter, it did not prove to be a significant distraction.

Development testing for the airlock and adapter continued through 1970 and 1971. Static tests at Huntsville subjected the components to the structural loads anticipated during launch. Internal pressurization and leakage tests ensured the integrity of hatches and seals under prelaunch and orbital conditions. Concurrently, NASA's Plum Brook Station in Ohio, along with contractors and NASA engineers, verified the systems for jettisoning the payload shroud. Successful tests demonstrated the effectiveness of the explosive system designed to separate the shroud into four segments, with only minor issues needing attention.

The final major development tests took place in 1971 and 1972. At Houston, high-fidelity mockups of the cluster modules were subjected to vibroacoustic tests. These tests aimed to simulate the vibration and sound pressure experienced during powered flight, assessing the modules' structural resilience and the adequacy of the qualification test criteria.

During this period, the workshop underwent testing in 4,500-cubic-meter chambers at MSC, and the payload assembly—including the airlock and adapter—was rigorously evaluated. These comprehensive tests ensured that the Skylab airlock and its associated systems were fully prepared for the demands of spaceflight, embodying the extensive engineering and design efforts invested in the Skylab Space Station program.

During the Apollo Applications era in 1967, the multiple docking adapter was still in its formative stages. Early images from this period, such as MSFC-67-IND 7200-021, show the adapter as a relatively simple structure. However, significant changes had occurred by the time it was adapted for Skylab, as illustrated by the 1971 photograph ML71-5280. This evolution was marked by enhancements that allowed it to better support the Skylab mission's needs.

Docking the command and service module

In December 1971, a critical phase in the adapter's development took place. The adapter was prepared for shipment from Martin Marietta's Denver facility to the McDonnell Douglas plant in St. Louis. There, it would be mated to the airlock, as sh in the archival image MSFC 026857. This process was essential for integrating the adapter into the Skylab configuration. At the same time, a backup adapter was being readied for pressure tests, a step documented in the image ML71-7637.

During the testing phase, the workshop's environmental conditions presented challenges. It was found that the specifications for the workshop's ecological zones were too stringent. Of 53 environmental zones, 33 were too high, while six were too low. Had these specifications remained unchanged, components tested at inadequate levels might have failed during launch despite passing initial qualification tests. Conversely, many components would have failed their qualification tests unnecessarily, leading to costly redesigns and retesting.

A notable issue arose with the control moment gyros, which initially could not pass the qualification tests due to excessively conservative vibration specifications. When a test gyro was subjected to vibration tests at MSC, engineers discovered that the specifications were overly restrictive. By relaxing these specifications, the gyros could pass the tests without redesigning.

Various mockups and trainers were developed to address the challenges of testing and training. Both Martin Marietta and McDonnell Douglas created zero-gravity trainers, neutral buoyancy trainers, and high-fidelity one-g mockups. Zero-gravity trainers, small enough to fit into a KC-135 aircraft, were used for testing critical features such as crew restraints and extravehicular aids in a weightless environment. These trainers provided valuable data during the developmental phase, helping to refine designs before manufacturing the final flight articles.

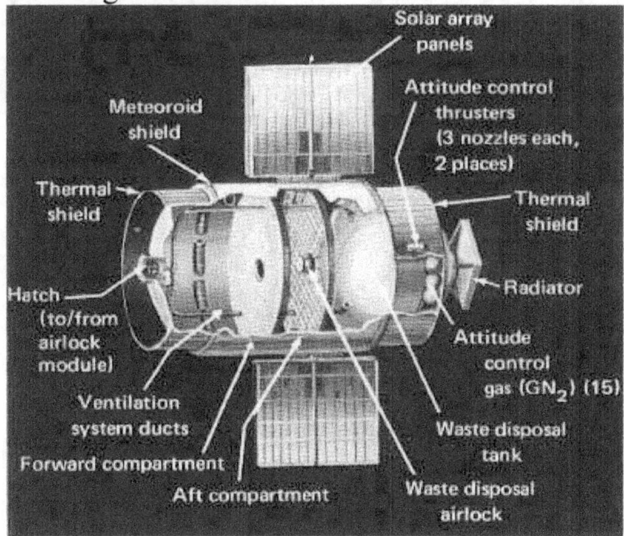

Neutral buoyancy trainers, featuring wire-mesh mockups of entire modules, were immersed in a large water tank at Huntsville. These trainers were used to verify astronauts' ability to move objects within the modules and develop extravehicular activity procedures. Later in the program, high-fidelity one-g trainers, which accurately replicated flight modules and contained the most precise equipment available, were employed as crews began to learn and practice flight procedures.

Overall, the progress in developing both the airlock and multiple docking adapter during 1971 was satisfactory, with significant strides in refining designs and preparing for the Skylab missions.

In December 1971, during a midterm program review, Leland Belew provided an optimistic update regarding the Skylab Space Station's development. He reported that neither of the space station's modules faced technical issues that could delay the program. However, there were concerns with the earth-resource experiments. Both the infrared spectrometer and the multispectral scanner encountered problems. Specifically, the coolers, essential for maintaining the proper operating temperature of their detectors, were malfunctioning. Additionally, the scanner experienced faults within its data-recording system.

Despite these setbacks, progress continued. The multiple docking adapter, an essential component for the station's airlock, had been accepted and dispatched to St. Louis for integration. The expectation was that the combined modules could be delivered to the Kennedy Space Center (KSC) by September 5, 1972, provided that the remainder of the program proceeded flawlessly. This timeline was based on an optimistic estimate of near-total success in the remaining tasks.

In St. Louis, engineers focused on completing other tests and checkouts while accommodating the missing earth sensors. They worked diligently for another six months to prepare the modules for the final stages of testing. By mid-1972, the modules were ready for their last evaluations, including a comprehensive fit-and-function review with astronauts rigorously verifying every on-orbit procedure. An altitude chamber test was also conducted to simulate the modules' performance in space conditions. Although no serious discrepancies emerged during these tests, some minor issues remained for technicians to address at KSC.

On October 5, 1972, the airlock module and the multiple docking adapter, the last of the flight modules to be shipped, were loaded onto a Super Guppy aircraft in St. Louis. This unique aircraft, built by Aero Spacelines, Inc., in the mid-1960s, was designed to transport oversized cargo, primarily for the Apollo program. Constructed from sections of four Boeing 377 Stratocruisers, the Super Guppy was, for a time, the world's largest aircraft in terms of cubic capacity. Its cargo for Skylab included the telescope mount, the instrument unit, and the Command and Service Module (CSM).

The Super Guppy arrived at KSC the following day, where the modules were unloaded and transported to the Vehicle Assembly Building. There, they were carefully stacked atop the workshop, marking a significant milestone in the preparations for Skylab's launch.

The workshop project at McDonnell Douglas's Huntington Beach, California, plant, which began in April 1969, had been a focal point of development and change during 1969 and 1970. The team there had modified the S-IVB stage 212, repurposing it for its new role in the Skylab program. This intense activity set the stage for the successful assembly and eventual launch of Skylab, marking a critical chapter in the history of space exploration.

The construction of the Skylab workshop at McDonnell Douglas's facility in Huntington Beach, California, was a focal point of the space station's development. By December 1970, a test version of the workshop was being prepared for shipment to the Manned Spacecraft Center in Houston. The workshop's design had evolved significantly, with a comprehensive redesign aimed at enhancing the habitability of the dry workshop module. Major changes included a complete overhaul of the crew quarters layout, the addition of a viewing window in the wardroom, and substantial upgrades to the food storage and preparation areas.

Despite these improvements, the project faced significant challenges. The waste management system's design remained unresolved until the end of 1970, and new requirements for the earth-resource experiments necessitated modifications to the attitude control system. These changes contributed to an increasingly complex engineering workload at McDonnell Douglas. The workshop, already the most intricate of the habitable modules, experienced delays due to these extensive redesigns.

By mid-1971, concerns about the progress of the workshop project prompted an evaluation by Headquarters. The review highlighted inefficiencies in management, questionable engineering practices, and inaccuracies in cost and schedule forecasts. It also pointed out the convoluted management structure involving Huntsville's Skylab office, the workshop project office, and the resident manager in California. The evaluation recommended strengthening Marshall Space Flight Center's management oversight, addressing the contractor's shortcomings, and instilling a greater sense of urgency.

In response, Marshall's program manager, Leland Belew, took decisive action. He appointed a 24-man Orbital Workshop Task Team led by William K. Simmons, Jr., the Marshall workshop project manager. This team was tasked with providing "timely on-site programmatic and technical interface" with McDonnell Douglas, a role commonly referred to as that of a "tiger team." Their mission was clear: to get the project back on track.

In August 1971, Simmons and most of his team, including James C. Shows from the Houston Skylab office and Richard H. Truly from the astronaut corps, relocated to California for a year. On the contractor's side, McDonnell Douglas assigned key officials to oversee the project: Walter Burke, the president of McDonnell Douglas Astronautics Company and a veteran of the Mercury and Gemini programs, and Fred J. Sanders, who had previously managed the airlock project before moving to California in 1969.

To streamline the project's progress, Simmons and Sanders instituted a weekly meeting schedule to review progress and plan future work. They also paired tiger-team members with their counterparts at McDonnell Douglas to address specific areas of responsibility. This collaborative approach was crucial in addressing the project's complexities and ensuring the Skylab workshop would be completed on schedule.

In addressing the myriad issues faced during Skylab's development, Houston's two representatives were primarily focused on resolving problems related to crew interfaces. Given that these concerns spanned nearly every system within the spacecraft, Richard H. Truly, one of the Houston team members, found himself in one of the most demanding roles. Truly's responsibility was to ensure that crew convenience and workload management were handled precisely, which required considerable negotiation and advocacy.

The core challenge for McDonnell Douglas lay in the sheer complexity of Skylab's systems. The project involved thousands of individual components sourced from both the company's facilities and various suppliers, including NASA. These parts needed to be delivered and assembled in a meticulously coordinated sequence. Any failure or redesign of parts, particularly after testing, could cause significant delays. Simmons discovered early on that McDonnell Douglas lacked an integrated schedule to manage the sequencing of these components. Additionally, information was moving sluggishly through the management structure, causing change orders to be delayed by weeks before reaching the production floor.

To address these issues, Simmons swiftly established a master schedule to prioritize tasks and streamline workflow. Recognizing the need for more immediate oversight, he arranged for the company's deputy operations manager to be stationed just off the shop floor, facilitating faster implementation of changes. While Simmons and Fred J. Sanders focused on the operational details, Walter Burke, the president of McDonnell Douglas Astronautics Company, played a crucial role in monitoring ongoing problems and ensuring that critical tasks received the necessary attention. Burke's presence positively influenced the organization, reinforcing the importance of resolving issues promptly.

During the fall of 1971, Simmons reported various technical problems to Leland Belew. These issues ranged from minor inconveniences, such as paint flaking off stowage lockers and scratches on the workshop window's electrically conductive coating, to more significant concerns. For instance, the reliability of brazed joints in hydraulic tubing was inconsistent. A particularly worrisome problem emerged when it was discovered that the iodine used to disinfect the drinking water was reacting with the brazing material, extracting nickel ions and rendering the water toxic. Although engineers introduced an ion-exchange resin to remove the harmful nickel, this solution inadvertently also removed the iodine, complicating the water treatment process further.

developing Skylab's workshop was a formidable challenge that required meticulous work and perseverance from the involved teams. Initially, the project faced significant hurdles related to the workshop's solar panels. By mid-October, despite ongoing efforts, the situation remained problematic. The proposal to address these issues demanded considerable effort from the crew, which met with resistance from key figures such as Truly.

Marshall Space Flight Center Director Eberhard Rees expressed growing concern as the May 15, 1972 delivery date approached. He was disheartened by the poor progress, particularly in contrast to the relatively high completion rates of the airlock and docking adapter—70% and 85%, respectively. In contrast, the workshop's development was a disappointing 25% completion. This discrepancy underscored the gravity of the situation and the need for immediate action to rectify the shortcomings.

The Skylab midterm review, held two months later, revealed that development and qualification testing were still significantly behind schedule. Persistent issues plagued several systems, including the thruster attitude-control system, the solar arrays, and the potable water system. The early months of 1972 showed little improvement, as old problems were addressed only to be replaced by new challenges.

However, by March 1972, there was a turning point. Simmons's weekly reports began to highlight progress as the checkout program gained momentum. By mid-May, the focus shifted to the crew compartment fit-and-function review. This critical evaluation involved the crew thoroughly inspecting the workshop from top to bottom. The review, which took four days to complete, wrapped up on May 27. Subsequently, the task team began assessing McDonnell Douglas's proposal for shipping the workshop on August 15.

In the ensuing months of June and July, preparations intensified for the final all-systems test of the workshop. The rigorous testing, which commenced on July 17 and concluded three weeks later, involved 510 hours of detailed examination, assessing every system within the workshop. Despite discovering a few anomalies, the overall performance was satisfactory, with only minor tasks remaining at the Cape. The majority of the work had been accomplished in California.

On September 7, 1972, NASA held a ceremony to mark the official acceptance of the completed workshop. Prominent officials, including NASA Administrator James C. Fletcher and Associate Administrator for Manned Spaceflight Dale D. Myers attended this significant event. The following day, the workshop, loaded aboard the U.S.N.S. Point Barrow, embarked on its 13-day journey via the Panama Canal to Florida.

The assembly of Skylab's workshop was a monumental endeavor, one that entailed far more than a simple checklist of tasks. The complexity of the workshop systems meant that every component had to function flawlessly before launch. Any partial success was unacceptable, as it could not be deferred to subsequent models. To draw a parallel, imagine if engineers were tasked with constructing the first Concorde supersonic jet and required it to be fault-free and ready for immediate commercial service. Such an expectation might have rendered supersonic travel a distant dream rather than a reality. By the end of September 1972, all flight hardware was in place at the Cape, prepared for final stacking and preflight testing, marking a significant milestone in the Skylab program's progress.

While Skylab's workshop awaited assembly and checkout at Cape Canaveral, attention turned to developing the launch system for the Space Shuttle. Per recommendations made the previous year, Deputy Administrator George Low mandated that the request for proposals for the Shuttle launch system include an analysis of the reentry hazards posed by the large fuel tanks. This requirement was to be a standard for all future projects.

Simultaneously, Low instructed the Office of Manned Space Flight to devise a method for deorbiting the S-IVB stages, which would be used to transport crews to Skylab. Unlike Apollo missions, where S-IVB stages were either left in space or sent to the lunar surface, this approach was unsuitable for Skylab missions. Studies indicated that the most straightforward method for deorbiting the empty upper stages was to vent excess propellants through the engine, a technique Low subsequently ordered to be adopted.

The discussion on orbital debris and its hazards raised concerns for Administrator James C. Fletcher, who had assumed office in 1971 after the decision to forgo controlled reentry for Skylab had already been made. Fletcher, unwilling to accept the inherent risks if there were feasible alternatives, instructed a reevaluation of the reentry issue. With just over four months before Skylab's launch, Program Director William Schneider directed both Marshall Space Flight Center and the Manned Spacecraft Center (MSC) to explore the feasibility of using the Apollo spacecraft's main engine to deorbit Skylab once the last crew had departed.

Initial responses from both centers were discouraging. Engineers at Houston identified numerous technical challenges and deemed the potential hazards to the crew unacceptable. The risk of the Apollo spacecraft encountering difficulties during undocking while the workshop was on a reentry trajectory was deemed too great. Marshall pointed out that implementing the necessary modifications to the launch vehicle and adjusting launch procedures would delay the launch and significantly increase costs. Conducting the requisite studies alone would take about six months, leaving insufficient time to address the issues before the last crew's launch.

Despite these challenges, Schneider remained determined. In April 1973, a team at Houston began investigating techniques and procedures for deorbiting Skylab using the Apollo spacecraft's service propulsion system. By the time the workshop was launched, the team had made substantial progress and identified many problems that needed resolution. However, their efforts were ultimately in vain. The loss of the micrometeoroid shield and damage to the workshop's solar arrays during launch introduced too many engineering uncertainties that could not be resolved in time. On July 13, 1973, Schneider decided to halt all studies on controlled deorbit, acknowledging that any reentry issues would need to be addressed in the future.

As Skylab evolved from an ambitious concept on paper to a tangible piece of hardware, the program office in Houston shifted its focus toward the operational intricacies of this pioneering space station. Unlike the Apollo missions, which adhered to a tightly regulated schedule dictated by various time-sensitive events, Skylab's operational framework necessitated a departure from this rigidity. Apollo missions were characterized by their precise, often minute-by-minute, timelines, with contingency plans meticulously crafted to handle potential spacecraft failures. Skylab, however, was designed with a different set of priorities.

The core of Skylab's mission was its focus on Earth and solar observations, which required a more flexible operational schedule. Unlike the relatively short Apollo missions, which lasted no more than two weeks, Skylab's missions were planned to extend over several months. This shift in mission duration introduced new challenges, particularly in data management. The space station's extended periods of isolation from ground stations meant that data could not be transmitted in real time. Instead, the information had to be stored onboard until Skylab passed over a ground station, at which point the telemetry data could be "dumped" into a ground receiver.

In addition to these technical and logistical adjustments, Skylab's operations necessitated a reevaluation of the relationships between Houston and Huntsville. The Marshall Space Flight Center (MSFC) in Huntsville, responsible for much of Skylab's hardware development, would need to collaborate closely with Houston, establishing in charge of overall mission operations. This new dynamic marked a significant shift from the earlier Apollo Applications Program (AAP) planning, fraught with conflict.

The early AAP planning was marred by disagreements, particularly concerning using the lunar module and developing a low-cost wet workshop. Houston had little faith in the proposed concept, while Huntsville was eager to advance the hardware development. This discord was compounded by Huntsville's desire for a more influential role in flight operations. Integrating Huntsville's support into Skylab operations required extensive negotiation and adjustment, which began in earnest in late 1966. Initial efforts concentrated on communication protocols and the role of Huntsville's Operations Support Center.

Tensions occasionally flared, with Huntsville officials suspecting their Houston counterparts of employing obstructionist tactics to undermine MSFC's involvement. The exclusion of Huntsville from Martin Marietta's integration contract, which omitted any reference to their support role, was a particular point of contention. Despite these challenges, it became increasingly clear that Houston would need to concede ground. As developing Skylab progressed, excluding Marshall from the operations support became technically challenging, highlighting the necessity of a more cohesive working relationship between the two centers.

Through these evolving dynamics, Houston and Huntsville would achieve a remarkable degree of cooperation, markedly improving upon the contentious interactions that had previously defined their collaborative efforts. This newfound synergy was crucial to the successful operation of Skylab, marking a significant milestone in the evolution of space station operations and inter-agency collaboration.

In 1967, as the Skylab project progressed, Huntsville sought to expand its role in flight operations. The Marshall Space Flight Center (MSFC) aimed to assume more significant responsibility, advocating for a supporting role in the Apollo Applications Program (AAP) flights and aspiring to establish itself as a spacecraft design and operations leader. Initially, Houston's Mission Control Center (MSC) resisted this change. However, a compromise began to take shape in June 1967 when Christopher Kraft, MSC's Director of Flight Operations, reluctantly agreed to incorporate Marshall engineers into AAP flight operations. This integration required that the engineers be part of MSC's organizational structure, established not satisfactory to Huntsville. Marshall preferred to maintain a separate team, with the lead engineer reporting directly to MSC for operational requirements.

In 1967, as the Skylab project progressed, Huntsville sought to expand its role in flight operations. The Marshall Space Flight Center (MSFC) aimed to assume more significant responsibility, advocating for a supporting role in the Apollo Applications Program (AAP) flights and aspiring to establish itself as a spacecraft design and operations leader. Initially, Houston's Mission Control Center (MSC) resisted this change. However, a compromise began to take shape in June 1967 when Christopher Kraft, MSC's Director of Flight Operations, reluctantly agreed to incorporate Marshall engineers into AAP flight operations. This integration required that the engineers be part of MSC's organizational structure, established not satisfactory to Huntsville. Marshall preferred to maintain a separate team, with the lead engineer reporting directly to MSC for operational requirements.

By November 1967, a tentative agreement was reached: Huntsville would establish a Systems and Experiments Section within MSC's Flight Control Division. This arrangement was a step toward recognizing Marshall's role while maintaining a degree of separation. As the project moved into 1969, the two centers appeared to be approaching a workable solution. In February, Robert Thompson, MSC's program manager, reassured Marshall's counterpart, Belew, that Huntsville would be informed of all developments. This would be achieved through coordinated efforts between the two offices and by MSFC's review and approval of the evolving operating procedures. Houston would submit change proposals through Marshall's program office to prevent interference with Marshall's contractors. Additionally, supplemental contracts would be established to formalize Houston's relationships with contractors such as McDonnell Douglas.

While these concessions were a positive step for Huntsville, Marshall officials sought a formal agreement to outline the "total operations interface." Such an agreement would acknowledge Houston's oversight of mission operations while recognizing Marshall's comprehensive responsibility for hardware, from design to operational procedures. This approach mirrored how the Kennedy Space Center (KSC), with its roots in the von Braun organization, recognized Marshall's role in Saturn test programs. Huntsville hoped for similar acknowledgment from MSC for Skylab operations, asserting its role in ensuring the performance and operational integrity of the space station hardware.

During his visit to the Marshall Space Flight Center in June 1967, Vice President Hubert H. Humphrey, as head of the National Aeronautics and Space Council, was briefed on the center's programs, including Skylab, by von Braun.

Preparations For Flight

In April, during a crucial planning meeting, the Marshall Space Flight Center (MSFC) staff approved a recommendation that the center seek a more proactive role in real-time operational decisions that affected individual hardware modules and the integrated Skylab space station cluster. This decision marked a pivotal shift in operational strategy, reflecting a desire for greater involvement in the ongoing management of Skylab's complex systems.

Two months later, center representatives convened in Houston to discuss operational strategies further. At this meeting, the Johnson Space Center (MSC) reviewed its specific requirements for flight planning, while Huntsville delineated the functions of its support center. A debate ensued over management philosophies, highlighting the contrasting approaches of the two centers. Despite the differences, they reached a consensus on several key points. First, the mission requirements document, jointly prepared by the two program offices, was established as the primary tool for mission planning. This document would serve as the formal communication link between the two centers. Second, Houston was tasked with preparing the operational data book, utilizing information from Huntsville and its contractors. Third, Marshall agreed to grant Houston access to its configuration control board, enhancing the collaborative effort between the centers.

However, not all aspects of Skylab operations were resolved smoothly. A significant point of contention was the upgrade of the workshop itself. Houston sought enhancements, while Huntsville adhered to a no-change policy. This disagreement arose during the telescope-mount design review in May 1970, where the discussions reached an impasse. An MSC official described this period as a challenging standoff, but it eventually led to improved relations. By the end of the year, the two centers had established a basic framework for Skylab operations, which included a flight management team composed of program managers and MSC's operations managers who would set policy. Despite Houston's majority on this team, Huntsville and the Kennedy Space Center (KSC) were guaranteed representation in all decisions. Daily operations would remain under the purview of MSC's flight control teams. In cases where hardware issues arose, the flight director could seek assistance from a Marshall liaison team stationed in the Flight Operations Management Room. This liaison team could call upon a larger group of engineers at Huntsville's operations center, ensuring a robust support system.

A sophisticated communications network was established to connect the two centers, providing Huntsville with detailed updates on Skylab's condition. This network was vital for ensuring smooth operational coordination.

Attitudes in Houston shifted significantly after the Apollo 11 mission and the decision to proceed with the dry-workshop approach for Skylab. Before July 1969, many MSC officials viewed Skylab as an unwelcome distraction from their primary focus on Apollo. However, Skylab emerged as the next major program following the lunar landing. In August 1969, the Flight Control Division began formulating an operations plan. Division Director Eugene Kranz aimed to incorporate many of the successful elements of the Apollo missions into Skylab operations. Nonetheless, adjustments were necessary to accommodate the extended duration of the Skylab missions and the increased number of flight systems.

Given the constraints, Houston could not maintain a full complement of staff in the Mission Control Center throughout the Skylab missions as it had for Apollo. An Earth-orbital mission required less extensive support. Consequently, during astronauts' working hours, a "high-level" shift would manage operations, while a skeleton crew would be on duty at night, with additional engineers available on call.

In the Apollo missions, two flight controllers typically shared the responsibility for managing spacecraft systems. However, Skylab presented a more complex challenge due to its five distinct units, including the telescope mount. To address this, a similar division of labor and additional requirements would necessitate nearly 30 engineers on the team. The Flight Control Division Director Eugene Kranz was concerned that such a large group could impede decision-making, leading to excessive discussions rather than prompt action. His preliminary plan aimed to streamline the process by designating one systems expert per spacecraft and consolidating various duties. This approach would reduce the number of flight controllers in the control room during peak activities to no more than 20, with normal operations requiring only 11 controllers.

In November, Kranz revisited Skylab planning as part of a broader review conducted by the Flight Operations Directorate. Manning requirements were a key focus, but several other issues were scrutinized, including the operational implications of Skylab's new 50-degree orbital inclination, Houston's interactions with principal investigators, and the demands of uncrewed operations. Kranz identified 11 aspects of Skylab operations that differed significantly from Apollo missions and called for a thorough review of these critical issues.

Over the next 30 months, the flight-control organization underwent significant restructuring. Several instructors were retrained as systems engineers and personnel assigned to experiments received extensive training, with some taking extended courses in solar physics. By October 1972, a dedicated flight-control team was assigned to Skylab full-time. This team was tasked with developing data processing procedures and conducting mission simulations with the flight crews. Following Apollo 17's splash down, the division focused on Skylab.

When Skylab missions commenced, the division's preparations generally proved effective. However, one notable issue emerged with the transmission of data. Skylab's signals were captured by one of thirteen tracking and data networkstations and relayed through the Goddard Space Flight Center to Houston. Although Skylab was often close enough to a station to transmit data in real time, there were frequent instances when data had to be recorded for later transmission. Skylab's telemetry system, which required only five minutes to transmit data collected over two hours, represented a significant departure from Apollo. This system posed challenges for Houston's flight-control teams, who were accustomed to having immediate access to only 10% of the data during lunar missions. The increased telemetry from Skylab and the long intervals between transmissions made immediate data access impractical. Instead, data were processed using a method known as "redundancy removal," which meant only changes to the data were sent to Mission Control, further complicating the task of real-time monitoring.

As Skylab, America's first space station, neared its launch, the transition from Earth-based preparation to space-bound execution revealed triumphs and trials. One of the pivotal challenges was the late installation of new equipment, which unfortunately led to a lack of proficiency among some flight controllers. This shortcoming became painfully evident during the workshop's post-launch crisis. Gene Kranz, a key figure in mission control, highlighted the gravity of the situation: "Because of the lack of proficiency in the data retrieval task, the flight controllers were generally inefficient in accomplishing contingency analyses."

The necessity for improvement was swiftly recognized. Following the first crewed mission, a specialized training program was established to address these gaps. Twelve individuals were trained explicitly in data retrieval, a crucial step in ensuring the team was adequately prepared for future missions. Meanwhile, as Kranz's division focused on refining operations, the Flight Crew Operations Directorate embarked on developing a comprehensive Skylab flight plan. This plan was designed to outline in detail the schedule of each crew's activities in space.

Initially, the drafts of this flight plan served multiple purposes beyond mere scheduling. They functioned as training tools for flight planners, whose Apollo-era experience proved less applicable to the unique challenges of Skylab. These preliminary drafts highlighted critical issues, shaped crew-training requirements, and exposed potential problems with experiment priorities. The mission requirements document provided a foundation for this planning, detailing objectives, experiment needs, extravehicular activities, recovery zones, and information regarding television and photography.

The Flight Crew Operations Directorate established general guidelines for scheduling crew activities, which were initially relatively rigid. For example, early protocols dictated that all crew members would dine together. However, as the complexity of scheduling increased, these guidelines became more flexible to accommodate the dynamic needs of the mission. Despite using computers, much of the scheduling was carried out manually. The goal was to fulfill all objectives outlined in the mission requirements document. When it became clear that meeting every objective was not feasible, program offices revised the document, often by reducing the frequency of specific experiments. The extensive planning process produced over 10,000 pages of books, checklists, cue cards, maps, and charts.

In parallel, Huntsville began preparing for mission support in mid-1970. The Mission Operations Office played a central role in this preparation, identifying seventeen major tasks and appointing managers. Coordination was primarily achieved through monthly meetings involving task managers, prime contractors, and representatives from Marshall Space Flight Center's (MSFC) major divisions. By the end of 1971, nineteen mission-support plans had been developed.

Marshall engineers frequently collaborated with their counterparts in Houston. A significant series of meetings in mid-1972 focused on reviewing hardware characteristics and operating procedures. In October of the same year, Huntsville tested its preparations with a mission simulation, setting the stage for future participation in Houston's dress rehearsals.

Marshall consolidated its support efforts in the Huntsville Operations Support Center, a facility that had proven its worth during the Apollo program. For Skylab, ten mission support groups were established, each tasked with managing a major system, such as attitude control. Initial manpower projections estimated over 400 engineers for these support groups, drawn from the program office, MSFC laboratories, and contractor teams. Saturn engineers were assigned to monitor launch vehicle operations during checkout and the early stages of flight. Additionally, personnel managed a sophisticated communications network, connecting Huntsville with Houston and the Cape through voice, television, and high-speed digital data lines.

The Skylab program introduced several innovations and adjustments compared to earlier crewed missions, notably the Mission Operations Planning System (MOPS). This advanced system, unavailable during previous space missions, allowed support personnel to access real-time data from Houston's computers. Through MOPS, immediate printouts of current flight and experiment data could be generated, greatly enhancing the efficiency and responsiveness of mission support.

The Skylab mission was strategically divided into five distinct phases, each with specific support requirements. Pre-launch support commenced in October 1972. Launch vehicles and workshop engineers were stationed at Kennedy Space Center (KSC) during this phase to ensure readiness. The second phase, involving workshop launch and deployment, spanned only a few critical hours but demanded peak operational activity from the support center. The crew's launch marked the third phase, during which peak operations were again required, particularly in the early stages of each launch.

The fourth phase encompassed crewed operations. During this time, the support center's coordinating staff operated at full strength while astronauts were actively working. Outside of active mission hours, support staff maintained a reduced on-call presence to handle emergent issues. The final phase, uncrewed operations, necessitated ongoing monitoring by the Marshall Space Flight Center (MSFC). Although the workshop was uncrewed, several systems, including the solar telescopes, continued functioning and required oversight.

The selection of Skylab crew members was a source of tension among Houston's astronauts. The astronaut corps had expanded significantly in the mid-1960s, but as NASA's fortunes shifted, it became clear that some astronauts could not fly until at least the 1980s. The selection process, managed by Deke Slayton, the director of flight crew operations, added to the frustration. Slayton's recommendations were typically approved by the head of the Manned Spacecraft Center, Chris Kraft, and then forwarded to NASA Headquarters. His selection criteria emphasized experience, favoring astronauts who progressed naturally from Gemini missions to Apollo flights. This policy particularly benefitted those who had joined the program by 1963 or the 1966 test pilots who received early assignments. In contrast, the scientist-astronauts who had joined in 1965 and 1967 were disadvantaged, as their flying opportunities were limited under this system.

When the initial group of Air Force pilots completed their year of rigorous flight training, they found themselves at a disadvantage. Dissatisfaction among scientist-astronauts emerged following the historic Apollo 11 lunar landing. Despite early expectations that subsequent Apollo missions would place a stronger emphasis on scientific exploration, Deke Slayton, the Director of Flight Crew Operations, opted to select only test pilots for the next three Apollo missions.

In October 1969, discontent among the scientist-astronauts grew more pronounced. They voiced their frustrations to NASA Headquarters, criticizing the selection criteria prioritizing operational experience over scientific expertise. Slayton responded by emphasizing the perilous nature of lunar missions, arguing that the priority was to ensure the crew's safety rather than solely focusing on scientific qualifications. His rebuttal underscored the critical nature of having highly skilled pilots who could manage the inherent risks of space travel, suggesting that scientific competence was secondary to the essential need for operational expertise.

The situation for scientist-astronauts deteriorated further in 1970, with NASA canceling one Apollo flight in January and two more in September. This series of cancellations raised concerns that no scientist would have the opportunity to participate in lunar exploration. As the year drew to a close, the Space Science Board intervened, seeking assurances from NASA that at least two scientists would be included in each Skylab mission. This move coincided with a resurgence of dissatisfaction among scientist-astronauts stationed in Houston.

In January 1971, NASA's chief scientist Homer Newell traveled to Houston to address the growing discontent among the scientist-astronauts. One by one, the scientists expressed their frustrations with the crew selection process, asserting that the preferences of test pilots like Slayton heavily influenced it. They felt that crew selections were based more on flight experience, special skills, and personal relationships than scientific merit. Many argued that those with a strong interest in science were disadvantaged. The astronaut group strongly advocated for a revised selection process to better account for scientific expertise and proposed that each Skylab mission should include two scientists. They emphasized that the demands of scientific work far outweighed the time required for flight operations, underscoring the need for a crew composition that reflected this imbalance.

Newell incorporated these concerns into his recommendations to NASA Administrator Thomas Paine. He proposed assigning Harrison Schmitt, the only astronaut with a Ph.D. in geology, to a lunar landing mission as soon as possible. He suggested that each Skylab mission should include two scientists. Additionally, Newell recommended a review of NASA's crew-selection process and restructuring the scientist-astronaut program to foster a stronger commitment to scientific careers. However, recognizing that he had only heard one side of the issue, Newell characterized his recommendations as tentative.

The recommendations sparked an extended debate about the composition of Skylab crews. Slayton and other NASA officials, including Robert Gilruth, argued against including more than one scientist per flight. They reasoned that the technical complexities of Skylab hardware necessitated a high level of systems expertise, which was established predominantly among test pilots. Gilruth, in particular, highlighted the likelihood of encountering systems problems during missions, noting that Skylab's systems could not be modified post-launch. As a result, the training focused heavily on systems management and malfunction procedures, emphasizing cross-training to ensure that each crew member had a balanced proficiency across major experiments.

As the debate over Skylab crew composition continued, the importance of an astronaut's specific academic background became a secondary concern. Dale Myers, who was open to accommodating the scientists 'requests, initially considered including a second scientist on at least one mission. However, the persuasive arguments put forth by Robert Gilruth shifted Myers's stance. The tragic deaths of three Soviet cosmonauts during reentry on June 29 underscored the perilous nature of space missions and reinforced the priority of operational considerations. Consequently, Myers endorsed a plan to see each Skylab mission staffed by two pilot-astronauts and one scientist-astronaut, with the first flight's scientist being a physician.

On July 6, 1971, Myers recommended approving this plan, leaving the selection of specific crew members to Houston. Despite some reservations from Homer Newell, the plan was ultimately adopted. The crew selections, announced on January 19, 1972, reflected a compromise between NASA's interests and the aspirations of the scientist-astronauts.

The first Skylab mission would be commanded by Charles "Pete" Conrad, a seasoned astronaut with three previous spaceflights, including the Apollo 12 lunar mission. He was joined by two astronauts new to spaceflight: Joe Kerwin, a physician who had joined NASA in 1965, and Paul Weitz, who entered the program a year later. The second mission was led by Alan Bean, a veteran who had flown to the moon with Conrad in 1969. His crew included Owen Garriott, an electrical engineer with a Ph.D. from Stanford, and Jack Lousma, a Marine major assigned as the pilot. The third crew was commanded by Gerald Carr, another Marine test pilot, and included Edward Gibson, a Ph.D. from Caltech, and Air Force Lt. Col. William Pogue.

The final crew selections balanced NASA's operational needs and the desire for scientific expertise, resulting in fewer veterans and scientists than originally hoped for. In retrospect, the significance of the crew composition was somewhat overstated. On all three Skylab missions, the test pilots performed scientific work admirably, while the scientist-astronauts demonstrated skill in maintaining and repairing spacecraft systems. Ultimately, the success of the missions hinged more on effective teamwork and the positive attitudes of the crew members than on their academic qualifications.

The Skylab Space Station program, an ambitious venture in space research, faced several challenges and adjustments as it progressed toward its missions. Among these was the decision to include a physician on the second mission, a proposal that the medical directorate championed fervently. Despite their advocacy, the fears surrounding a prolonged 56-day mission were ultimately unfounded. The Apollo telescope mount experimenters, crucial to the mission's success, were ably supported by astronauts Garriott and Gibson. Their presence proved sufficient for the tasks, reflecting the crew's capability to handle the specialized equipment and conduct scientific research.

In an ideal scenario, the second or third crew would have included an earth-resources specialist to leverage the new experiments added late in the program fully. However, the late addition of these earth-resources experiments posed a problem. None of NASA's scientist-astronauts had extensive qualifications in the relevant hardware, and given the experimental nature of the instruments, additional expertise might not have been utilized effectively. This concern was compounded by Slayton's assertion that the flight plan would offer limited time for independent research, a prediction that proved largely accurate.

Preparation for the mission began in earnest in October 1970, driven largely by the efforts of the Apollo telescope mount investigators. Richard Tousey from the Naval Research Laboratory, who had initially approached Houston in 1967 with a proposal for a solar physics course for astronauts, renewed his request in February 1970. In his letter to the program office, Tousey emphasized the need for scientific training for the crew. He noted that while astronauts could operate the telescope mount without a deep understanding of solar physics, the data quality would suffer. He expressed concern that a last-minute cram course before the launch—when operational training would be critical—would be inadequate. Ideally, the crew should have begun training 24 months before the launch. With the mission scheduled for July 1972, there was little time to spare.

Despite these pressing needs, Houston's enthusiasm for initiating crew training was initially lukewarm. The astronauts were deeply involved in design reviews, and training chief John Von Bockel was preoccupied with Apollo missions. Nevertheless, by June 1971, steps were taken to address the concerns of the telescope-mount investigators. A meeting held in Denver resulted in the decision to start a comprehensive 10-week, 60-hour course in solar physics that fall. This course, designed by Dr. Frank Orrall of the University of Hawaii, would involve active participation from principal investigators, ensuring that all crew members received the same level of training, irrespective of their backgrounds.

The training facilities themselves were meticulously designed to simulate the space environment. At the Manned Spacecraft Center, the one-g trainer included a detailed workshop and an Apollo command module. The upper deck of the workshop featured a scientific airlock, complete with a coiled metallic hose for access. The double ring of storage lockers and water tanks was arranged for easy accessibility in zero gravity. The lower deck of the trainer was organized into compartments clearly labeled for various functions, providing a comprehensive simulation of the modules that astronauts would encounter in space.

The one-g trainer at the Manned Spacecraft Center was designed to provide astronauts with a realistic simulation of space operations. The top-left view of the trainer features the airlock module, established and mounted to allow for lateral rotation. This airlock was crucial for managing atmospheric gases under high pressure, with six cylindrical oxygen tanks and six spherical nitrogen tanks ensuring a stable environment for the crew. Below this, the airlock, docking adapter, and telescope mount were visible. Notably, the black ring on the left was the fixed shroud, and the telescope mount, positioned at the head of the stairs, was designed for deployment in its flight attitude. Unlike the other modules, the telescope mount lacked an interior workspace, requiring astronauts to conduct their tasks from its exterior. Additionally, the bottom-left view reveals the power supply and circuit breaker panels inside the airlock, essential for maintaining the trainer's operational integrity.

As the Skylab program progressed, there were calls for enhancements to the training regimen. Suggestions included incorporating real-time solar observations into the course rather than scheduling them for later. Dr. Frank Orrall's curriculum was supplemented with lectures on the broader implications of solar physics, aiming to demonstrate its relevance beyond mere solar study. This approach was intended to increase the astronauts' engagement by highlighting the broader applications of solar physics in scientific research.

Many astronauts found the material challenging when the training course commenced in late October. One astronaut candidly described the experience as overwhelming, struggling to keep pace and grasp the complexities of solar physics. Communication difficulties also arose between the astronauts and the solar physicists, who specialized in a highly technical field. Jerry Carr found that focusing on becoming a proficient observer rather than attempting to master solar physics as a subject helped him navigate the course more effectively.

Simultaneously, the Manned Spacecraft Center's training office developed a comprehensive Skylab training program led by Robert Kohler and supported by a team from Martin Marietta. The training syllabus was extensive, totaling 2,200 hours over 18 months, with a schedule structured around a 28-hour training week. This rigorous program encompassed 450 hours of briefings and reviews, 450 hours dedicated to experiment work, and nearly 700 hours of simulator training. The demanding nature of this schedule marked a significant increase compared to the Apollo missions, which had averaged around 1,200 hours of training. The inclusion of extensive briefings was a notable feature of the 1971 training regimen, underscoring the program's commitment to thorough preparation.

In this demanding environment, astronauts like Charles Conrad, Jr. engaged with the display and control panels of the telescope mount, exemplifying the hands-on approach necessary to master the equipment and ensure mission success.

The early phases of Skylab Space Station training focused on rigorous preparation to ensure mission success. The initial crew training sessions centered around mastering the complex medical experiments to be integral to the mission. One notable scene captured during this training involved astronaut Kerwin, positioned in a rotating chair designed for the human vestibular function experiment, designated M 13 1. As Kerwin navigated the rotating chair, astronaut Weitz diligently recorded data. At the same time, Conrad was seen pedaling a bicycle ergometer in the background, contributing to the study of physical performance in varying conditions.

Training for Skylab missions was conducted in two distinct phases. Initially, principal investigators provided in-depth lectures covering the theoretical underpinnings, objectives, and data collection methodologies pertinent to the experiments. This was followed by a comprehensive operational training phase led by Martin Marietta instructors, who detailed the practical aspects of equipment use, maintenance, safety procedures, and support systems. Much of the training was dedicated to familiarizing astronauts with the Apollo spacecraft, a critical component of the Skylab missions. Over 130 hours of briefing sessions and extensive simulator training were conducted by North American Rockwell to ensure astronauts were adept in handling the spacecraft. Despite the limited time Skylab crews would spend in the Apollo spacecraft, these hours were crucial due to the potential risks associated with any operational errors.

The most substantial instructional time was allocated to the Skylab workshop itself. Martin Marietta focused on the intricacies of the telescope mount, while McDonnell Douglas covered the remaining systems essential for mission operations. 1971, the training office's schedule was frequently adjusted to accommodate spacecraft testing. Traditionally involved in testing flight hardware, astronauts were allotted 100 hours in the Skylab syllabus. However, the actual time spent on testing often doubled this estimate, and the schedule did not fully account for the delays. Testing frequently occurred at contractor facilities or NASA centers, requiring astronauts to travel to locations such as Huntsville or St. Louis, only to face postponements. The most significant disruptions came from Huntington Beach schedule slips, significantly impacting the training office's planning.

Additional training requirements kept the astronauts engaged in a variety of tasks. They familiarized themselves with navigational stars and stellar experiments at the Morehead Planetarium in Chapel Hill, North Carolina. They worked with astronaut maneuvering units in Denver and Langley Research Center in Virginia. Extravehicular activity training, crucial for operations outside the spacecraft, took place in Marshall's large water tank. The Neutral Buoyancy Simulator, extensively utilized during Skylab's design phase, proved to be an effective environment for practicing these activities, allowing for meticulous planning and timing of each maneuver.

Training transitioned from theoretical to practical in early 1972 as crews began utilizing Skylab simulators. These simulators featured computer systems that replicated the visual and operational conditions astronauts would encounter in space. The telescope mount console, a central component, was prominently featured, and astronauts spent up to 200 hours monitoring solar activity through its video screens. The computer system could simulate normal and abnormal conditions across various control panels, providing comprehensive training experiences. Often, while one crew trained in the workshop simulator, another crew practiced flight operations in the command-module simulator, honing their skills for docking with Skylab. Additional Apollo simulators were used to prepare for launch aborts and rendezvous procedures, allowing astronauts to operate independently or in conjunction with Mission Control.

Houston's training philosophy emphasized that while all crew members needed proficiency with major experiments, specialization was also essential. The commander was responsible for the Apollo spacecraft, and the scientist took charge of extravehicular activities, the solar telescope, and medical experiments. The pilot focused on workshop systems and earth-resource equipment. This division of responsibilities was evident in the training sessions, ensuring that each crew member was well-prepared for their specific role in the Skylab missions.

In the meticulous preparation for Skylab missions, each astronaut's training regimen was tailored to their specific responsibilities, revealing a notable variation in time spent on different aspects of their preparation. Despite his extensive flight experience in the command module, astronaut Charles Conrad dedicated 400 hours to Apollo simulator training, surpassing his colleague Paul Weitz by 55 hours. Weitz, conversely, invested nearly twice as much time in earth resources training as his crewmates. Astronaut Kerwin, who focused on the medical experiments, accumulated 181 hours of preparation, far exceeding the time spent by his partners. This approach underscored a flexible training schedule prioritizing practical competence over rigid adherence to set hours.

The training syllabus served as a guide rather than a strict framework, allowing for adjustments by crew commanders and mission training coordinators. Commanders wielded considerable authority over their training schedules. For instance, Conrad, believing that 20 hours was insufficient for workshop activation, insisted on extending this period to 125 hours to ensure thorough mastery of the task. Instructors evaluated the crews based on operational competence rather than mere hours of exposure, reflecting the program's focus on practical readiness.

By September 1972, the training shifted from individual to team-focused sessions by introducing "mini-sims." These intensive training sessions in the workshop simulator mirrored the demands of a mission day, beginning at 6:00 a.m. and concluding at 10:00 p.m., simulating the astronauts' daily schedule in space. During these sessions, crews received instructions via teleprinter as they would in flight, with limited voice contact with the ground only when the simulated flight passed over a ground station. This format allowed instructors to address specific questions but maintained a high degree of realism. The mini-sims proved to be a valuable investment, enhancing the integration of training and exposing flight planners to various scheduling constraints.

As the launch date approached, pressures increased due to disruptions caused by simulator breakdowns, reviews, and last-minute demands. By January 1973, the first crew was behind schedule, and their work weeks extended to 60 hours. The launch date was pushed from April 30 to May 14 to accommodate delays at the Cape, granting the training office some respite, though the pace remained intense.

Upon completing the missions, Von Bockel expressed general satisfaction with the training program despite acknowledging areas for improvement. He had initially advocated for training only one backup crew but ultimately complied with the need for two, as one prime crew included a doctor and the other two included physical scientists. This approach ensured readiness for any potential crew replacement. Von Bockel recognized that while astronauts were highly motivated and engineers frequently addressed their questions, instructors often struggled to keep pace with the crew's knowledge acquisition. He recommended that future programs prepare training materials well in advance to address this issue.

One significant challenge noted was the extended interval between training and actual performance of critical tasks. For instance, the last crew's deactivation and reentry occurred 13 weeks after their training, resulting in a procedural error—though promptly corrected—that highlighted the need for refresher training during the flight. Von Bockel suggested that future missions incorporate time for such refresher training to mitigate the effects of this delay.

The launch of Skylab in August 1972 evoked a sense of nostalgia reminiscent of the Apollo program's glory days at the Kennedy Space Center (KSC). At that time, the Vehicle Assembly Building was a hive of activity, with Apollo 17—the final mission of the lunar landing program—undergoing its final tests in one high bay. Adjacent to it, in a second bay, the Saturn V booster designated for Skylab was being assembled, while a third bay housed a new 39-meter pedestal, soon to serve as the launch platform for the Skylab missions. This scene underscored the deep connection between Skylab and the Apollo program; both initiatives utilized the same facilities, operations, and hardware. By 1970, a single office managed both programs, reflecting their intertwined histories.

Despite these similarities, Skylab marked a significant departure from its predecessor. Initially conducted from Cape Canaveral, the Saturn IB launches were relocated to NASA's Complex 39, designed specifically for the Skylab missions. This change necessitated modifications to the launch facilities and new equipment for the workshop's unique payload. The introduction of experimental hardware added complexity to the checkout process, requiring new procedures and equipment.

The transition was not without challenges. Initial launch preparations faced extensive debate and logistical adjustments, but the changes were implemented smoothly and economically once finalized. Launch operations, however, proved more problematic. The new hardware revealed numerous defects, which were anticipated given the complexities of new flight equipment. The schedule was designed to accommodate these delays, reflecting a preparedness for such issues.

One of George Mueller's primary objectives for the Apollo Applications program was to ensure continued employment for the Saturn industrial team. Unfortunately, reductions in NASA's budget thwarted these plans. By mid-1968, KSC officials grappled with the challenge of maintaining a Saturn IB launch team during an extended period of inactivity. The team, which numbered nearly 3,000 people—90% of whom were contractor personnel—had previously overseen the Saturn IB program's impressive record of 14 successful launches. Chrysler Corporation's Space Division, McDonnell Douglas, and IBM were key contractors responsible for the various stages and components of the Saturn IB rockets.

With Apollo missions scheduled to transition from Saturn IB to the Saturn V rockets after Apollo 7 in October 1968 and the Saturn V launches occurring at complex 39 on Merritt Island, a substantial gap loomed in Saturn IB operations. The first Apollo Applications mission's postponement to late 1970 extended this hiatus, prompting Mueller to approve a plan drastically reducing the Saturn IB launch team by 87%. This plan left only 350 personnel in a standby capacity. The two launch complexes were maintained minimally, with removable equipment stored and primary structures periodically sandblasted and repainted. However, the retained staff, while skilled, were not optimally suited for maintenance tasks, and retaining key personnel proved challenging. The alternative—assembling a new

Saturn team in 1970—was deemed even less practical.

As Kennedy Space Center (KSC) officials wrestled with maintaining the Saturn IB launch team, another study was underway to explore alternative launch sites. Both launch complexes on Cape Canaveral presented significant drawbacks. Launch Complex 34 (LC-34), an aging facility originally constructed for Army use had suffered from years of exposure to salt air, leading to corrosion and wear. Its design, constrained by inadequate funding, had undergone numerous modifications over its seven years of operation to accommodate crewed flights, yet it remained undersized and outdated.

Conversely, Launch Complex 37 (LC-37), designed by NASA engineers in 1961, was better suited to the requirements of Saturn rockets. It featured a service structure, launch umbilical tower, and blockhouse that was more appropriately scaled for the Saturn IB operations. However, LC-37 had not yet been adapted for crewed missions, and retrofitting the complex for this purpose would require nearly two years.

In contrast, the Advanced Programs Office at KSC advocated for using the newer Launch Complex 39 (LC-39) on Merritt Island for Apollo Applications Program (AAP) missions. Consolidating manpower and equipment at this state-of-the-art facility promised to save money and streamline operations. LC-39, constructed to support the massive Saturn V rockets, was designed on a larger scale and featured the innovative mobile launch concept.

Unlike the older Cape Canaveral complexes, where rockets were assembled directly on the pad, LC-39 utilized the Vehicle Assembly Building (VAB) for rocket assembly. With its enormous dimensions, the Saturn V was assembled in this controlled environment and then transported to the launch pad five kilometers away by a crawler. The mobile launcher, equipped with a 136-meter tower, replaced the traditional stationary umbilical towers. This tower included eight service arms that provided essential electrical, pneumatic, and propellant services to the various stages and modules of the space vehicle.

In October 1968, von Braun, right, escorted Homer E. Newell through the Skylab mockup at the Marshall Space Flight Center.

This modern approach, emphasizing mobile assembly and support infrastructure, represented a significant advancement over the older facilities, offering a more efficient and adaptable solution for the evolving needs of the Apollo Applications Program.

Launching Skylab

Several logistical challenges emerged as the Kennedy Space Center (KSC) contemplated the transition to Launch Complex 39 (LC-39) for Saturn IB launches. LC-39, designed to support the colossal Saturn V rockets, featured a mobile service structure with a ninth arm specifically for astronaut entry into the command module. Positioned opposite the tower at the pad, this mobile structure was essential for crewed missions. However, the complex had only one such structure, which created issues for accommodating the smaller Saturn IB rocket.

The primary complication arose from the need to adapt LC-39's infrastructure to the Saturn IB, establishing significantly smaller than the Saturn V. An Apollo spacecraft mounted atop a Saturn IB was 43 meters shorter than one on a Saturn V. Consequently, much of the existing supporting equipment would be misaligned for the smaller rocket. Service arms 7 through 9, connected with the Apollo spacecraft on a Saturn V, would be positioned far above the spacecraft on a Saturn IB. These arms, known for their ability to swing clear in 2-5 seconds, were essentially mechanical bridges measuring 18 meters long and weighing up to 25 tons. Relocating these arms presented a major challenge.

Additionally, work platforms in the Vehicle Assembly Building (VAB) and on the mobile service structure faced similar issues. Although these platforms did not need to swing, their large dimensions—up to 18 meters square and three stories high—complicated their repositioning. The launch team also needed to adjust the propellant, pneumatic, and electrical lines nearly covering the mobile launcher's back side.

In February 1969, Boeing proposed a solution to minimize modifications by using a 39-meter pedestal to elevate the Saturn IB. This would align the second stage and instrument unit, as well as the Apollo spacecraft, to match the height of the Saturn V configuration. This approach would allow using the existing launcher's upper service arms and work platforms, reducing the need for extensive reconfiguration. The estimated cost for this modification was around $5 million, significantly less than the cost of constructing a new launcher.

The major concern with this proposal was the dynamic interaction between the rocket and the pedestal during liftoff. The pedestal needed to absorb the forces exerted by the rocket without causing dangerous oscillations. During ignition, hold arms on the launcher restrained the vehicle for four seconds, allowing the launch control team to verify that all systems were functioning correctly. If the engines shutdown suddenly, the vehicle could rebound with significant force. Boeing recommended further studies on the dynamics between the rocket and the pedestal to address these concerns.

In 1970, as NASA decided to complete the lunar landings before focusing on Skylab, the debate over using LC-39 for Saturn IB launches resurfaced. Grady Williams, chief of design engineering, generally supported Boeing's report. His office undertook a geometric evaluation and preliminary layout of the pedestal, including stress and weight analyses. Although Huntsville had some concerns regarding vehicle-pedestal dynamics and wind loads at liftoff, Saturn officials appeared willing to proceed with the proposed modifications.

As the Skylab Space Station project advanced, a crucial decision loomed over the launch preparations, with significant implications for budget and schedule. Williams, a key figure in the decision-making process, concluded that the proposed modifications would not delay Skylab's mission. However, Walter Kapryan, the director of launch operations, raised several concerns regarding the changes. The principal issue was the reliance on a single pedestal for the Interstage Booster (IB) launches. This arrangement would constrain the Kennedy Space Center's (KSC) schedule, necessitating weekend work and limited operational flexibility. Should the pedestal suffer severe damage during a launch mishap, repairs could delay the final crew's departure beyond the planned eight-month life of the Skylab workshop.

Despite these concerns, Kapryan acknowledged that transitioning operations to Launch Complex 39 (LC-39) would save costs. This was particularly relevant if NASA deactivated Launch Complexes 34 (LC-34) and 37 (LC-37). Ray Clark, the director of technical support, believed that the estimated savings of $10 million might be understated, suggesting the actual figure could be as much as 50% higher. Nevertheless, Clark cautioned that dual operations on LC-39 could present challenges, particularly during hurricane season. The center had only two crawlers to move three substantial structures: the two launchers and the mobile service structure. Since each transfer required seven hours, managing these operations during a hurricane could strain the launch team significantly.

From Huntsville, Roy Godfrey, Saturn manager, echoed the call for an early decision. He highlighted the urgent need to address modifications at LC-34, where ground support equipment adjustments were incurring costs of nearly $4,000 per day. Godfrey also stressed the importance of allowing adequate time for the necessary changes at LC-39. To accommodate a six-month study of the pedestal design followed by a year of wind-tunnel tests and data analysis, Huntsville needed to commence design work by mid-July. While Godfrey did not insist on an uncrewed test launch of the pedestal, he emphasized the necessity of close coordination between centers to make a well-informed decision. He argued that the advantages of the proposed change should encompass not only the identified cost impacts and program risks but also the potential for additional costs and risks revealed through detailed analysis and testing.

In Houston, the prevailing sentiment mirrored that of Huntsville. The anticipated savings from operating LC-39 outweighed concerns about conducting a crewed launch from an untried pedestal. Transitioning to LC-39 would benefit the principal contractor, North American Rockwell, by eliminating the need to transfer Apollo equipment from Merritt Island to the Cape and reducing the manpower required for launch operations. However, if a decision were delayed beyond May 15, much of the anticipated savings might be eroded. Houston was already advancing with the design work for LC-34 equipment and planned to finalize material contracts by June.

On April 23, the Skylab Office presented its case to Debus, emphasizing that operations from LC-39 would yield substantial cost savings and demonstrate the versatility of the Merritt Island complex. The presentation sparked a range of questions, including whether the cost estimates for LC-34 encompassed rehabilitation costs. The response was negative. Debus also inquired about the purpose of the wind tunnel tests and the potential for disputes arising from the involvement of non-union workers from Chrysler alongside union personnel at LC-39. Following the presentation, Debus consulted with his staff and found a consensus supporting the proposal.

On the same day, a meeting in Huntsville revealed lingering concerns about the proposed move to Launch Complex 39 (LC-39). The Marshall Space Flight Center staff expressed apprehension about the technical risks of launching a vehicle from the new pedestal. They noted that simulations and dynamic analyses could not fully address these risks, leaving uncertainties to persist until the first launch. Huntsville's support for the transition was contingent upon several key requirements: a pedestal load test to confirm its rigidity, a pull test to assess the vehicle's stiffness, and an additional three months of checkout time to address any unforeseen issues.

The urgency of reaching a decision was underscored by the need to finalize plans before May 15 to maximize savings. A delay beyond June 1 was deemed unacceptable due to the risks it would pose to both cost and schedule. On April 27, during a meeting of officials from the four program offices, Program Director Bill Schneider questioned the feasibility of meeting the May 15 deadline, given the necessity of obtaining administrator approval. Schneider's primary concern was ensuring the safety of the pedestal. He questioned, "How do we prove we can safely launch from LC-39?" Despite this, the prevailing opinion at Kennedy Space Center (KSC) was that testing and data analysis would ensure the pedestal's reliability. The deputy Saturn manager at Huntsville considered the cost savings a compelling argument for using LC-39, particularly given NASA's pressures to limit operating costs. After thoroughly debating the need for a trial launch, Schneider concluded that operational advantages should take precedence over cost considerations.

Decisions from Washington came sooner than expected. On April 29, 1970, Myers tentatively authorized the transition to LC-39, although he prohibited any irreversible actions until further confirmation. Administrator Paine provided verbal approval on May 11, and four days later, NASA informed the Congressional Space Committees of its intention to use LC-39. In June, Schneider requested KSC to provide substantiating data to ensure that flight-crew safety standards would remain uncompromised. Morgan subsequently submitted a plan that included design reviews, dynamic and stress analyses, a wind tunnel program, and several pull tests to measure the deflection of the vehicle and the pedestal.

Despite these preparations, doubts about the pedestal persisted. In November 1970, program offices revisited the idea of a trial launch to train the crew and validate the system, with Chrysler officials suggesting a static firing as a potential training exercise. After reviewing the proposal, Schneider concluded that KSC's existing plan was robust. The following month, the Management Council endorsed his recommendation against a trial launch.

The pedestal, colloquially known as the "milkstool" among locals, was a standout feature of Skylab's setup at LC-39. This massive structure weighed 250 tons and was designed to support the launch table with four steel pipe legs, each more than half a meter in diameter. Its sheer size and robust construction were emblematic of the scale and complexity of the Skylab mission.

The engineering and construction challenges involved in developing the Skylab Space Station's launch pedestal were intricate and critical to the mission's success. The pedestal's design began in July 1970, and several key considerations and disputes marked its construction.

The pedestal was a monumental structure with columns that stood 15 meters apart at their base but leaned inward, narrowing to less than half that width at the top. This design was supported by a network of horizontal and diagonal pipes that braced the structure. From an aerial perspective, the launcher table appeared as a massive doughnut featuring an 8.5-meter exhaust hole at its center. On this deck, essential components such as hold and support arms, fuel pipes, and electrical lines were meticulously arranged. A removable platform was positioned above the exhaust hole, providing technicians access to service the eight engines of the Saturn IB's first stage.

The decision-making process surrounding the pedestal's construction was marked by scrutiny. Initially overseeing the project, Buchanan rejected Chrysler's bid to build the pedestal under a sole-source contract. Chrysler's proposal was deemed problematic due to concerns about its complexity and the potential for distortion from the intense exhaust of the Saturn rockets. Despite Chrysler's assertion that their proposal would expedite construction, the Kennedy Space Center (KSC) preferred to adhere to a competitive bidding process. Consequently, Reynolds, Smith, and Hills, the architects for the mobile launcher, were awarded the pedestal contract, while KSC chose to design the pedestal's support systems in-house.

Designing the pedestal posed significant challenges, primarily in minimizing vertical and horizontal vibrations. The specifications set forth by Huntsville were stringent, allowing only minimal sag under substantial loads. However, designers were constrained by the need to balance weight and stiffness, as the Saturn V rocket, coupled with the crawler, was near the maximum capacity for the launch system. The weight of the pedestal was capped at 225 metric tons, a figure carefully calculated to ensure that it could support the Saturn V without surpassing the crawler's weight limits.

Additional factors included the intense heat from the rocket's exhaust, with flame temperatures approaching 2700 K, and the impact of wind loads. During operations at the pad, the service structure would absorb much of the wind's force, and an arm connected to the top of the rocket would help dampen vibrations. However, this protection would not be available in the final hours of the countdown. Wind tunnel tests determined a maximum permissible wind speed of 32 knots for a safe launch. While designers considered strengthening the pedestal by connecting it to the launcher tower, studies showed that it alone would be sufficiently stiff.

The pedestal's construction also led to the only major contractual dispute concerning Skylab's launch facilities. In the fall of 1970, the Small Business Administration requested that the contract be reassigned to one of its affiliated firms. KSC rejected this request, emphasizing the need for an experienced team to adhere to the tight six-month schedule. The pedestal was a critical component of Skylab's launch system, and any delay could significantly impact the entire program. KSC insisted on open bidding due to the high precision required in alignment and elevation, which exceeded typical industry standards. Despite attempts by the Small Business Administration to influence the decision through congressional and NASA Headquarters channels, the matter remained unresolved for over a month, causing delays in the overall schedule.

By late December, a decisive ruling from NASA Headquarters favored the Kennedy Space Center (KSC) in the ongoing contractual dispute. However, when bids were finally opened the following month, Holloway Corporation, a modest electrical firm based in nearby Titusville, submitted the lowest bid of $917,000. This bid presented another challenge for KSC, including a proposal to complete fabrication work by a smaller Jacksonville firm. Despite these initial concerns, Holloway Corporation and its partners secured the necessary steel pipe and completed the pedestal construction on time and according to specifications. The successful outcome led the Small Business Administration to send a letter to Congress reprimanding NASA for its reluctance to engage small firms.

Throughout the development phase, KSC experienced the same frustrations as other Apollo Applications offices, primarily due to frequent revisions of schedules. The decision to proceed with the dry workshop configuration provided a more solid foundation for planning, and by December 1969, KSC had outlined a preliminary launch plan. A key objective was to minimize the time the spacecraft spent on the launch pad. While Apollo missions typically required eight weeks on the pad, Skylab operations aimed to reduce this to 24 workdays. This reduction was intended to limit exposure to adverse weather conditions and cut down the high costs associated with launch operations, which averaged around $100,000 per day in the final month.

KSC planned to perform as much work as possible inside the assembly building to achieve this tighter schedule, including removing work platforms from the workshop's interior. Pad access would only be restricted to critical tasks such as testing the water supply, checking instrument readings, or installing late experiments. Despite these efforts, veterans in the Launch Operations Office were skeptical about the feasibility of maintaining such a strict timeline. Over the next year, debates persisted regarding pad time and access.

Charles Mars, the Skylab project leader for the operations group, anticipated that principal investigators would insist on late access to their experiments, ultimately forcing adjustments to the schedule. Mars aimed to accommodate these needs by planning to leave access platforms in the workshop during rollout and allocating pad time for scientists. However, at a review in September 1970, Debus, the program manager, supported the program office's stance that pad access would be granted only by exception. To Mars's surprise, KSC adhered to this policy for the next 30 months, maintaining a rigid schedule despite the limitations it imposed on the project. While the workshop remained off-limits, additional pad requirements extended the timeline beyond the initial projections.

By June 1970, Skylab's pad time planning had been extended to six weeks, allowing two weeks for contingencies. This decision came in response to concerns from Huntsville, which led the Kennedy Space Center (KSC) to eliminate the contingency cushion, though estimates for the schedule continued to rise. During the December program review, Paul Donnelly, the associate director for operations, presented a revised schedule of 44 days, which included 30 workdays. The primary increase in time—nine days—was due to the complex task of filling and testing the oxygen and nitrogen tanks necessary for maintaining the workshop's atmosphere. Donnelly committed to reviewing the schedule further to identify which requirements could be compressed. By early 1971, the operations office had reduced the time allowed and scheduled several tests to be conducted in parallel. Consequently, the planned pad time was adjusted to a more manageable 30 days.

The operations plan devised for Skylab in 1971 followed a methodical, building-block approach. Each major module's components and systems were individually checked out before technicians assembled the space vehicle and conducted integrated systems tests midway through the eight-month schedule. These integrated tests were crucial, as the major modules had not been previously assembled, either mechanically or electrically. Before rollout, the launch team would store food, film, and other consumables. Given the experience from previous crewed programs, which revealed numerous unforeseen issues during the first launch, Skylab's schedule was extended several months beyond that of a typical Apollo mission. This extension also accommodated increased launch activity: starting August 1972, three vehicles would be concurrently prepared at Launch Complex 39. Although the launch of Apollo 77 in December would alleviate some of the workload, KSC was soon confronted with its first dual countdown, resulting in Skylab launches being spaced just 24 hours apart. The scale of these operations necessitated an early start.

Launching a Skylab crew required less planning than earlier Apollo missions, as it essentially followed the same procedures. However, extensive operations in Earth orbit introduced new stowage plans and test procedures. Moreover, the change of launch sites necessitated a trial run of the LC-39 facilities. Key milestones included the unique mating of the Apollo spacecraft with the docking adapter before liftoff and testing the pedestal in January 1973.

Facility modifications were integral to the operations discussions, with a significant focus on a new "contingency" arm for accessing the workshop. The initial December 1969 plan had proposed entry through a side door, a new feature advocated by KSC. Technicians would access this door via service platforms in the assembly building, while at the pad, a new swing arm would provide contingency access. By 1970, this arm had become the principal means of access to the workshop. The launcher's uppermost service arm, previously used by Apollo astronauts to board the command module, was relocated adjacent to the workshop's side hatch. An airlock, designed to safeguard the workshop's interior from contamination, replaced the Apollo white room at the end of the arm. To avoid constructing a second airlock for the assembly building, the engineering office also recommended using the new arm there.

By the end of the year, access plans for the remainder of the space vehicle were finalized. There was a lot of traffic to the airlock, and the multiple docking adapter was routed through the new swing arm. Inside, technicians accessed the workshop's forward dome hatch and could reach the telescope mount from access ramps on the top work platform, which was equipped with an additional clean room. Although KSC had initially not planned to service the telescope mount at the pad, mid-1970 saw Huntsville identifying several service requirements, leading to the selection of arm 8.

A significant focus of the Skylab operations debate was the mobile service structure, establishing the only major item at LC-39 without a backup. The structure could be moved, but the five-kilometer trip between pads took approximately six hours. Consequently, if operations required the service structure at pad A, pad B would remain unsupported for at least a day. This constraint led Kennedy Space Center planners to consider the implications of relying on a single mobile service structure.

Hans Gruene, director of launch vehicle operations, challenged the decision to proceed with the Skylab mission as planned. He had noted that loading cryogenic propellants into the S-II stages of Saturn V rockets had previously caused insulation cracks, necessitating inspections and repairs on the pad. Given this history, Gruene anticipated similar issues with Skylab and questioned whether the service structure would be available when needed. If the service structure were unavailable, it would be necessary to devise an alternative method for accessing the S-II stage or return the rocket to the assembly building, which would be impractical.

Events that summer validated Gruene's concerns. In July, Huntsville mandated that the S-II insulation be inspected on the pad, further underscoring the need for the service structure. While workers could access the Saturn V from a bosun's rig, their ability to perform necessary tasks was severely constrained. Using the service structure for both Skylab vehicles presented clear scheduling and design challenges. Notably, the payload shroud on the Skylab workshop was nearly three meters larger in diameter than that of the Apollo spacecraft. The service structure's bottom platform would need to be extended for effective servicing to accommodate this difference.

The debate over this issue continued for several months before resolving. By October, Kapryan agreed to modify the lowest platform to facilitate servicing of the Skylab workshop. However, this modification would leave only one platform available to service the lower half of the IB rocket. Kapryan recommended restoring the platform to its original configuration after the launch of the first crew to ensure all workstations were available for the subsequent missions. He emphasized that the cost of the modification, $85,000, would be outweighed by the operational benefits. His proposal was approved, allowing the modifications to proceed.

Additional adjustments were required to adapt the Saturn V facilities for the smaller IB booster. The five swing arms used to service the lower stages of the Saturn V were replaced with a single arm, being extended by three meters to reach the IB booster. Umbilical lines and a withdrawal mechanism were relocated from Launch Complex 34. A new work stand was constructed in the assembly building to access the structural section between the two stages. 19 firing panels were installed at the launch control center to support IB operations.

The propellants team at KSC faced a significant challenge on the pad. The liquid oxygen system, which pumped 37,850 liters per minute into the Saturn V, operated at four times the rate the IB could accept. Instead of altering the system, they adapted the Saturn V's replenishment system, which pumped 4,540 liters per minute, about half the desired rate for the IB.

Payload testing, excluding the workshop, occurred at the Operations and Checkout Building eight kilometers south of Complex 39. One notable modification was adding a clean room for the telescope mount. Situated in the building's high bay, this room was designed to support the calibration of the experiment optics, with specifications requiring the floor to move less than five seconds of arc in 24 hours. Adjacent rooms housed the air conditioning unit and ground support equipment for telescope testing. Another modification involved the integrated test stand used for systems testing on the Apollo spacecraft, which was established and reconfigured to place the command module at the bottom level, facilitating a crucial mating test between the spacecraft and the docking adapter. The Apollo laboratories were also adapted to accommodate Skylab experiments, reflecting the program's evolving requirements.

In April 1971, the initial components of the pedestal for the Skylab Space Station began arriving at the construction site adjacent to the assembly building. The assembly process commenced with the sandblasting and painting of pipes, which were then meticulously welded into six-meter sections. By early May, the pedestal's framework was visibly taking shape. The launch table, comprised of eight distinct segments, was delivered in mid-June. By early July, this launch table was securely mounted atop the pedestal, and an access bridge was subsequently added to facilitate entry.

During the fall of 1971, contractors undertook the task of outfitting the pedestal with essential infrastructure. This included installing engine service platforms, routing new fuel and power lines, and integrating a quench system designed to cool the rocket exhaust. Concurrently, construction on the clean rooms within the checkout building began, although this was delayed due to issues concerning a partition that separated Apollo 76 operations from Skylab activities. Despite these setbacks, progress was on track by Christmas, and modifications to the checkout building proceeded smoothly.

Testing and preparing experiment hardware was complex and demanding due to numerous interfaces and interactions among components. Skylab was equipped with over 70 experiments, many interconnected with other experiments and flight hardware. For instance, the ultraviolet panorama telescope, developed in France, was a sophisticated instrument with eight separate parts that interacted with each other and seven other flight hardware items. The telescope relied on 41 interfaces for proper functionality, more than half of which needed to be tested at the launch site. Despite being categorized as a less complex experiment, the telescope's numerous interfaces were meticulously tracked using a fit-check matrix—a detailed chart listing all hardware connections and their verification status.

Experiment hardware was divided into three distinct groups based on complexity. The first group comprised the simplest experiments, which did not require continuous support from the development center or contractors. These experiments were generally installed before arriving at Kennedy Space Center and remained untouched during testing. The second group included experiments necessitating ongoing support from their developers, such as earth-resource instruments and various corollaries, representing approximately 40% of the experiments. These required off-module testing and were subject to rigorous oversight. The third group, involving preflight and postflight medical experiments, consisted solely of non-functional hardware, with the development centers retaining full responsibility for their preparation.

The hardware testing was largely managed by module contractors, with McDonnell Douglas handling experiments mounted in the workshop. Responsibilities included inspection, bonded storage, handling, installation and removal of hardware, documentation, planning, and coordination of the checkout process. The contractors addressed any anomalies encountered during testing, with oversight from the Kennedy Space Center.

Although not directly involved in prelaunch checkout, principal investigators played a crucial role. Each experiment had a dedicated Kennedy engineer assigned as a point of contact, and investigators were encouraged to review test procedures and data. However, to avoid disrupting the test schedule, the responsible centers carefully coordinated investigators' activities in advance. Investigators with political connections were handled with particular care to mitigate potential complaints. Generally, those who visited the Cape and witnessed the intricate nature of the operations were more understanding of the constraints. At the same time, those who did not were often less tolerant of the restrictions imposed by the testing process.

The coordination between NASA's various centers during the Skylab program presented its share of challenges, particularly when it came to defining test procedures for the spacecraft. While the launch team at Kennedy Space Center found it relatively straightforward to align test procedures with Houston—given the similarities between Skylab's command and service modules and their Apollo predecessors—relations with the Marshall Space Flight Center in Huntsville proved more contentious.

A significant point of friction emerged over the workshop test procedures. The conflict stemmed from a longstanding rapport between Huntsville's Saturn engineers and Hans Gruene's launch vehicle operations team, a connection forged over years of collaboration. However, the checkout of the Skylab workshop fell under Ted Sasseen's spacecraft operations office, which had a limited history of interaction with Huntsville. This lack of established rapport meant that creating new working relationships was a slow and arduous process.

Compounding the issue was Marshall's perceived overemphasis on safeguarding its Skylab hardware, a sentiment that did not sit well with those outside of Huntsville. NASA's typical procedure involved design centers defining test requirements, which Kennedy then used to craft detailed test procedures. The design centers reviewed these procedures, resolved disagreements, and the launch team conducted the tests. However, Huntsville's insistence on controlling the operation—especially the critical first integrated-systems test—led to a protracted negotiation process over a year.

Another point of contention arose regarding the preflight tests of the telescope mount. This situation marked the first occasion that a crewed spaceflight center was tasked with performing tests directly at the launch site, previously handled by contractors. This disagreement became evident in December 1970 during a review of the telescope mount's flight procedures. Gene Cagle, the engineering manager responsible for the telescope mount, strongly objected to Kennedy's position that his team would operate as a contractor. Cagle argued that his team, already stretched thin, lacked the resources to meet Kennedy's stringent requirements, which included maintaining 73 forms with multiple signatures and adhering to a rigorous quality assurance process. He estimated that fulfilling these requirements would necessitate 700 personnel, three times the number available to him. Cagle also contested the application of Kennedy's quality control standards, arguing that the testing organization ensured the quality of its work at Huntsville.

Kennedy officials, however, remained unmoved by Cagle's objections. Their procedures were grounded in years of experience. They were designed to manage the intense pressures of launch operations, which differed markedly from the more relaxed environment of Huntsville's checkout processes. The Cape's launch atmosphere, characterized by the urgency of a looming deadline and the need for thorough documentation, left little room for shortcuts. Contractors were accustomed to working within this framework, and Kennedy's refusal to grant additional manpower to Cagle's team aligned with its checks-and-balances philosophy.

The impasse persisted for nearly a year before a resolution was achieved. Spacecraft operations at Kennedy extended assistance by providing Cagle with systems engineers from its liaison team. These engineers accompanied the telescope mount through its journey from Huntsville to Houston and finally to the Cape, integrating with Huntsville's test team. In turn, Kennedy agreed to perform quality checks similar to those Houston conducted for thermal vacuum tests, and Marshall worked to accommodate Kennedy's other requirements. This compromise ultimately allowed the testing and preparation for Skylab's launch to proceed more smoothly.

In the summer of 1972, preparations for Skylab's launch entered a critical phase. The checkout of the telescope mount proceeded without major issues, culminating in a commendation letter from Dr. Kurt Debus acknowledging the exceptional efforts of the test team. The arrival of flight hardware marked the beginning of a pivotal period for the Skylab project, with the launch team assuming a central role that persisted for the next nine months.

On July 19, 1972, the first spacecraft, designated CSM-116, arrived at the Operations and Checkout Building aboard NASA's Super Guppy, a specially designed transport aircraft known for its ability to carry oversized cargo. The spacecraft was promptly inspected in an altitude chamber during the following week. The subsequent two months of checkout procedures were carefully coordinated around the Apollo 17 mission requirements, highlighting the intricate scheduling needed to balance overlapping space programs.

The arrival of the S-IC booster, the 13th flight article in the Saturn IC stage production, on July 26, further accelerated the pace of operations. Transported from New Orleans via the barge Orion, the booster was an essential component of the Saturn V rocket. By August 22, all four propulsion stages for the initial Skylab missions were in place, setting the stage for the final assembly.

Skylab's activity intensified after the Apollo 17 rollout and the Labor Day break. Within two weeks, the stages were assembled atop their respective launch vehicles. On September 22, the workshop and payload shroud embarked on a two-week journey from Huntington Beach, California, aboard the Point Barrow, a vessel specially equipped by the Navy's Military Sealift Command for such delicate operations. The telescope mount arrived shortly after that and was transported by the Super Guppy.

Once in high bay 2, Skylab's components were integrated with the Saturn V rocket. Early operations went smoothly, benefiting from using proven equipment and established procedures. However, the process encountered significant challenges with deploying the meteoroid shield. Scheduled for October 3-7, this task was crucial as technicians needed to verify the shield's deployment before accessing the workshop for further work.

The meteoroid shield, a critical component designed to protect the workshop from space debris, required precise installation. McDonnell Douglas, responsible for rigging the shield in its launch configuration, faced considerable difficulties. The shield, weighing 545 kilograms, had to be positioned around the workshop like a corset on a sleeping giant. Despite the efforts of 32 technicians, the initial fit was problematic. Several bulges and gaps were evident along the shield's edges, complicating the deployment process.

The discrepancy arose because the flight shield differed from the static-test article in earlier deployment tests. Faced with persistent issues, the launch team attempted various adjustments, including loosening and re-tightening the shield's trunnion bolts. Despite these efforts, the gaps remained. McDonnell Douglas eventually halted the procedure and employed ultrasonic scanning to assess the shield's contact with the workshop surface. The results revealed that only 62% of the shield's surface was aligned correctly with the workshop, underscoring the challenges of this complex installation task.

As preparations for Skylab's launch progressed, the workshop underwent various crucial tests and adjustments. It was pressurized, and the contact areas between the meteoroid shield and the workshop were meticulously re-mapped. This time, the results showed 95% contact between the two surfaces. Given the substantial pressure differential expected during flight, the team in Huntsville deemed the configuration acceptable.

Many of the Saturn and Skylab large components were transported to the launch center aboard special barges or in special modified aircraft.

Once in orbit, the meteoroid shield was designed to deploy and stand 13 centimeters from the workshop. The verification of this deployment, however, proved to be another challenge. During initial tests, two of the latches intended to secure the shield in place during flight failed to engage. Additionally, three of the sixteen torsion rods responsible for rotating the shield outward were over-torqued, with one needing replacement. Although the upper latch failed again on the second attempt, the lower latch was sufficient to hold the shield in place, and Huntsville ultimately accepted the condition.

By late October, final preparations for the flight had begun, albeit several weeks behind schedule. Testing of the workshop launch vehicle started in early November, running concurrently with the workshop's checkout. In mid-November, the solar arrays, with their wings folded, were mounted onto the workshop. Thanksgiving completed testing of the refrigeration system, and the waste-management system was ready by Christmas.

The Saturn IB rocket was rolled out on January 9. The airlock and docking adapter, arriving on October 6, were the last major components to reach the launch center. Over the next four months, all modules underwent exhaustive examinations in the Operations and Checkout Building. Testing of the telescope mount revealed a few significant issues, and by mid-January, the Huntsville team had successfully attached the thermal shield and solar arrays.

Other hardware, however, presented more difficulties. The earth-resource experiments, among the latest additions to the Skylab program, were particularly troublesome. As late as January, Martin Marietta, responsible for these instruments, reported ongoing issues with signal conditioners, videotape recorders, and heat control for the multispectral camera's window. End-to-end tests, where technicians simulated subject matter for the cameras, repeatedly failed to record significant portions of data.

To resolve these issues, the Martin team, advised by a Kennedy Space Center (KSC) employee, employed two rudimentary procedures—cable wiggling and pin probing—previously banned at the Denver plant. During testing, a technician wiggled each cable at specific intervals, and the data output was compared with the movement. This revealed several erratic channels. Further probing of cable connector pins identified defective joints. With the installation of new connectors, the instruments finally performed satisfactorily, resolving the persistent issues.

The challenges faced with the earth-resource experiments aboard Skylab were emblematic of the broader issues encountered throughout the space station's development and deployment. Skylab's preparation period was marked by extensive testing and modifications, highlighting the inherent difficulties of pioneering space technology.

In the eight months preceding Skylab's launch, most of the hardware encountered problems. Approximately one-third of the equipment required repairs while still on the ground, and an additional one-fifth suffered mechanical issues during installation. These issues were not merely inconveniences but critical setbacks significantly impacting the mission's timeline. Most notably, 61% of the experiments had to be removed from Skylab due to either test failures or late design changes, necessitating extensive additional work to correct and retest the systems. This added workload extended the checkout period considerably and illustrated the challenges of integrating and validating cutting-edge technology.

The Kennedy Space Center (KSC) project officer reflected on these difficulties, noting that the experiences with Skylab did not align with the optimistic belief that increased industry experience would naturally lead to fewer failures. Instead, the high incidence of test failures and design changes underscored that much of Skylab's hardware was at the forefront of technological advancement, making it particularly vulnerable to problems during testing. The officer estimated that about one-third of Skylab's experiments might have failed once in space without the rigorous prelaunch checkout.

On January 19, 1973, program officials convened at Merritt Island for a critical design certification review of the launch complex. This meeting marked the culmination of various reviews that had been underway since June 1972. These reviews, conducted by the Manned Spaceflight Management Council, scrutinized Skylab's hardware, experiments, and mission operations. During the KSC review, the focus was on identifying single-point failures, such as issues with the mobile service structure, and addressing elements of the launch complex that had undergone significant modifications from previous Apollo missions.

The review did not reveal any major shortcomings. However, it concluded with requests for KSC and the Marshall Space Flight Center to address several specific areas, including dynamic analysis of the launch pedestal and a re-evaluation of previous Saturn IB launch issues.

As the review took place, additional delays were becoming apparent. Schneider, a key figure in the program, observed that testing on the airlock-docking adapter was lagging behind schedule. By early January, testing had fallen four days behind, casting doubt on whether the launch team could complete the module stacking by January 19. This delay became inevitable when the team had to remove the control and display panel from the earth-resource experiments. The test office, confronted with the prospect of another week's delay, rescheduled the mating of the modules to January 29.

These events underscored the complexities and challenges of preparing Skylab for its mission, reflecting both the ambitious nature of the project and the persistent difficulties encountered in advancing space exploration technology.

Upon evaluating the persistent test problems, Schneider determined that the entire Skylab schedule would need to be adjusted, projecting a delay of at least two weeks. Although it was hoped that this lost time might be recovered, there was a substantial risk of further delays. A NASA spokesperson addressed the situation, emphasizing that the delays were not attributable to any single factor but were the result of the complexities involved in the first-time testing of the Skylab modules and their associated experiments. As a result, tentative launch dates were set for May 14 and 15, with firm dates expected to be confirmed by late March.

Over the following two months, progress improved as fewer issues arose. On February 9, 1973, an integrated systems test was conducted, marking the first comprehensive evaluation of the Skylab workshop and its launch vehicle as a unified system. This ten-day test was largely successful, though minor problems with the refrigeration system, mostly related to ground support equipment, were noted.

By February 20, Rockwell had transported the Apollo command and service modules to the assembly building for integration with the Saturn IB rocket. The process was completed quickly, and the Saturn IB was positioned on the launch pad within a week. March was devoted to extensive testing and preparation. On March 7, Martin Marietta completed the final simulated passes with the earth-resource cameras, an essential step in verifying their functionality. Two weeks later, the entire launch team conducted a simulated countdown and liftoff of the workshop during the flight readiness test, establishing the final major milestone before the vehicle departed the assembly building. This test spanned four additional days, focusing on initial workshop operations.

Concurrently, technicians worked on loading provisions, with the task reaching 70% completion by the end of March. During the final two weeks in the assembly building, various crew compartment fit and function tests were performed

to ensure all components were properly installed and functioning. A report on April 12 confirmed that the internal Orbiting Workshop (OWS) was prepared for flight. Final preparations included the installation of the payload shroud, which protected the telescope mount during launch. On April 14, the ordnance necessary for stage separation and emergency destruction of the vehicle was added, and the workshop rolled out two days later.

On the launch pad, the initial tasks involved connecting and testing support systems, including fuel, water, electricity, environmental control, and high-pressure gas lines. On April 25, the countdown demonstration test began—a crucial rehearsal that simulated the final week before launch. This exercise, a longstanding component of Saturn prelaunch operations, was particularly significant for Skylab as it tested integrated operations for two space vehicles. Months of meticulous planning were validated when the dual countdown proceeded smoothly. Following simulated liftoffs on May 2 and 3, fuel tanks were drained, and insulation was inspected. The terminal count for the Saturn IB-Apollo, featuring a dry run with the crew aboard, commenced shortly after that.

In a deliberate double exposure photograph, the first two space vehicles of the Skylab program were depicted side by side, illustrating their significant but often overlooked spatial separation of 3 kilometers. On the left, the Saturn IB rocket, poised on its launch platform, was designated to carry the first Skylab crew. On the right, the Saturn V rocket's third stage has been replaced by the Skylab space station, marking a distinct difference in their configurations.

As the time approached for Skylab's maiden launch, anticipation and apprehension were palpable. While the laboratory, its launch vehicle, and Launch Complex 39 were prepared, the stakes were particularly high. Unlike many programs where the maiden flight was just one of several opportunities to perfect the mission, Skylab's success hinged significantly on this initial launch. Any failure could jeopardize future funding, with the $250 million required for a second attempt being a considerable barrier.

The final days before the launch were suspenseful, primarily due to weather-related issues. Following a heavy rain on May 4, it was discovered that the payload shroud was leaking. Subsequent high winds and more rainfall hampered efforts to seal the shroud. On May 9, the first day of the final countdown, a lightning strike to the mobile launcher necessitated a swift retest of the vehicle's systems. Fortunately, the thunderstorms subsided, allowing the final countdown to proceed without major issues.

Just before liftoff, an oversight was corrected by Martin Marietta technicians, who affixed a metal United States flag to the docking adapter, symbolizing the national pride and significance of the mission. At 1:30 p.m. on May 14, the Skylab workshop cleared the launch tower, and control of the mission was transferred from the Kennedy Space Center (KSC) to the Johnson Space Center (JSC) in Houston, reflecting the transition from launch to mission operations

Chapter 3 - Skylab 1

Skylab 1 (SL-1) marked a historic milestone as it embarked on its uncrewed mission, initiating Skylab's journey into the cosmos. Launched on May 14, 1973, aboard the final Apollo-era Saturn V rocket, Skylab emerged as America's first space station. However, the mission encountered significant challenges almost immediately after liftoff.

Launch of the Skylab 1 space station on May 14, 1973, the final flight of a Saturn V rocket.

The launch vehicle roared off from Kennedy Space Center's Launch Complex 39A, a site previously the departure point for many Apollo missions to the Moon. On this particular day, the Saturn V carried a payload that was more than just a scientific curiosity—it was an orbiting laboratory and future home for astronauts. Skylab's design was an evolution from the Apollo command and service module. While the module was adept at carrying scientific instruments, it lacked the capacity for prolonged manned operations. Skylab aimed to provide a spacious, self-sufficient environment where astronauts could live and work for extended periods, independent of frequent resupply missions.

The Saturn V rocket stands poised on Pad A of Launch Complex 39 at the Kennedy Space Center. Its passenger was Skylab, our first space station.

Despite its advanced design, Skylab faced immediate difficulties. The launch and deployment phases were fraught with problems. During liftoff, vibrations caused severe damage, including the detachment of a critical meteoroid shield and one of the station's two primary solar panels. The loss of the meteoroid shield, which also functioned as a sun shade, resulted in overheating within the station's interior, where temperatures soared to a sweltering 126°F. This issue was compounded by debris from the detached shield becoming entangled with the remaining solar panel, preventing its full deployment and leading to a significant power deficit.

In response to these unforeseen challenges, NASA engineers undertook an intense ten-day effort to develop procedures to make Skylab habitable. They meticulously trained future crew members and engineered a solution to mitigate the high temperatures. They stabilized the situation by rotating Skylab to reduce internal temperatures and deploying its secondary solar panels to maximize power generation. Despite these heroic efforts, the station's operational conditions were far from ideal.

Following Skylab's launch, Launch Complex 39A was deactivated as it underwent modifications to support the Space Shuttle program, which was initially set to begin in March 1979. The complex would remain dormant for decades until February 19, 2017, when SpaceX's CRS-10 mission marked its return to active use, highlighting the pad's enduring legacy in space exploration.

Skylab's inception can be traced back over a decade before its launch. The concept evolved from an initial idea of converting an empty rocket stage into a habitable volume in orbit. Skylab was transformed into a sophisticated space laboratory equipped with advanced power, communication, instrumentation, and control systems by its launch. Designed for crew health and comfort, it featured state-of-the-art equipment for scientific measurements and experiments. This meticulous development process culminated in a space station of unparalleled sophistication, poised for its historic journey into orbit aboard the Saturn V rocket.

The Skylab program yielded a wealth of scientific data, providing invaluable insights for researchers. Equally significant were the lessons the crews and flight planners learned about the complexities and unpredictability of human adaptability in space. Each crew brought a unique approach to their mission. The first crew, known for their resourceful "can-do" attitude, was succeeded by a driven and overachieving second

crew. The third crew, characterized by their methodical and sometimes stubborn approach, set new standards in space endurance. These experiences underscored an essential truth: astronauts were not interchangeable; their traits and abilities significantly influence mission outcomes.

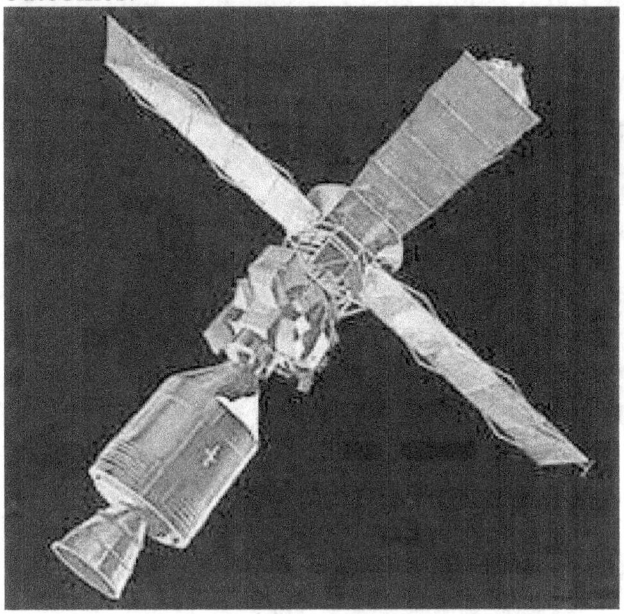

An intermediate concept envisioned a cluster which included a lunar module, the solar observatory with its windmill-like solar arrays, and a command and service module.

Skylab's scientific achievements were remarkable. The program generated overwhelming data, sometimes nearly overloading the scientists responsible for analyzing it. The missions also provided crucial insights into astronauts' physical adaptation to prolonged weightlessness. Before Skylab, extended exposure to microgravity was a looming concern for future long-duration missions. The third crew's record-breaking 84-day mission, which remained unbeaten for four years, proved that humans could endure extended periods in space without significant physical deterioration.

Despite the success of Skylab, its story ended in an uncertain and, at times, dramatic manner. After the final crewed mission, Skylab remained in orbit, a silent sentinel in space, for five more years. Meanwhile, the next phase of American crewed spaceflight, the Space Shuttle program, faced numerous technical and financial setbacks, delaying its maiden flight. NASA had originally planned to use the Shuttle to boost Skylab into a higher, more stable orbit, extending its operational life. However, as delays mounted, this plan became increasingly unlikely. By 1979, Skylab's orbit was decaying, and its uncontrolled reentry into Earth's atmosphere became inevitable.

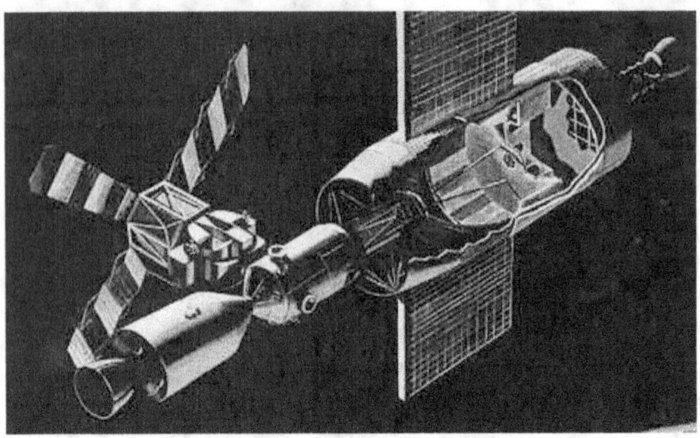

In a "wet workshop" concept, the Skylab cluster consisted of a lunar module with the solar observatory mounted above it, a command and service module, a docking adapter, and a Saturn upper stage, emptied of its propellants in achieving orbit. This spent stage would be outfitted in orbit for occupancy.

For three months in 1979, Skylab captured public attention as it had not since the success of its first crewed mission. Media coverage was intense, with widespread speculation and concern over where the spacecraft would reenter and what damage it might cause. In the end, Skylab's reentry was more spectacle than catastrophe. The workshop disintegrated in the atmosphere, with debris falling harmlessly over the Indian Ocean and sparsely populated Western Australia regions, leaving only memories of its pioneering achievements behind.

Both triumph and challenge marked Skylab's legacy. In its final mission, the Saturn V rocket had placed Skylab into a perfect orbit, 436 kilometers above Earth. The initial hours of the mission saw the successful deployment of key systems, with the laboratory beginning to come to life as planned. The telescope mount, a critical component of Skylab, deployed smoothly to its operational position, allowing the Apollo spacecraft to dock later. However, the meteoroid shield's premature

deployment and the solar arrays' subsequent failure quickly turned triumph into crisis.

As Skylab orbited out of range of the tracking stations, tension grew among NASA officials. Erratic telemetry signals indicated that something was wrong. The workshop's solar arrays, which were supposed to deploy smoothly, seemed malfunctioning. When Skylab passed over the tracking station in Carnarvon, Australia, the data was confusing and alarming. Telemetry suggested that one array had partially deployed but was not generating power, while temperature readings hinted at the possibility that both arrays were gone.

The next two weeks were a race against time. Engineers on the ground devised makeshift solutions to stabilize Skylab's temperature and energy systems. When they eventually arrived, the crew would have to undertake risky spacewalks to manually deploy a replacement solar shield and free the jammed solar panels. Through their combined efforts, the mission was saved, and Skylab's story transitioned from near-failure to remarkable recovery and success.

Skylab laid the groundwork for future space stations and long-duration spaceflight in its short operational life. The lessons from Skylab's trials and triumphs would resonate through the decades, influencing the design and operation of subsequent space habitats, including the International Space Station. Skylab's story was a testament to human ingenuity, resilience, and the relentless pursuit of knowledge beyond our home planet.

The final spacewalk

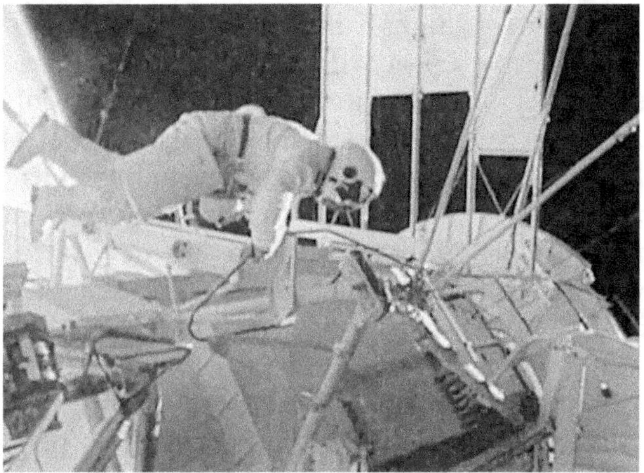

Saving Skylab

The failure of backup commands from the Goldstone Tracking Station in California and the Madrid Tracking Station in Spain intensified the mounting concerns about Skylab's condition. The post-launch briefing held at Kennedy Space Center soon focused on the crucial issue of the solar panels. By late afternoon, it became apparent that Skylab was facing at least two significant problems.

The most pressing issue was losing the workshop's solar panels, which severely reduced the station's electrical power. The solar panels on the workshop and the Apollo Telescope Mount (ATM) array each provided approximately 5 kilowatts of usable power. Nearly half of Skylab's power was lost with the workshop panels gone. While Apollo fuel cells could supply an additional 1.2 kilowatts of power for up to 20 days, this was only a temporary solution. Beyond that period, the command-service module would be forced to draw electricity directly from Skylab, further straining the station's power resources. Despite the system's initial design to accommodate some power loss, officials anticipated that the mission would remain viable until the Apollo fuel cells were exhausted. However, if the situation persisted, the crew would be required to curtail most of their experiments during the final week of the mission. The subsequent crews would face even more severe limitations due to the prolonged power shortage.

While the power shortage dominated discussions at an evening press conference, an even graver issue was quietly looming—the apparent loss of the micrometeoroid shield. Although the risk of damage from a meteoroid strike was relatively low, the shield's secondary thermal control function became a critical concern. The shield was designed to maintain the workshop within a comfortable temperature range, as cooling was more challenging than heating. Its exterior featured a black-and-white pattern for optimal heat absorption. At the same time, the interior of the shield and the workshop's outer surface were coated with gold foil to regulate heat transfer.

The failure of the shield to deploy as scheduled, combined with the ineffectiveness of subsequent ground commands, meant that the gold coating on the workshop would rapidly absorb excessive heat, potentially rendering the interior uninhabitable. Initial flight data revealed an anomalous lateral acceleration approximately one minute after liftoff, suggesting possible structural failure. This was shortly before the space vehicle reached its maximum dynamic pressure. As a result, workshop temperatures began to rise significantly, with many external sensors recording temperatures exceeding 82°C—above the maximum scale reading. Internal temperatures also climbed above 38°C. Based on thermal models, engineers in Huntsville predicted that internal temperatures could soar to 77°C and external temperatures could reach as high as 165°C, jeopardizing the food, film, and possibly the structural integrity of the workshop itself.

In response, Mission Control began to maneuver Skylab to reduce the exposure of the affected areas to direct sunlight, leading to some cooling of the station. The micrometeoroid shield, added to the wet workshop design in March 1967 amid uncertainties about meteoroid hazards, was expected to mitigate these risks. NASA had initially estimated the probability of a meteoroid strike at about 1 in 100. While a puncture from such a strike would not necessarily end the mission—since the crew could patch holes up to 5 millimeters in diameter and replenish the workshop's atmosphere—losing the shield's thermal control function severely challenged the mission's success.

The mission team faced various daunting challenges as night fell on the evening of Skylab's launch. The situation was bleak: the station faced severe overheating, a critical power shortage, and now, complications with its attitude-control system. The responses from the rate gyroscopes, essential for maintaining Skylab's orientation, were failing to average correctly. Additionally, the initial maneuvers had consumed an excessive amount of nitrogen gas, further compounding the problem. The notion of returning Skylab for repairs was, of course, out of the question. Repairing it in space seemed highly improbable, yet efforts to address the issues had to be made.

The immediate response was to postpone the crew's launch by five days. Engineers at the Marshall Space Flight Center in Huntsville began rigorous analytical studies to predict the likely temperatures within the workshop and evaluate their potential impact. Huntsville and Houston began investigating methods to deploy a thermal shield to mitigate the overheating. Simultaneously, NASA contractors and other centers were encouraged to conduct independent studies to find solutions.

Rocco Petrone, Director of the Marshall Space Flight Center, acted decisively. He authorized a special task force, led by the deputy directors of the Astronautics and Astrionics Laboratories, to tackle the crisis with all available resources. "Whatever you need at the center is yours," he instructed. Operating from the Huntsville Operations Support Center, the task force drew personnel from mission support groups and received additional aid from Marshall's laboratories and contractors' plants. Despite the urgent need, computer time had become a scarce resource. Much of the critical work was eventually carried out on Martin computers in Denver, and sometimes, resources had to be sourced from elsewhere.

The accident drastically disrupted Huntsville's usual operations. What was originally planned as a standard 40-hour workweek for operations personnel quickly transformed into a round-the-clock effort. The operations center was reinforced with Skylab design engineers, and the support teams for electrical power systems, attitude control, and environmental control were expanded significantly. The overall staff at the operations center grew from 400 to 600.

Initially, Eugene Kranz, Chief of the Johnson Space Center's Flight Control Division, attempted to manage the crisis with his four flight-control teams. Each team was assigned specific problems to address when not manning the consoles, allowing those who developed plans to implement them. However, by the 15th, this approach proved unfeasible. The complexity and urgency of the problems required continuous, dedicated attention. Consequently, two teams were assigned to man the consoles around the clock. In comparison, the remaining two teams focused on contingency planning, which included altering the flight plan, updating the activation checklist, supporting developing a sunshade, and reducing the workshop's power requirements.

During the intense period leading up to and following the launch of the Skylab Space Station, the entire Skylab team was fully engaged in addressing numerous challenges. From Huntsville to Houston and extending from Huntington Beach to Cape Canaveral, team members worked 16- to 18-hour days as the mission reached its critical stages. Such long hours became the norm, blurring the lines of day and night, yet tempers remained notably calm despite the exhaustion. The collaborative spirit between the Marshall Space Flight Center (MSFC) and the Johnson Space Center (JSC) was exemplary, a testament to the strong working relationships forged during Skylab's design and development.

Healthy competition among the various teams, particularly those working on sunshades, was prevalent; however, in retrospect, the focus was on the immense teamwork and the remarkable achievements realized in a concise timeframe. This period was often referred to by Huntsville officials as "the 11 years in May," reflecting the frenetic pace and intense effort involved.

One of the pressing issues facing the Skylab team was the electrical power situation. While it was bothersome, it did not pose an immediate threat. The real challenge lay in rapidly lowering the temperature inside the workshop. The situation was further complicated by the simultaneous loss of the solar panels and the meteoroid shield. This dual malfunction created a predicament where the negative impact of one issue exacerbated the other. Skylab's power generation depended on maintaining a solar inertial attitude, which required the sun's rays to strike the solar panels perpendicularly. However, this alignment exposed the full length of the workshop to solar radiation, increasing internal temperatures.

To mitigate this, Mission Control initially pointed the forward end of Skylab directly at the sun, which somewhat reduced internal temperatures and decreased power generation. Various attitudes were tested to find an optimal balance, ultimately determining that a 45-degree pitch toward the sun offered the best compromise. During daylight portions of each orbit, sufficient sunlight would hit the solar panels to charge the batteries for the subsequent dark period while internal temperatures stabilized around 42°C.

Steering issues complicated the search for a viable attitude. Skylab relied on nine rate gyroscopes to control its orientation by measuring rotation around three axes. Early in the mission, several gyroscopes overheated, causing erroneous readings and prompting flight controllers to cease averaging data from multiple gyroscopes. Fortunately, at least one functional gyroscope per axis remained operational. Despite this, the gyroscopes accumulated excessive and erratic errors, making it difficult for ground controllers to compensate effectively. As Skylab moved out of the solar inertial plane, random errors generated spurious signals that often caused the gyroscopes to reach saturation. Desaturation required a daylight pass in the solar inertial attitude, leading to further complications.

To address these issues and minimize the need for maneuvering, Mission Control implemented makeshift procedures. These included measuring roll attitude by monitoring temperatures on opposite sides of the workshop, determining pitch angle based on the electrical output of the solar wings, and calculating Skylab's momentum to ensure it was in the correct orbital plane. These ad hoc solutions consumed substantial amounts of attitude-control propellant, which could not be replenished. Despite the array of possible solutions to the various malfunctions, the inability to replace the propellant remained a significant concern, exacerbating the challenges faced by the Skylab team.

Skylab's launch had included a surplus of supplies, but within the first three days, the compressed nitrogen that powered the attitude-control thrusters was being consumed at an alarming rate. By May 17, 23% of the nitrogen had been depleted—twice the anticipated amount. Fortunately, as flight controllers gained proficiency in maneuvering the station, the situation improved, though the nitrogen expenditure remained higher than ideal. The consumption rate was tolerable until the first crew's arrival, which, ironically, was delayed by another five days on the 17th.

While much of Skylab's interior struggled with excessive heat, the airlock presented a different problem: it was too cold. By the 18th, temperatures in the airlock had dropped below 4°C. This was a critical issue because the suit umbilical system, which used water to transfer heat from the astronauts' suits during extravehicular activities, was at risk. Despite efforts to warm the airlock with heaters, its temperature continued to fall, nearing freezing on the 21st. If the water lines in the umbilical system froze, they could potentially damage the heat exchanger connected to the airlock's primary coolant loop.

The partially deployed workshop solar wing was made up of panels which were hinged together. Each was composed of hundreds of solar cells.

To address this, on the 20th, flight controllers adjusted Skylab's orientation slightly to expose the airlock to more sunlight. When this did not produce the desired result, they decreased the station's pitch to 40° the following day and rolled the workshop to place the water loops directly in the sun for one pass. These maneuvers succeeded in warming the airlock and generating additional electricity, but they also caused internal temperatures to rise, reaching approximately 54°C by the end of the 21st. The flight controllers had to carefully balance Skylab's position throughout the second week to maintain temperatures and power within safe limits. Although they struggled to achieve the expected stable condition, they managed to prevent significant damage to the station.

Despite losing the solar array, Skylab still had enough power to meet its needs until the crew's arrival, provided it remained perpendicular to the sun's rays. Power generation decreased sharply when sunlight struck the solar panels at less than a 90° angle. The estimated power requirement for the uncrewed Skylab was 4.5 kilowatts, just below the maximum output of the Apollo Telescope Mount (ATM) power system. As it became clear that maneuvers were necessary, engineers turned off non-essential heaters and transmitters, reducing power requirements to 3 kilowatts. This adjustment proved sufficient until the second week when high-angle maneuvers caused Skylab's electrical output to fall below this level. On the 24th, eight of the ATM's eighteen batteries failed due to excessive electrical demands. Only seven batteries were revived when Skylab was returned to a solar inertial attitude, highlighting the risks associated with further high-angle maneuvers.

The rapid accumulation of heat led to concerns about Skylab's provisions. The day after the launch, controllers began planning to restock the station's supplies, anticipating that the high temperatures would likely spoil all non-frozen food items.

In the food laboratories, prior tests had revealed that the canned food intended for Skylab could endure temperatures of up to 54°C for at least two weeks without degradation, while dehydrated items were expected to last even longer under similar conditions. To validate these findings, new tests were conducted by baking one batch of food at 54°C and another at the temperature typical of the food storage area. Periodic sampling confirmed that the heat did not affect the mineral content or taste of the food. By the 22nd of the month, officials in Houston concluded that the food was suitable for consumption, leading to the decision to forgo any additional food resupply for the workshop.

Similarly, the initial outlook for Skylab's medical supplies was fraught with concern. It was feared that the heat might compromise half of the 62 medications aboard. In response, Houston's medical team reassessed the resupply needs, considering heat tests and insights from pharmaceutical companies. Concurrently, Huntsville officials scrutinized the condition of film stored aboard Skylab. While the film intended for solar telescopes was safeguarded in the docking adapter, the film for earth-resource cameras and other experiments was kept in workshop vaults. The primary issues were the heat and low humidity, which threatened to dry out the film's emulsion. Salt packs placed in the vaults to maintain moisture were expected to be effective for only about four days. Kodak engineers suggested that the crew could potentially restore the film by rehumidifying the vaults, a process that might take up to 20 days. Consequently, arrangements were made to carry additional film aboard the Apollo spacecraft.

As temperatures on the exterior of Skylab climbed, reaching as high as 150°C, Huntsville engineers grew concerned about the spacecraft's structural integrity. Despite these fears, the spacecraft was successfully pressurized without any issues. Another pressing concern was the potential release of toxic gases into the workshop. The aluminum walls of the S-IVB tank were insulated with polyurethane foam, which, while effective at temperatures far below freezing, could emit harmful gases such as carbon monoxide, hydrogen cyanide, and toluene diisocyanate when exposed to 150°C. Of these, toluene diisocyanate was the most hazardous and potentially lethal, even in small concentrations. Chemical experts from industry and academia deemed the risk minimal, and tests by McDonnell Douglas indicated that the concentration of toxic gases in the workshop's large volume would be insufficient to pose a severe threat. Nonetheless, as a precaution, the workshop was vented and repressurized four times. Upon first entering, the crew was equipped with gas masks and instructed to sample the air to ensure safety.

Devising a Sunshade

In the aftermath of Skylab's malfunction, the immediate challenge was to find a solution to shade the workshop and prevent further deterioration. While the situation was urgent, there was optimism about finding an effective remedy despite some skepticism from the media. The key to resolving the issue lay in protecting only a portion of the exposed surface, which could potentially reduce temperatures to manageable levels. Unlike on Earth, where rigid supports and strong materials would be necessary due to wind and weather, space presented a different set of conditions where the lack of wind made it possible to use less robust solutions.

In the week following the incident, Skylab officials explored many potential fixes. These ranged from spray paint and wallpaper to balloons, window curtains, and extensible metal panels. While many ideas were considered, only ten were deemed promising enough to warrant further design and development. Huntsville officials, ever proactive, began contemplating a replacement for the meteorite shield shortly after the launch. Initial proposals varied widely, and while some ideas were unconventional, none were dismissed outright if they met the criteria of being lightweight and simple to deploy.

Some concepts were quickly ruled out. The astronaut maneuvering equipment, a new and untested system, was rejected due to the crew's lack of confidence. Thermal engineers also opposed deploying a weather balloon through the scientific airlock, fearing it could reflect heat and potentially damage the ATM solar panels. They favored a flat shade with sufficient distance from the workshop wall to mitigate this risk.

Max Faget's engineering directorate convened on launch night to brainstorm solutions in Houston. After extensive discussions, several proposals were assigned to teams for further study. Spray paint and wallpaper were eliminated as viable options. Although spray paint performed well in vacuum tests, it posed logistical challenges and risks of contamination. Wallpaper was discarded due to uncertainties about its performance under the harsh space conditions.

Three viable solutions emerged from the discussions: extending a shade from a long pole attached to the telescope mount, deploying a shade from the maneuvering Apollo spacecraft, or using a device extended through the scientific airlock on the solar side of the workshop. Each option had its pros and cons. Extending a shade from the telescope mount would require extravehicular activity (EVA), a process NASA meticulously trains for. However, the crew was already familiar with working on the telescope mount, and with the aid of a portable foot restraint, they could manage the task.

Deploying a shade from the Apollo spacecraft offered a potentially quicker fix with a simpler design. However, it faced the challenge of maneuvering around the workshop in space. The most straightforward approach was using the scientific airlock to extend a shade from inside the workshop, a method for which procedures were already prepared. The challenge here was designing a device that could fit through an opening just 20 centimeters square and then expand to cover an area of 7 square meters.

Given the tight timeline, with a launch date set for May 20, Faget's team at Johnson Space Center focused on the spacecraft-based shading solution, which seemed the most promising for timely implementation.

Left: View of the two sunshades deployed over the Skylab workshop during the first and second missions to help cool the station, as seen during the third crew's approach. Right: Gibson emerging from the Skylab Airlock during the mission's first EVA, photographed by Pogue. NASA

Amid the intricate preparations for Skylab's operations, one significant task involved deploying a crucial piece of equipment known as the SEVA sail. As the mission approached, astronauts faced the challenge of properly installing this sunshade. Standing in the Apollo spacecraft's hatch, an astronaut would begin the process by attaching one corner of the shade to the aft end of Skylab's workshop module. The spacecraft would then be carefully maneuvered laterally to secure the opposite corner of the shade. With precision, the Command and Service Module (CSM) pilot would guide the spacecraft toward the forward end of Skylab, allowing the shade to unfurl gradually. The astronaut would make a final attachment at the telescope mount, completing the setup.

This sunshade, eventually referred to as the SEVA sail, was a product of meticulous planning and design. Caldwell Johnson, the chief of the spacecraft design division, was tasked with overseeing its development. Johnson assembled a dedicated team to tackle this challenge. The team worked within the confines of the centrifuge building, where the public observed them through a mezzanine window. The presence of onlookers added an unusual element to their task, akin to feeling like "goldfish in a bowl."

Normally, Skylab flew in the solar inertial attitude, with the solar arrays directly facing the Sun. After the micrometeoroid shield was lost, ground controllers maneuvered the space station so that the workshop centerline was at a 45-degree angle to the Sun. The solar arrays then operated with less effectiveness, but temperatures within the workshop were kept at a tolerable level.

The creation of the SEVA sail involved a collaborative effort. Seamstresses stitched the bright orange fabric while parachute packers carefully folded the sail to ensure its proper deployment. Design engineers focused on integrating the various fasteners essential for its attachment. One of the most significant hurdles faced by the team was obtaining accurate and up-to-date data on Skylab's configuration, as some of the available drawings were outdated. In some instances, the engineers had to rely on photographs provided by McDonnell Douglas to fill in the gaps.

Johnson also encountered the challenge of managing suggestions from other NASA officials. While these suggestions were well-intentioned, they risked complicating the design and jeopardizing the project's timeline. Despite these obstacles and some minor delays, the SEVA sail progressed rapidly. By the Management Council meeting on the 16th, it was tentatively selected as the first shade for deployment.

The NASA teams debated the feasibility of extravehicular activity (EVA). The Johnson Space Center (JSC) leaned toward avoiding EVA due to the complications encountered during Gemini missions. In contrast, the Marshall Space Flight Center preferred conducting EVA from the telescope mount, driven by concerns that debris might obstruct the scientific airlock.

On the evening of the Skylab launch, Huntsville engineers began designing an alternative sunshade under the pressure of impending deadlines. This new design resembled a window blind and was created continuously, overnight. By the 15th, the design was complete, and fabrication commenced immediately. Testing began the following evening at the neutral buoyancy simulator, where Russell Schweickart, the backup crew commander, and Joe Kerwin, the scientist-pilot of the prime crew, tested various devices and assessed visibility from the telescope mount. The atmosphere was charged with media presence as news reporters watched the underwater activities.

The testing revealed that the initial design needed refinement. After the session, Schweickart and Kerwin joined 75 Marshall engineers for a debriefing. The astronauts were still in quarantine, and the setting, with participants wearing blue masks, resembled a surgical ward. Schweickart sketched new ideas on a blackboard, emphasizing the need for simplicity given the imminent launch. The group eventually agreed on a revised design: two cantilevered poles, each 17 meters long, would be assembled from 11 smaller sections. A continuous rope loop, threaded through eyelets at the poles' ends, would allow the shade to be deployed much like hoisting a flag. By early morning, this new configuration was set, ready to be implemented in the upcoming mission.

As the Skylab mission approached its launch date, meticulous planning and development continued for the sunshade, a critical component of the space station's equipment. Among the innovations under consideration was the development of the twin-pole sail by the Huntsville team. This design featured poles whose height above the workshop could be adjusted to avoid potential debris, adding a layer of flexibility to the deployment process.

Meanwhile, in Houston, another team led by Jack Kinzler was working on a different sunshade solution. Although Kinzler's Technical Services Division was known for its quick turnaround on flight items, it was not initially consulted for this project. Kinzler, motivated by a desire to support his friend and astronaut Pete Conrad, took the initiative to develop a practical sunshade design. Drawing from his extensive experience with the Apollo spacecraft, he was acutely aware of the constraints related to weight and size.

Kinzler's design process began the morning after the launch. He sought to create a sunshade that could be deployed through the scientific airlock, simplifying the crew's operations. His solution involved a combination of coiled springs and telescoping rods, which allowed a large cover to be deployed from a compact porthole. By the 13th, Kinzler's concept had materialized into a working prototype. He attached a parachute canopy to telescoping fishing rods fitted with hub-mounted springs. The entire assembly was compact enough to fit into a container roughly the size of the airlock canister. When deployed, the fishing rods extended and locked into a horizontal position, stretching the parachute into a smooth canopy.

The effectiveness of Kinzler's parasol design quickly became apparent through demonstrations, convincing Houston management of its viability. Encouraged by this positive feedback, Kinzler continued refining his design. The selection of the prime sunshade was a significant topic during a telephone conference among Skylab officials on the 19th. The decision to delay the crew's launch a second time had diminished the SEVA sail's initial advantage. Concerns were raised about the practicality of deploying the SEVA sail, especially given the demanding 22-hour launch day ahead.

Additionally, there were fears that the Apollo thrusters might contaminate the telescope mount and its solar panels.

Medical representatives preferred the parasol, as they wished to avoid the risks associated with extravehicular activity (EVA) before the crew had acclimated to the space environment. Deke Slayton supported using the scientific airlock, describing it as "the most direct approach and the least difficult operation for the crew." While some officials, like Schneider, believed the Huntsville twin-pole sail had the best chance of success, others, such as Kraft, wanted to eliminate it due to its 25 kg overweight issue. In a subsequent status briefing that night, JSC's director recommended continuing to develop the SEVA sail as a contingency should Huntsville's design fail the critical neutral-buoyancy tests.

The selection and development of the parasol for Skylab were complex and iterative processes marked by meticulous design and testing. The decision to approve Kinzler's parasol, establishing Conrad's preferred choice, marked a significant milestone. Initially considered ahead of the twin-pole sail, this parasol was designed to provide shade efficiently in the unique space environment.

The Huntsville team, confident in their design, prepared the parasol for deployment in the neutral-buoyancy tank, a crucial step in verifying its functionality. As Schweickart noted, the primary challenge was convincing NASA management of the design's viability. Despite a tight schedule, the Marshall Space Flight Center adhered to traditional design and development processes, including preliminary and critical design reviews, rigorous bench checks, and structural tests.

Huntsville aimed to complete the parasol by the 22nd, in time for NASA management's review in the neutral-buoyancy simulator. On the 18th, a tank test confirmed the parasol's feasibility but revealed potential issues, such as the possibility of pole sections separating under stress. Adjustments were promptly made, including modifications to the locking nut, reducing the shade's weight, and incorporating Teflon inserts to minimize friction. These changes led to a successful dress rehearsal in the tank.

In Houston, the parasol underwent significant revisions. The initial design, which used fiberglass fishing poles, was upgraded to stronger aluminum rods, and coiled springs were replaced with a "rat-trap" spring mechanism. The canopy was enlarged after Huntsville's thermal engineers recalculated the necessary dimensions. A major design alteration involved reshaping the frame to address the misalignment of the airlock with the shaded area. Kinzler's redesign included four arms of equal length—6.5 meters—allowing the rods to extend beyond the off-center canopy, improving symmetry in packing and deployment.

By the 17th, Director Chris Kraft had redirected most of Johnson Space Center's resources toward the parasol project. Faget's engineering division provided crucial design support while the program operations manager, Donald Arabian, oversaw configuration control and testing. Arabian expanded the parasol team beyond Houston, collaborating with North American and Grumman to address specific requirements. During the second week, he and Kinzler supervised the development closely, exercising joint veto power over changes. This approach led to what both men described as "engineering after the fact," where initial concepts were quickly built and detailed later.

Selecting the right shade material was governed by strict criteria, including thermal performance, lightweight construction, compactness, and stability across a broad temperature range. The material also needed to deploy reliably and avoid contamination. Initially, a spacesuit fabric composed of nylon, Mylar, and aluminum was chosen for its thinness—less than 0.1 millimeters—and suitability. However, nylon's susceptibility to deterioration under ultraviolet rays posed a challenge. Applying thermal paint could mitigate this issue, but added thickness was problematic given the parasol's tight fit within its container.

With no extensive data on nylon's long-term exposure in space, Houston decided to proceed without the paint, relying on the possibility that the second crew could replace the parasol if necessary. Huntsville, however, remained cautious about the unpainted nylon's durability. This decision underscored the ongoing challenges in developing equipment for the harsh space environment, where each design choice could significantly impact mission success.

Several days after the accident, Robert Schwinghamer's office initiated a comprehensive testing program at Johnson Space Center (JSC), evaluating the effectiveness of the sunshades under consideration. This program involved a rigorous assessment of a dozen materials through 49 tests. The testing, which included 100 hours of exposure to solar and vacuum conditions, revealed that nylon lost half its pull strength. Despite these results, Houston officials remained relatively unconcerned. They anticipated that the aluminum inner surface of the shade would effectively reflect most of the heat.

Contrastingly, studies conducted in Huntsville indicated a significant decline in the nylon's strength and shielding performance. On the 23rd, during the design certification review in Huntsville, all critical aspects of Skylab's sunshade problem were discussed extensively. The review focused particularly on the candidates for sunshades and materials testing. The findings were clear: every test result demonstrated that nylon deteriorated over time when exposed to ultraviolet rays. In the executive session, Skylab's top officials decided to retain the parasol as the primary choice but insisted on adding a protective covering for the nylon to enhance its durability.

Houston had anticipated this decision and selected Kapton, a highly ultraviolet-resistant tape, for the protective covering. Additionally, the twin-pole and SEVA sails, made from the same nylon-reinforced material, would be coated with thermal paint to improve their resilience. Langley Research Center was tasked with continuing work on an inflatable device as a backup in case the parasol encountered unforeseen issues.

The decision in Huntsville left JSC with less than a day to implement modifications to its two shades. Crews began applying Kapton to the parasol and spray-painting a SEVA sail on Wednesday evening. However, Caldwell Johnson's team faced several challenges. Contaminants in the paint required a lengthy straining process, and the drying time in the oven was longer than anticipated. By Thursday morning, there was considerable uncertainty about whether the SEVA sail would dry for the launch.

The modifications to the parasol proved even more problematic. The added bulk of the Kapton made stowage complicated the release mechanism, raising serious doubts about its functionality in space. Morale at the Houston center, which had been high the previous day, plummeted. During a final review at Kennedy Space Center, the parasol was reaffirmed as the primary device with its unprotected nylon. Materials experts estimated that the nylon would last at least 28 days in space. If signs of deterioration appeared, Marshall's twin-pole shade would be deployed as a backup.

Packing the parasol for the mission presented further difficulties, even without the Kapton. The final configuration revealed that the extension rod was recessed more than expected, complicating the astronauts' task of connecting the rod sections. To address this, the parasol team decided to add a 5-cm sleeve to facilitate the assembly. Manufacturing of this new component began as the parasol was delivered to Ellington Air Force Base, with the sleeve arriving on a separate flight just before the spacecraft's final closeout.

NASA's immediate focus was reducing power requirements, but there was also a longer-term need to increase power availability to ensure mission success.

The development and operational challenges the Skylab Space Station faced were significant, particularly when addressing the limitations of the Apollo electrical systems and the Advanced Technology Module (ATM) during the station's extended missions.

Initially, the ATM and Apollo's electrical systems were deemed sufficient for the early stages of Skylab's mission. However, as the station prepared for its more demanding 56-day flights, it became evident that these systems would not suffice. This realization prompted John Schneider to enlist the efforts of teams at both Houston and Huntsville to explore viable solutions.

In their quest to enhance Skylab's operational capabilities, Johnson Space Center (JSC) considered adding a solar-winged module designed to dock at the side port of the docking adapter. Concurrently, the Marshall Space Flight Center examined various designs for a portable solar array. Unfortunately, the complexity and time required for these hardware modifications made it impractical to implement them for the initial crew.

Telemetry data revealed a more immediate solution: Remains of the meteoroid shield were still holding one of the workshop's solar arrays. The hope was that removing this debris during a standup extravehicular activity (EVA) from the Apollo hatch could swiftly resolve the issue. Schneider, however, tempered expectations, acknowledging that the situation was uncertain. The team anticipated that, at best, they might secure photographic evidence to aid in the array's eventual deployment.

To address the debris removal problem, the Chief of Marshall's Auxiliary Equipment Section was tasked with developing specialized tools. The initial approach involved adapting tree-trimming shears from a local hardware store. This was soon followed by a collaboration with the A. B. Chance Company of Centralia, Missouri, known for their tools designed for power companies. The company agreed to showcase their full range of tools, leading to the selection of a cable cutter and a universal tool with prongs for prying and pulling. Both tools were modified for attachment to a three-meter pole.

While these tools were being developed, the Space Simulation Branch in Huntsville prepared a Skylab mockup in a neutral buoyancy tank. This simulation included loose wires, twisted bolts, and fragments of a meteoroid shield, alongside a model of the command module fl in from Houston. NASA officials evaluated the tools on the 21st of May, and astronaut Paul Weitz practiced freeing a solar array. Although the tools had already been dispatched to the Kennedy Space Center, a certification review raised concerns about the pointed tips of the cutters. Consequently, new heads with blunt tips were quickly fabricated and installed at the launch site.

The final preparations for launch faced an unexpected setback when a lightning strike struck the service structure's mast, causing a spacecraft gyroscope to go offline. Despite this, the guidance and navigation system was promptly retested, and the countdown resumed. Adjustments to the schedule were required when issues with the parasol delivery arose; propellants were loaded three hours ahead of schedule, and final stowage was delayed until 3:00 a.m., just as the crew was preparing for boarding.

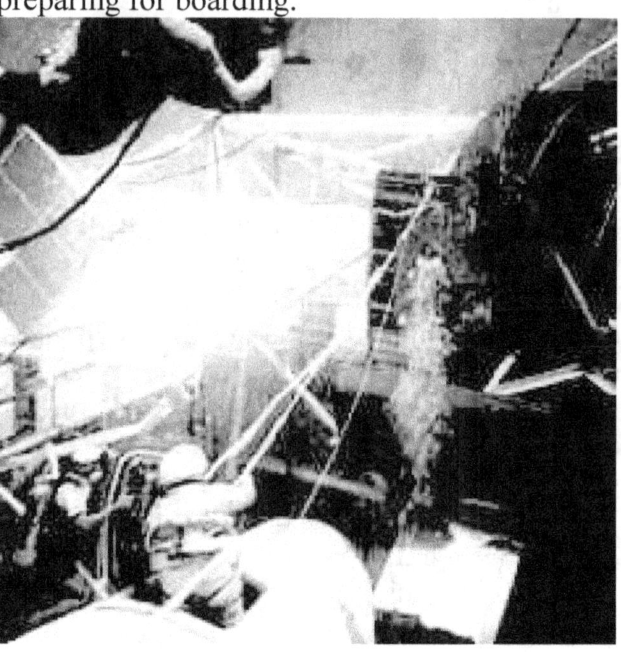

Extensive testing and many hours of practice in simulators such as the neutral buoyancy tank helped prepare the Skylab crewmen for spacesuited performance in the weightless environment. This huge water tank simulated the weightless environment which the astronauts would encounter in space.

Chapter 4 - Skylab 2

The launch on the morning of May 25, 1973, proceeded flawlessly. Skylab Crews

NASA's original vision for Skylab's crew composition was diverse, including an astronomer, a medical doctor, and other scientists to ensure a broad range of expertise. However, as the missions progressed, practical considerations led to a revised strategy. The tragic loss of three Soviet cosmonauts aboard Soyuz 11 highlighted the critical need for highly experienced pilots on long-duration space missions. Consequently, NASA included two trained pilots on each Skylab mission alongside one scientist-astronaut. This adjustment was a testament to the evolving understanding of the demands of space travel and the importance of having seasoned professionals aboard.

The inaugural Skylab mission, launched in 1973, was marked by the assembly of a highly skilled and diverse crew, each member bringing a wealth of experience and expertise to the historic endeavor. Commander Charles Conrad, Jr., a seasoned U.S. Navy captain with an illustrious career as an astronaut, led the crew. Conrad had been selected as part of NASA's second astronaut group in 1962 and had already proven his capabilities in space during several notable missions. He first ventured into space aboard Gemini 5 in 1965, followed by commanding Gemini 11 in 1966. However, his most celebrated achievement came in 1969 when he became the third person to walk on the Moon as the commander of Apollo 12, a mission that solidified his reputation as a consummate professional in the realm of space exploration.

Launch of Skylab 3 on a Saturn IB rocket.

Alongside Conrad was Pilot Paul J. Weitz, also a U.S. Navy commander. Weitz had joined the astronaut corps in 1966, but unlike Conrad, he had yet to experience spaceflight before the Skylab mission. His naval background gave him the skills and discipline required for the challenging space environment. Scientist Pilot Dr. Joseph P. Kerwin, another U.S. Navy commander, completed the trio. Selected as an astronaut in 1965, Kerwin was a medical doctor with a deep understanding of emergency procedures, making him an indispensable member of the crew. Though he had not flown in space before Skylab, his expertise was critical in studying the effects of spaceflight on the human body—a primary research focus of the mission.

Dr. Kerwin's role as the first scientist-astronaut to board Skylab was particularly significant. With his extensive medical training, Kerwin was tasked with monitoring the health and safety of the crew during their extended stay in orbit. His work involved pioneering research into the physiological impacts of space travel, contributing valuable insights to inform future missions and long-duration spaceflights.

(May 1973) --- These three astronauts named by the National Aeronautics and Space Administration as the prime crew of the first manned Skylab mission are, left to right, Joseph P. Kerwin, science pilot; Charles Conrad Jr., commander; and Paul J. Weitz, pilot.

The structure of each Skylab mission crew was meticulously planned, comprising a commander, a pilot, and a scientist pilot. The commander and pilot were typically individuals with extensive experience as spacecraft and aircraft pilots, often with robust engineering backgrounds that equipped them to handle the complexities of space missions. The scientist pilot, a relatively new and vital role within the astronaut corps, brought a unique combination of scientific acumen and practical flying skills. This blend of expertise ensured that the crew could conduct advanced scientific research while maintaining the operational integrity of the mission.

Supporting the primary crew were two backup crews, ready to step in should the need arise. For the first Skylab mission, the backup crew included Russell L. Schweickart, a civilian astronaut previously serving as the lunar module pilot for Apollo 9 in 1969. Schweickart's experience with lunar missions made him a valuable asset to the Skylab team. Alongside him were Bruce McCandless II, a lieutenant commander in the U.S. Navy, and Story Musgrave, a civilian physician and astronaut. McCandless and Musgrave, both selected as astronauts in the 1960s, had not yet

flown in space but were rigorously trained and prepared to undertake the mission if necessary.

The Skylab 2 backup crew of Bruce McCandless, left, Russell L. Schweickart, and Dr. F. Story Musgrave.

Between September 6 and September 20, 1973, the Skylab 2 crew—astronauts Charles Conrad, Joseph Kerwin, and Paul Weitz—and their backup team—Russell L. Schweickart, Dr. F. Story Musgrave, and Bruce McCandless—engaged in a rigorous program of mission simulations designed to prepare the astronauts for their upcoming mission aboard the Skylab space station.

The training took place in the Mission Training and Simulation Facility in Building 5 of the Manned Spacecraft Center (MSC), now known as NASA's Johnson Space Center in Houston. The facility housed Skylab and Command Module simulators and ground-based trainers tailored to mimic the Skylab environment. Over 11 days, the crews alternated between these simulators, immersing themselves in the scenarios they would encounter during their 28-day mission.

The simulations included 11 scenarios, each representing a unique day of the planned mission. This intensive preparation was crucial for ensuring that the astronauts could effectively manage the space station's systems, conduct experiments, and handle any unexpected challenges in space.

By mid-afternoon, the crew had reached Skylab and encountered a situation that was somewhat in line with their expectations. Commander Charles Conrad reported that "Solar wing two was gone completely off the bird," while "Solar wing one was partially deployed." A bulge of meteoroid shield material was found underneath it, impeding its full deployment. Sunlight had tarnished the gold foil on the workshop's exterior, but the scientific airlock appeared to be nearly clear of debris.

Despite the challenging conditions, the astronauts were determined. During their initial inspection, Weitz struggled to broadcast the damaged area due to the confined space within the spacecraft, though Houston received some clear views. Conrad completed a flyaround of Skylab and remained hopeful that the crew could resolve the issue during a standup EVA.

Following a meal, the astronauts began their attempt to extend the solar array. Weitz used the tools while standing in the open hatch, with Kerwin holding his legs and Conrad maneuvering the spacecraft. However, as Apollo passed over the California tracking station about 40 minutes later, it became apparent that the crew was experiencing significant difficulties. Frustration was evident as

the astronauts expressed their struggles with four-letter expletives, despite Houston's efforts to reestablish communication.

During the flyaround inspection of Skylab, Conrad reported: "Solar wing 2 was completely gone, and solar wing 1 was partially deployed." He also reported that debris from the micrometeoroid shield was jammed around the partially deployed wing, holding it in place and preventing full deployment.

Conrad's report was disheartening; the metal strip wrapped around the array beam, though only a centimeter wide, was riveted in place by bolts that had attached themselves as the shield was torn away. Despite Weitz's strenuous efforts, the panel remained stubbornly in place. Conrad's assessment of the situation highlighted the complexity and gravity of the challenge they faced.

During their training at Houston, Commander Charles Conrad had humorously suggested to his colleague, Paul Kerwin, that if they ever had to resort to a particular emergency procedure, it would mean it was time to return home. Yet, when faced with a real issue aboard Skylab, Conrad pragmatically communicated to Mission Control, "We might as well try the EVA. Because if we ain't docked after that, I think you guys have run out of ideas."

The emergency procedure was intricate and demanding. It required the crew first to depressurize the spacecraft and then open the forward tunnel hatch. They had to remove the probe's back plate to bypass certain electrical connections. Once these steps were completed, the crew carefully centered the probe and drogue. They maneuvered the spacecraft using the Apollo's thrusters to align with the docking adapter. The precision of their approach was evident as the two docking surfaces met, and all twelve latches engaged successfully.

View of Skylab 2 from the Skylab Command/Service Module during the final flyaround inspection. Damage can be seen on both a solar panel and the parasol solar shield.

As the program managers held a midnight press briefing to announce the successful docking, the crew took the opportunity to tidy up the Apollo cabin. They concluded a grueling 22-hour day with a sense of accomplishment and relief, knowing they had overcome a significant challenge in the mission's history.

First Workshop Visit

When astronauts Conrad, Weitz, and Kerwin awoke aboard Skylab, their immediate concern was ensuring the workshop's atmosphere was safe for habitation. Engineers from various NASA centers had taken meticulous precautions to prevent any harmful gaseous residues from the insulation used in the spacecraft. The spacecraft's maneuvering had maintained temperatures that prevented the insulation from degrading, while a thorough pressurization and depressurization cycle over several days had effectively purged any undesirable atmospheric elements.

As Weitz donned a gas mask and began sampling the atmosphere, he found no traces of noxious or toxic gases. Confident in the environment's safety, he and Kerwin entered the workshop and initiated its activation. Despite the internal temperature of 130°F, the remarkably low humidity allowed them to work comfortably for up to five hours.

The crew then focused on deploying the parasol thermal shield, a crucial step in regulating the workshop's temperature. They began by attaching the canister to the scientific airlock. Conrad and Weitz worked meticulously, ensuring that they did not distort any structural components as they extended the folded shield through the airlock and into space. Slowly, the struts of the shield extended, and the sunshade began to unfurl. Although one corner of the trapezoidal shield did not extend fully, it was successfully positioned over the workshop's outer surface. The deployment of the parasol shield led to an immediate drop in the internal temperature of the workshop.

A Sense of Up and Down

Scientist Pilot Kerwin provided a fascinating insight into the adaptation to microgravity: "You do have a sense of up and down," he explained. "And you can change it in two seconds whenever it's convenient. If you go from one module into another and you're upside down, you tell your brain, 'Brain, I want that way to be up.' And your brain responds, 'Okay, then that way was up.' Your brain will follow if you need to rotate 90 degrees to work. I don't think it's a matter of the vestibular system. It's strictly about your eyes and brain, and it works remarkably well."

As the crew settled into their new environment, they established a routine that balanced experimentation, housekeeping, eating, sleeping, and exercising. This routine was carefully coordinated between the flight and ground teams to ensure that routine maintenance tasks and scientific experiments were performed efficiently.

Each day was structured around a 16-hour work period followed by 8 hours of sleep, with all crew members adhering to the same sleep schedule. This disciplined approach helped the astronauts maintain their physical and mental well-being while adapting to the unique conditions of space.

The solar observatory and workshop solar arrays drew their energy through panels containing thousands of small solar cells.

Orientation in the space station was never a problem. As the crew adjusted to its new environment, Scientist Pilot Joe Kerwin reported that there was, indeed, an up and down in space. It was simply a matter of telling yourself which was which, he observed. In this photo taken by Astronaut Weitz on the final day of the mission, Kerwin was shown in the hatch between the Multiple Docking Adapter and the Apollo spacecraft.

In coordinating their daily schedules, the Skylab crew meticulously allocated time for crucial Earth and Sun observations, carefully considering Skylab's orbital position. Once these observation periods were established, other activities were assigned the remaining time.

The solar observatory's power system supplied most of Skylab's electrical power, necessitating stretching the experiment program over a longer period. Despite this adjustment, the experiments were conducted almost as initially planned. However, six experiments faced challenges because they were designed to use the Sun-facing scientific airlock, now covered by the parasol shield. To address this, the flight crew adapted the program by performing three experiments using the Earth-facing scientific airlock. The remaining two experiments were mounted on the solar observatory's truss and one on the Sun shield during extravehicular activities.

Accomplishing The Repair

Despite the myriad challenges faced on Skylab's first day in orbit, NASA officials maintained steadfast optimism about deploying the space station's parasol. On the afternoon of May 26, the crew commenced their work by first activating the docking adapter and airlock. The heat inside the workshop was stifling, with astronaut Weitz describing it as a "dry heat, like the desert." The crew worked diligently to manage the oppressive temperature, periodically stepping outside the workshop for relief. This careful, methodical approach extended the operation to approximately two hours.

The deployment process began with connecting the parasol canister to the scientific airlock. Once the port was opened, astronauts threaded extension rods and gradually extended the parasol. There was a palpable sense of disappointment as the folded arms swung outward and the fabric began to spread. The parasol, designed to shield the station from the sun's intense radiation, did not deploy as expected. Conrad reported that the canopy was misaligned, covering only about two-thirds of its intended area.

Despite the crew's concerns, Mission Control in Houston immediately greeted the parasol deployment news with cautious optimism. They attributed the wrinkles observed in the parasol to the extreme cold of space, where the shade had been extended but remained unopened during the dark phase of the orbit. NASA officials hoped that exposure to sunlight would allow the material to stretch and smooth out.

Over the next three days, internal temperatures in the workshop began to stabilize. The external temperature of the parasol plummeted by 55°C overnight, while internal temperatures fell by 11°C on the first day. The uneven coverage of the parasol led to noticeable hot spots within the workshop, including a hot area near Joe Kerwin's sleeping compartment. By May 29, engineers projected that the workshop's temperature would stabilize around 26°C, which, though 5°C above the desired level, was still deemed acceptable.

Full-scale operations commenced on May 29, including medical tests, solar observations, and preparations for the initial Earth-resources pass. Power consumption quickly became a significant concern, running close to Skylab's output of 4.5 kilowatts. This issue was particularly pronounced when operating the telescope mount, which drew 750 watts. During an evening news briefing, Flight Director Neil Hutchinson acknowledged that power limitations were among several factors affecting early mission planning.

The situation worsened on May 30. Four batteries failed during the Earth-resource maneuver, which required retracting Skylab's solar panels and relying on battery power. Despite numerous attempts to restore functionality, only three batteries were reactivated when the spacecraft resumed its solar inertial attitude. This failure, Skylab's second in a week, further reduced power capacity by 250 watts and raised serious concerns about the reliability of the electrical systems.

In response to these escalating issues, the Management Council decided to advance the launch date of the second crew by two weeks. Discussions were held to explore solutions for the solar array problem, with a decision scheduled for June 4. Rusty Schweickart led a team investigating the issue since the day after launch. By May 29, discussions with the crew and analysis of televised footage enabled engineers at Huntsville to develop a feasible plan to address the jammed solar array.

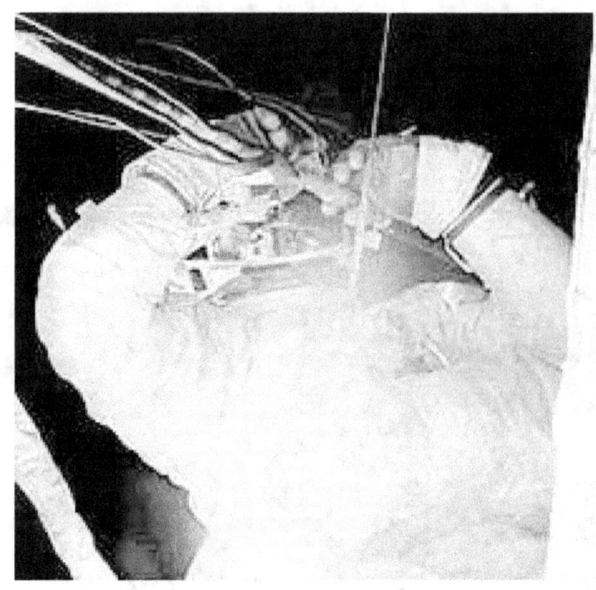

Kerwin extended the long-handled cable cutter while Conrad affixed the cutter jaws to the strap holding the solar wing. Together, they cut the strap, which freed the wing and then they extended it to its full position.

Over the following four days, a challenging yet viable procedure was devised. Two crew members were to exit through the airlock port and navigate the airlock trusses to reach the solar array at the forward edge of the workshop. Equipped with an eight-meter cable cutter, one astronaut would use a pole as a handrail to approach the solar array. They would attach a beam-erection tether—a nylon rope with hooks—between the solar wing and the airlock shroud. This tether was intended to break a frozen hydraulic mechanism and restore the array's functionality.

The Skylab Space Station's operational challenges were considerable, particularly regarding tasks that required precise, hands-on work in a zero-gravity environment. One of the most pressing issues was the absence of footholds, significantly hampered the astronauts' ability to perform tasks that necessitated using both hands. Despite this challenge, progress was made. By June 2, astronauts Russell Schweickart and Ed Gibson had successfully demonstrated the procedure for managing the solar array in a water tank. This crucial step proved that similar operations could be achieved in space. This demonstration was pivotal, as actions that could be performed in the controlled environment of the tank were typically translatable to the actual conditions of space.

The material to be used in the protective shields was carefully selected. It had to be light, very compact when folded, tough and resistant to harmful radiation. Above all it had to stand up in the space environment for half a year.

However, as June 4 approached, the Management Council in Huntsville received a sobering update on Skylab's condition. The situation was critical: if no additional batteries failed, the first crew might complete the planned experiments, but the lack of sufficient power would likely impede the work of subsequent crews. Schweickart, tasked with reviewing the procedure for freeing the solar array, presented films of his practice sessions to the council. Despite some concerns—particularly about the difficulty of attaching a cutter to debris eight meters away and the absence of an alternative method to secure the pole—the council decided to proceed. The potential benefits of successfully deploying the solar array, which promised a significant increase in power, outweighed the risks. Even in the event of failure, valuable information could be gained for future attempts.

Solar wing 1 was held firmly in its partially deployed position by a strap, apparently from a portion of the micrometeoroid shield.

Rigging of the protective shields for Skylab was rehearsed many times in the neutral buoyancy simulator. Here, a shield made of netting (for simulation only) was being rigged underwater to determine the problems to be encountered in space.

On the evening of June 6, Schweickart gave the crew a detailed briefing on the operation. A list of tools, assembly instructions, and procedural steps was transmitted the following day over the teleprinter. The astronauts used their free time to review the procedures, and any remaining questions were addressed during an hour-long session with Mission Control. On June 8, the crew practiced the operation inside the workshop, using television and radio to communicate with Mission

Control. Kerwin donned his pressure suit for a more realistic simulation, and Conrad made several minor adjustments to the beam-erection tether. Despite their preparations, neither astronaut was particularly optimistic about the outcome.

As Skylab began its dark pass on the morning of June 7, the crew opened the airlock hatch. Conrad assembled the necessary tools under the light of the airlock shroud, and the astronauts moved to the antenna boom. When daylight allowed, Kerwin attempted to fasten the cutters, but his initial efforts were unsuccessful. Unlike the water tank model, the actual flight model presented new obstacles; cable connectors obstructed his footing, making it challenging to stabilize himself. As Kerwin struggled, he found that one hand was immobilized, wrapped around the antenna, while the other struggled to control the pole. Every movement of the pole caused his body to react, complicating the task further. Despite several attempts to position the cutter close to the restraining strap, the pole's movement made it challenging to open the cutters properly.

The situation grew more tense when communications with Houston were lost at 11:42 a.m., after Kerwin had been working intensively for over half an hour, his pulse rising to 150. In a breakthrough moment, Kerwin had an idea to prove crucial. He stabilized his position by shortening the tether, connecting his suit to the antenna. This adjustment allowed him to complete the task, securing the cutters and making the necessary adjustments to free the solar array.

The efforts of astronauts and engineers in the water tank at Marshall Space Flight Center in late May and early June 1973, experimenting with various cutting tools and techniques, had culminated in a successful resolution to a critical problem. Schweickart's and Gibson's practice, coupled with the ingenuity demonstrated during the spacewalk, underscored space missions' collaborative and adaptive nature, where preparation and innovation combined to overcome formidable challenges.

After successfully cutting the debris strap, astronauts Conrad and Kerwin worked together to free the frozen actuator by pulling on the line. Ten minutes into their efforts, they informed Houston that the pole was securely fastened to the solar array. Though the immediate crisis was averted, the crew still faced several challenges.

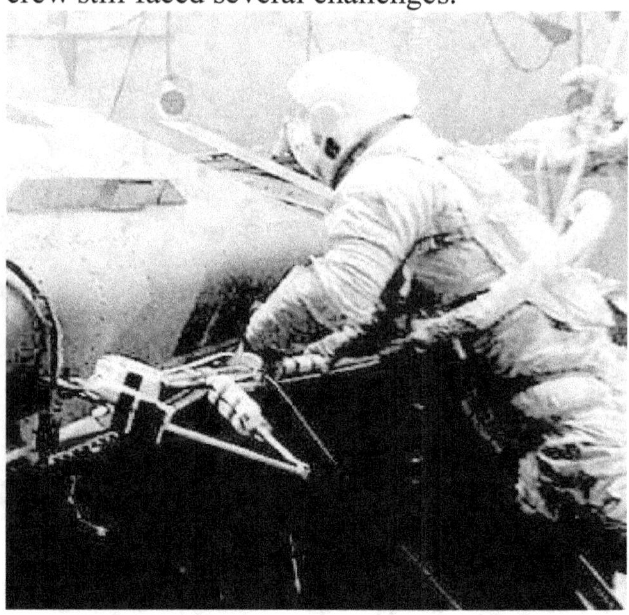

Engineers or astronauts carefully tested each of the tools in the neutral buoyancy tank. Techniques for operating the tools were devised.

Conrad encountered difficulty at the solar array. The holes intended for the erection tether were slightly smaller than those on the ground model, preventing him from attaching both hooks. After struggling with the second hook, he proceeded with just one. Meanwhile, Kerwin cut the restraining strap without much trouble, but the frozen damper proved more resistant. The two astronauts worked in unison to overcome this obstacle, eventually succeeding in their task.

When Houston inquired about the solar array's deployment, Conrad humorously recounted the experience: "I'm sorry you asked that question. I was facing away from it, heaving with all my might, and Joe was also heaving with all his might when it let go, and both of us took off. By the time we settled down and looked at it, those panels were out as far as they would go." Fortunately, the array was fully extended by the following day, generating nearly 7 kilowatts of power.

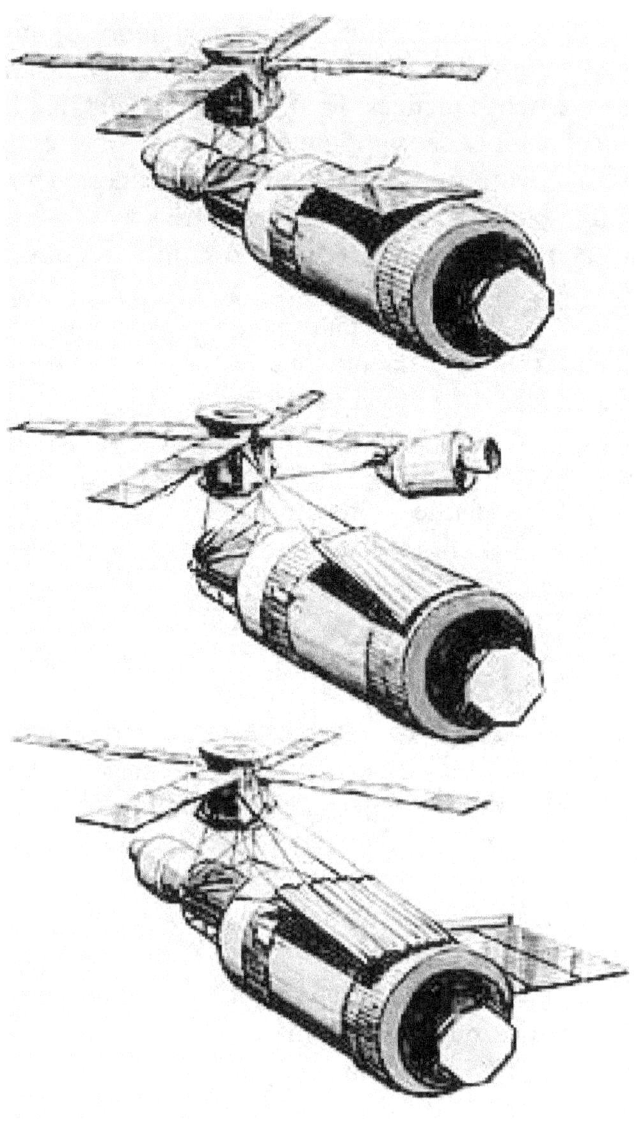

With the protective micrometeoroid shield missing, it became necessary to shield the workshop from the Sun by other means. Several methods were devised, and two were ultimately used. The three sketches show the schemes selected. The upper sketch depicts shading by the rectangular parasol, which was deployed from inside the workshop by the first crew. The second concept called for the astronauts to rig a shield from the command and service module maneuvering alongside Skylab. This second concept was not used. The lower sketch shows the twin-pole sail, which was deployed by the second crew during a spacewalk.

Despite the successful deployment, the incident attracted criticism from Congressional figures. Senator Frank Moss, chairman of the Senate space committee, demanded a thorough investigation into the accident, a requirement of NASA's agency policy. Bruce Lundin, director of the Lewis Research Center, was appointed to lead the inquiry on May 22. Lundin's committee began by scrutinizing the flight data to reconstruct the sequence of events. They identified the failure of the meteoroid shield as the primary cause of the accident and meticulously reviewed the shield's development, focusing on design, fabrication, and testing management.

Left: In the Mission Control Center at NASA's Johnson Space Center in Houston, capsule communicator and backup Skylab 2 commander Russell L. Schweickart, right, and Skylab 4 astronaut Edward G. Gibson react with joy at the news of the deployment of the jammed solar array wing. Right: The freed solar array was visible in this photo taken by the departing Skylab 2 crew who deployed it. NASA

The committee considered ten potential failure modes for the shield but concluded that only two were plausible. The first involved a gap between the shield's edge and the workshop wall. Although NASA's design specified a tight fit, the shield had gaps exceeding the design specifications by half a centimeter at launch. Wind tunnel tests supported the idea that pressure buildup in these gaps could have contributed to the accident. However, flight data pointed more convincingly to the shield's auxiliary tunnel as the likely culprit. This tunnel, intended to vent pressure as the launch vehicle ascended, had not been constructed according to design specifications. Lundin's team theorized that pressure buildup might have been sufficient to lift the shield into the airstream one minute after launch. As the shield detached, it wrapped around one solar array and damaged the latches on the other. Gravity and aerodynamic drag kept the array close to the workshop for over eight minutes until the S-II stage separated from the workshop. The stage's retrorockets then fired, causing the solar array to be ripped from its hinge.

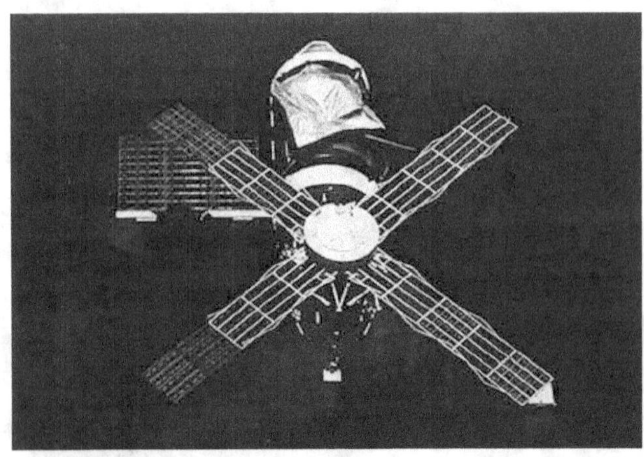

The workshop's shading and deployment of the solar wing made Skylab fully operable. The workshop solar array's addition provided much-needed electrical power, and the parasol shielded the workshop from high temperatures.

The board's investigation revealed a critical oversight: NASA and McDonnell Douglas failed to detect the defect despite six years of development and testing. The error was partly attributed to an overreliance on the assumption that the shield would fit as specified in the design criteria, which proved flawed.

developing the Skylab Space Station faced significant challenges, particularly regarding its shield system. The shield, intended to protect the station from the harsh space environment, was a "large, flexible, limp system" that could not be rigged according to the original design specifications. This unexpected outcome highlighted a critical oversight in the planning and execution stages of the project.

A committee evaluating the project strongly criticized NASA's handling of the shield system. The primary issue identified was NASA's failure to treat the shield as an independent subsystem with a dedicated project engineer overseeing its development. Instead of having a focused expert manage the intricate details, the project lacked cohesive engineering leadership and judgment.

The committee's investigation revealed that the design deficiencies and the communication breakdown within the project team were not due to constraints of time, budget, or technical expertise. Rather, these issues stemmed from a systemic failure to apply sound engineering principles and maintain vigilant leadership over an extended period.

Three of the board's recommendations stood out for their broad implications for NASA's management practices. First, it was advised that a project engineer be appointed for complex systems involving multiple engineering disciplines. This recommendation aimed to ensure that each intricate component received the required attention and expertise.

Second, the board cautioned against excessive emphasis on documentation and formal procedures. It emphasized the importance of engineers gaining hands-on experience with actual hardware, developing an intuitive grasp of computer-generated results, and effectively utilizing flight data to enhance their understanding.

Lastly, the board recommended assigning an experienced chief engineer to oversee major projects, such as the Skylab workshop or airlock. By relieving this engineer of administrative and managerial duties, the focus could remain on the subtle integration of all system elements, ensuring that the overall system functioned harmoniously and effectively.

These recommendations were designed to address the Skylab shield project's shortcomings and improve NASA's approach to managing complex engineering systems in future endeavors.

The NASA Dispute

A significant dispute emerged between NASA's Office of Public Affairs and the Office of Manned Space Flight (OMSF) regarding the handling of private communications. The crux of the debate centered on whether American press access to air-to-ground discussions, particularly those concerning operational and medical issues, should be restricted. The Office of Public Affairs was concerned that excluding the press from such conversations might damage relations with the media. Conversely, OMSF worried that prohibiting private communications could jeopardize the mission's success. This internal conflict highlighted the broader struggle to balance transparency with operational security.

Meanwhile, the crew faced difficulties with Skylab's ergometer, the space station's primary exercise equipment. Although they eventually found a workable solution for using the machine, the intense physical activity led to concerns about their health, further complicating the situation. In the final two weeks of the mission, with additional power provided by the repaired solar array, the astronauts managed to complete most of their remaining tasks.

Since its inception, NASA had prided itself on the transparency of its programs, distinguishing itself from the Soviet Union's secretive approach. The agency had traditionally kept the media well-informed with transcripts of air-to-ground transmissions and regular briefings. This openness, however, faced scrutiny starting in the late 1960s when the press began to criticize a perceived "credibility gap." Concerns were raised when private conversations from the Apollo 9 mission were seen as a departure from NASA's open policy. Publications like the Washington Post and the Houston Post debated whether the public had a right to access detailed medical information about the astronauts.

In response, NASA Administrator Thomas Paine affirmed in March 1969 that while the agency supported an open policy, private communications should be reserved for special medical or operational emergencies. Any such private discussions would be summarized for the press to maintain transparency.

For Skylab, the Johnson Space Center (JSC) proposed modifications to this policy, citing the mission's duration and unique medical demands as justifications. The new policy allowed daily private medical conversations, though flight surgeons would inform the press of significant medical updates without releasing or transcribing the full tapes. Operational communications could also be private if there was a real need, with a public affairs officer summarizing these discussions for the media. Additionally, weekly unmonitored calls between astronauts and their families were permitted.

OMSF supported the Houston plan, arguing that doctor-patient confidentiality was standard practice and that astronauts, like many individuals, had a right to private medical consultations. The need for private communications was underscored by past experiences, such as the Apollo 15 mission, where there was reluctance to disclose information about astronaut James Irwin's irregular heartbeat due to concerns about potential negative publicity.

Overall, the initial Skylab mission highlighted the complexities of managing media relations, maintaining transparency, and addressing the practical needs of astronauts in space, setting the stage for future missions and their associated challenges.

In January 1973, a significant debate arose within NASA regarding handling private communications during the Skylab missions. John Donnelly, the assistant administrator for public affairs, moved decisively to prevent any policy shifts that might affect the agency's commitment to transparency. He informed Administrator James Fletcher that there was a fundamental disagreement between his office and the Office of Manned Space Flight (OMSF) regarding using private lines for routine medical discussions.

Donnelly argued strongly against using private channels for regular medical updates. He believed that the condition of astronauts was as crucial a news element as the status of the spacecraft itself, particularly for a high-profile mission like Skylab. Donnelly was skeptical that private medical communications would foster a more honest reporting of astronauts' health. He noted that pilots typically hesitated to disclose issues that could potentially shorten their mission or increase their dependence on ground support. He feared that adopting a private line might undermine NASA's long-standing policy of openness, which had earned the agency global respect and admiration.

Despite these concerns, a compromise was reached in March 1973. Fletcher approved a policy allowing routine medical discussions over a private channel. While these conversations would not be directly paraphrased for the press, the flight surgeon would provide summary bulletins. Private operational communications were also permitted, but only in extreme emergencies. These communications could be initiated by either flight controllers or the crew, and would be paraphrased for the press by the public affairs office.

The new policy was tested less than a week after the first Skylab crew's launch. On May 28, 1973, Commander Pete Conrad, who shared many

Houston officials' dissatisfaction with the open policy, requested a private conversation for the following morning. Conrad indicated that it was not an emergency but preferred to discuss specific issues away from public scrutiny. Despite objections from the public affairs office, NASA officials approved Conrad's request.

During the private conversation on May 29, Conrad addressed several topics, including difficulties with the bicycle ergometer, the solar array, the docking probe, and workshop temperatures. He expressed surprise at the mission's smooth progress and reported that the crew was in good condition. However, when newsmen received a summary of the conversation an hour later, they questioned whether an actual emergency had existed. Donnelly publicly criticized the decision to grant a private communication, arguing that the situation did not warrant it. He sought to release the tape of the conversation and proposed that Conrad be formally reprimanded, though neither action was taken. Instead, Dale Myers agreed to implement a new procedure where the capsule communicator would confirm the presence of an emergency before arranging any further private communications.

The same day, a second controversy emerged regarding "channel B dump data," which included information recorded on board and relayed to Earth regularly. This data was distinct from live air-to-ground conversations broadcast over the primary channel. In February, Conrad had mistakenly informed the press that transcripts of these recordings would be provided, although NASA had intended to keep them confidential. The issue resurfaced on May 29 when the availability of the tapes became a contentious topic. At Donnelly's insistence, Myers initially agreed to release all channel B data but later excluded information about medical experiments.

As the Skylab Space Station program progressed, tensions arose concerning releasing sensitive medical data and crew communication policies. The initial concerns about public access to medical information were rooted in fears that laypersons might misinterpret or misrepresent the data, potentially undermining public trust in the program. Donnelly, a key figure in mission management, sought to address these concerns by advocating releasing all Channel B material to counter a growing "climate of mistrust." He argued that withholding such information could imply a lack of transparency and suggested that failure to disclose fully could jeopardize the credibility of the entire mission.

Despite these arguments, George Low, the program manager, upheld the doctors' request to restrict the release of sensitive medical data, pending further discussion with Fletcher, the administrator. On the 31st, Fletcher rescinded an earlier directive that required Mission Control to interrogate the crew about the seriousness of operational problems. This decision stemmed from concerns that such inquiries might deter astronauts from reporting genuine issues if they felt pressured by the questioning. Fletcher emphasized the importance of maintaining an open line of communication, asserting, "We do not want to risk the safety of the flight by having the astronauts infer, from our questions, that they should not use the private communications loop when a real need might exist."

This policy adjustment reflected a broader perspective that extreme operational emergencies should encompass any significant concern the crew might have that could not be adequately addressed over open lines. Fletcher also endorsed Myers's decision to withhold medical data from Channel B tapes, a practice aligned with NASA's tradition of protecting proprietary information for medical investigators.

The debate over private communications persisted despite Fletcher's clarifying press release on NASA's policy. Donnelly expressed concern over NASA's credibility, especially after the launch of the second crew. He reported that journalists from Time, the New York Times, and the Chicago Tribune were discontented with the shift from traditional open policies. These reporters were apprehensive that the increasing reliance on private communications might weaken their negotiating leverage with the Soviet Union for the Apollo-Soyuz flight. Although Donnelly's vigorous defense of journalistic interests might have strained his standing within NASA, the press showed limited concern about these issues, primarily due to the overshadowing Watergate scandal.

Following the launch of the second crew, Skylab missions largely avoided private operational conversations, likely due to the policy's controversy. Astronaut Conrad, for instance, felt that the absence of a private communication channel hindered effective communication, leaving him "in the dark" about Houston's plans. He recounted a particular instance where he learned about EVA plans to free the solar array from his wife during a birthday greeting, rather than through official channels. While he recognized that open lines allowed more information dissemination, he believed a private channel would have facilitated more candid discussions.

The lack of a private operational channel did not significantly impact the first two missions. Still, it became apparent that the open communication policy contributed to communication difficulties during the final flight. The experience highlighted the complex balance NASA had to maintain between transparency, operational effectiveness, and crew safety.

During the Skylab program, the issue of private communications within NASA revealed a significant tension between political interests and operational needs. The agency struggled to reconcile its desire for transparency with the practical demands of maintaining effective communication between the astronauts and mission control. Despite various attempts, a satisfactory compromise remained elusive.

The disagreement over medical experiments was a long-standing issue that dated back to the Gemini missions. Astronauts often felt they were being subjected to unnecessary experiments, viewing themselves as mere subjects rather than active participants in their missions. Mike Collins, for example, criticized the inflight sleep analysis experiment on Gemini 7, describing it as a scenario where "the tail wagged the dog," with decisions made by medics on the ground with inadequate information, such as brain wave data.

From the perspective of the medical directorate, these experiments were deemed essential for understanding the effects of spaceflight on human physiology. Dr. Charles Berry, a prominent figure in the medical team, acknowledged that many within NASA, including astronauts, regarded his staff as "over cautious." Nevertheless, he defended this approach, arguing that the caution was necessary due to the substantial pressure from critics of crewed spaceflight. Accurate data on astronauts' adaptation to space was crucial for the success of future missions.

One area of contention was the control and regulation of physical exercise. Medical experts wanted to impose strict guidelines on exercise routines due to their importance in monitoring cardiovascular adaptation and physical performance in space. They aimed to measure all physical activity to gather comprehensive data. Conversely, astronauts objected to these rigid controls, citing personal inconvenience and believing they were best suited to judge their exercise needs.

The compromise reached for Skylab involved a more flexible approach. Instead of regulating every aspect of exercise, the medical office agreed to periodic measurements of physical condition while leaving daily exercise to the discretion of the astronauts. Crew members were required to report the duration and intensity of their workouts. The ergometer, a primary exercise machine, was used to gauge the workload.

During the missions, the flight plan allotted the first crew 30 minutes per day for exercise, which could be performed on the ergometer or an isometric device. Additionally, twice a week, each astronaut underwent a metabolic experiment, known as M I71, to test physiological responses to exercise. This experiment measured changes in metabolic response, including blood pressure, heart rate, and oxygen consumption, providing valuable insights into the effects of exercise in a microgravity environment.

The metabolic experiment M I71, conducted during the Skylab missions, was designed with a structured protocol to assess the physiological impacts of exercise in space. The test comprised five distinct phases within a 25-minute timeframe: a rest phase to establish a baseline metabolic rate, followed by three exercise phases at increasing intensities—25%, 50%, and 75% of the crewman's maximum capacity, as determined by preflight tests. The final phase was a recovery period.

A secondary objective of M I71 was to evaluate the effectiveness of the ergometer, the primary exercise machine used during the mission. During the lunar missions, astronauts experienced a decline in physical conditioning due to the limited space available for exercise within the Apollo spacecraft. Although this decline was generally temporary, with fitness levels typically returning to preflight conditions within 36 hours, it highlighted potential risks for long-duration flights.

Problems with the ergometer became evident during the first run of the metabolic experiment by astronaut Weitz on May 28. Due to the elevated temperatures in the workshop, Kerwin had suggested shortening the exercise schedule to mitigate heat stress. However, Houston advised completing the full regimen to maintain experiment controls. Midway through the third exercise level, Weitz was forced to terminate the session. He reported that the waist and shoulder harness, establishing supposed to secure him to the bicycle, restricted his movement. As a result, he relied too heavily on his hands rather than engaging his leg muscles, which impeded his performance.

On May 29, during a private communication, Conrad reported difficulties with the ergometer, noting that its operation in space differed significantly from its use on Earth and questioning whether the crew could complete the entire exercise protocol. Later that day, Kerwin also failed to finish the exercise, while Conrad struggled through the final stages, describing the third exercise level as equivalent to "20 minutes of a full workload." He conveyed to Mission Control that he was "running out of gas" and felt he was using muscles not typically engaged on the ground.

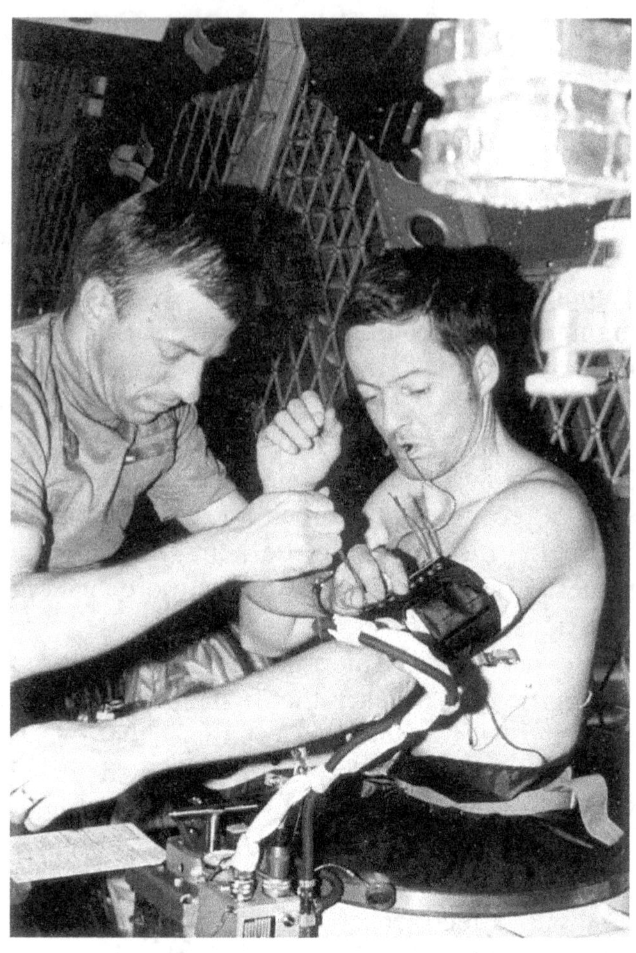

Weitz aids Kerwin with a blood pressure cuff.

In response to these challenges, the crew recommended reducing the workload on the ergometer by 10-20% to account for the difficulties encountered while exercising in the microgravity environment. This adjustment was proposed to help improve the practicality and effectiveness of the exercise regimen for future missions.

The Skylab Space Station, an ambitious step in human space exploration, faced various challenges that impacted its mission effectiveness and crew health. One of the initial problems encountered was related to the station's exercise regimen. The crew's physical activity was severely constrained by a demanding schedule that prioritized other tasks, effectively eliminating their designated periods for physical training.

On May 31, astronaut Don Kerwin voiced concerns over the impact of this scheduling on their physical fitness. He noted that the flight plan, managed by Houston Mission Control, had inadvertently minimized the time allocated for exercise. Kerwin pointed out that other critical tasks had overshadowed the planned physical training periods, leaving little to no time for exercise. He hoped Mission Control would reconsider and prioritize physical training, given its importance to crew health and mission performance.

As the mission progressed into its second week, the crew began to adapt to the constraints imposed by the space environment. They experimented with various positions on the ergometer, a stationary bicycle designed to maintain cardiovascular fitness in microgravity. Initially, they used a harness to secure themselves, but this method proved cumbersome. The astronauts discovered they could stabilize themselves more effectively by locking their feet into the pedals and bracing their hands against the ceiling or handlebars. Astronaut Ed Weitz described this adaptation as a significant improvement, likening it to a revelation compared to the cumbersome harness.

During these adjustments, astronaut Pete Conrad demonstrated an unconventional technique called "arm ergometry," where he pedaled with his hands while his feet were braced against the ceiling. This innovative approach allowed the crew to return to preflight levels of exercise once the workshop had cooled, addressing the initial issues caused by excessive heat.

However, not all aspects of the mission proceeded smoothly. On June 4, the crew shifted focus to freeing the solar array, confident that they had resolved the earlier ergometer issues. Yet, concerns arose regarding the health data from the M171 experiment. The initial results indicated abnormally high pulse rates, and Conrad experienced various heart palpitations. The medical team at Houston was alarmed by these findings, though they had not communicated these concerns to the press or the crew due to delays in data transmission.

The doctors attributed the elevated pulse rates to the heat and harness during the initial exercise sessions. However, there was some speculation about the potential early effects of weightlessness. With extravehicular activity planned, the medical team was particularly concerned about Conrad's

health. They recommended reducing his maximum workload and advising him to avoid strenuous activities if further palpitations occurred.

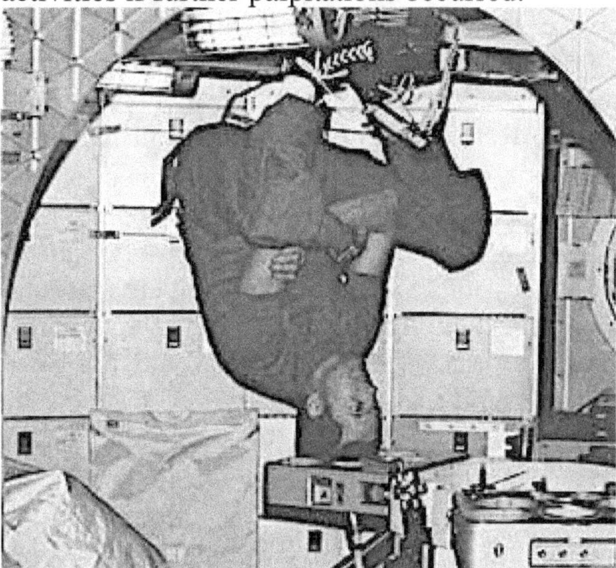

The weightlessness of space transformed Skylab's crewmen into skilled acrobats. They found that, within the roomy workshop, and without the impediment of gravity's pull, they could perform feats that even the most talented acrobats on Earth could not duplicate.

By June 5, Dr. Royce Hawkins addressed the situation in a press conference, describing Conrad's condition as a premature ventricular contraction. Although he acknowledged that this condition was common, the medical team remained vigilant, making special arrangements to monitor Conrad's health more closely during the M171 experiment.

During Skylab's second mission, the crew encountered significant challenges related to their physical regimen and mission scheduling, highlighting the difficulties of operating in space.

One critical issue emerged with the exercise routine, which established crucial steps for maintaining astronaut health in microgravity. Dr. Don Kerwin, concerned about the impact of the tight schedule on their physical training, found himself at odds with Houston Mission Control. On May 31, Kerwin expressed frustration that other scheduled tasks eroded the crew's allocated exercise periods. The rigorous timetable left little time for physical activity, and the crew often found exercise sessions scheduled inconveniently around major activities, diminishing their effectiveness. Kerwin hoped Mission Control would prioritize physical training to prevent adverse effects on the crew's health.

As the mission progressed, the astronauts experimented with various solutions to the exercise issue. During the second week, they modified their use of the ergometer, a stationary bicycle designed for space. Initially, the crew relied on a harness for stability, but this method proved cumbersome. They discovered a more effective technique by locking their feet into the pedals and using their hands to stabilize themselves against the ceiling or handlebars. Ed Weitz described this adjustment as a significant improvement, noting that it simplified the exercise process. Pete Conrad even developed a unique technique called "arm ergometry," where he pedaled with his hands while bracing his feet against the ceiling. This innovation allowed the crew to resume their preflight exercise levels once the workshop cooled.

Despite these adjustments, issues persisted. On June 4, the crew shifted their focus to freeing the solar array, confident that they had resolved the ergometer issues. However, concerns arose regarding the health data from the M171 experiment. The initial results indicated unusually high pulse rates, and Conrad experienced heart palpitations. The medical team at Houston, troubled by these findings, had not yet informed the crew or the press due to delays in data transmission. They initially attributed the abnormal pulse rates to the heat and harness used during exercise but were also considering the possibility of early effects of weightlessness.

The medical team's anxiety heightened as Conrad was scheduled for extravehicular activity.

That evening, the crew's physician, Charles Ross, informed Conrad about the medical concerns and advised that Houston made special arrangements for the M171 experiment the following day. Ross recommended that Conrad reduce his maximum workload and avoid strenuous activities if further palpitations occurred. The medical team's recommendations underscored the need for careful monitoring and adjustment to ensure crew safety.

On June 5, during a press conference, Dr. Royce Hawkins described Conrad's condition as a premature ventricular contraction but assured that it was not uncommon. By then, Mission Control had revised its approach based on the crew's feedback and adjusted the exercise protocols. The revised plan allowed the crew to exercise at their discretion, and Conrad completed the M171 protocol without further issues. Conrad later described the incident as a pivotal moment in the mission, emphasizing the critical role of physical exercise in space.

Throughout the mission, the crew struggled with a demanding schedule that included activating the hot workshop and adjusting to space operations. They often found that tasks took longer in microgravity than anticipated. Despite some initial difficulties, such as handling small items or locating equipment in stowage, the crew adapted quickly. By June 1, a holiday provided a welcome break, allowing the astronauts to catch up on housekeeping tasks and relax.

Conrad's feedback to Mission Control highlighted the need for better planning and time management. He suggested allocating more time for housekeeping and individual experiments, having one crew member perform complete procedures, and reducing the time lost between experiments. These suggestions aimed to streamline operations and improve overall efficiency. As a result of these experiences, subsequent missions benefited from increased physical activity schedules and improved operational planning, leading to more effective and healthier missions.

During the Skylab Space Station's early missions, the astronauts performed remarkable feats that captured the public's imagination. In a captivating 15-minute telecast, the crew showcased their ability to adapt to the unique conditions of space. Among the highlights was a demonstration of the "Skylab 500," an acrobatic routine that illustrated the effects of centrifugal force in microgravity. Astronaut Charles Conrad had bet friends in Houston before the mission that he could use centrifugal force to counteract weightlessness and walk upright on the storage lockers mounted around the upper deck of the workshop. The crew began performing on their hands and knees, gradually accelerating until they could walk rapidly along the lockers. This impressive feat, captured live on television, vividly demonstrated the astronauts' adaptability to the zero-gravity environment.

The mission's second week was marked by significant technical challenges and adjustments. Much of this period was devoted to addressing issues with the solar wing, which had failed to deploy properly. After resolving these problems, the crew resumed their schedule of scientific experiments on the 9th day, following a day of maintenance and relaxation. The first two weeks of high-intensity activities were followed by a more routine phase, where the crew adjusted to their new environment. Astronauts, including Kerwin, often found themselves ahead of schedule, a situation that seemed mundane compared to the earlier excitement. As Kerwin noted, there were moments when "it seemed like it had been day 18 for a week."

As the astronauts adapted to the station's conditions, they chose not to request additional tasks, understanding that the upcoming preparations for their return to Earth would occupy much of their final week. This decision was made to avoid setting unrealistic expectations for subsequent crews.

The initial three weeks in space presented considerable stress for the principal investigators. They faced the daunting prospect of losing years of painstaking research due to the station's power shortages. The absence of sufficient electricity exacerbated their anxiety, making each day a critical opportunity to gather valuable data. John Disher, one of the principal investigators, recalled the sense of urgency and fear that the mission might be cut short. The quality of the science being conducted was a mixed blessing; while the initial data was promising, the investigators felt an intense need for more information, a sentiment echoed by Robert Parker, the program scientist, who noted that "they felt starved for it."

Many scientists struggled with the perception that their experiments were being unfairly prioritized. Some believed their specific experiments were being neglected, leading to frustration and a lack of trust in the newly appointed scheduling coordinator, Parker. Communication issues and the mangling of their requirements compounded this frustration, leading to feelings of paranoia and dissatisfaction among the team.

When electrical power was finally restored, it did little to alleviate the pressure of the mission's tight schedule. The investigators were eager to compensate for the lost time, particularly in deploying the parasol and solar array, which had consumed about 15% of the science time originally allotted. At a news briefing on June 8, Parker likened the challenge to fitting a size 10 foot into a size eight shoe—something had to give. Medical experiments retained their top priority, with the crew increasing the frequency of cardiovascular and metabolic tests in the final weeks. Despite the challenges, the crew completed nearly 100% of the medical requirements and 80% of the solar observations. However, the earth-resources experiments, most affected by the initial power shortage, only reached 60% of their planned goals.

NASA officials highlighted the substantial amount of work achieved rather than the shortfall. The first crew managed to take 29,000 pictures of the sun and collect 14 kilometers of magnetic tape for earth resources.

While many scientists were generally pleased with the results during Skylab's first mission, some expressed significant dissatisfaction. According to Robert Parker, a program scientist, a few investigators felt they had been shortchanged, having invested considerable time and NASA's resources with minimal return in data. This sentiment was echoed in the belief that the mission had not yielded enough data, prompting a demand for more comprehensive results in future missions. To address these concerns, Parker organized periodic planning sessions during subsequent missions. These meetings helped investigators understand each other's challenges and alleviated some of the discontent. However, the most effective remedy was the continuous data flow from Skylab itself.

The console for the solar telescopes was notably complex, with numerous switches, monitors, and checklists. One astronaut, reflecting on the experience, remarked that it was almost inevitable to make mistakes, though the goal was to avoid significant errors. The crew, particularly on the first mission, faced frequent interruptions that hindered their ability to establish a consistent routine with the Atmospheric Telescope Module (ATM). This constant disruption contributed to various mistakes, with many attributed to the challenge of adapting to the console's intricacies.

A major issue with the ATM console was its flare detection system, designed to alert the crew of solar flares while they were away from the console. This system often malfunctioned, especially when the workshop passed over the eastern part of South America, where the Earth's radiation belt dips lower. The magnetic field in this region triggered frequent false alarms, leading to interruptions as crew members had to turn off the system repeatedly. Kerwin, in particular, experienced frustration with this detector, often finding it more of a hindrance than a help. An early mishap occurred on May 30, when Kerwin, reacting to an alarm, began recording a flare without realizing it was due to the South Atlantic anomaly. Although he quickly corrected his error, the incident did little to endear him to the flare detection system, which he later deemed "absolutely worthless."

Despite these challenges, the crew's persistence paid off on June 15 when they successfully observed a significant solar flare. The astronauts had sacrificed their free day to catch up on experiments despite a pessimistic forecast for solar activity. Houston reported a few subnormal flares, and Kerwin humorously requested "supernormal flares." Hours later, a sizable flare was detected. Kerwin proudly informed Houston, "I'd like you to be the first to know that the pilot [Paul Weitz] was the proud father of a genuine flare."

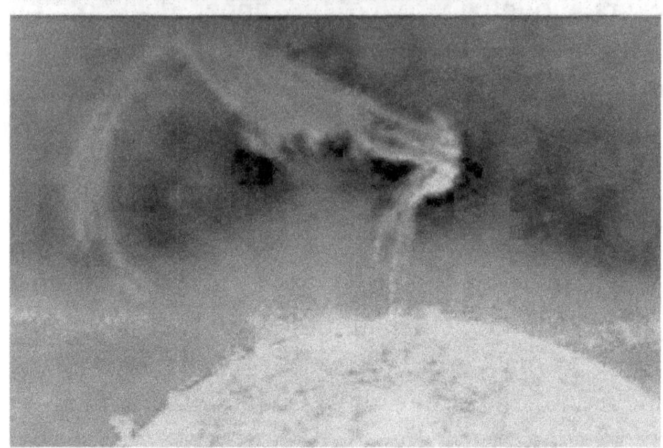

Typical of the photographs obtained by Skylab's solar instruments were these photographs of a solar eruption (top) and a solar prominence (bottom). (Naval Research Laboratory).

Weitz's efforts in tracking the flare through its rise and fall, involving multiple displays and flare programs, were highly commended by solar scientists.

Overall, the crew's performance was highly praised in postflight evaluations. Most of the experimental hardware functioned satisfactorily, and while some disappointment remained regarding the ATM work, the principal investigators 'perspectives changed positively with the arrival of the valuable data. This shift in attitude highlighted the scientific value of Skylab, which had previously been questioned by some astronomers who doubted the necessity and cost of crewed solar telescopes in space.

During the Skylab missions, the crew faced various challenges and discoveries that painted a complex picture of life aboard the space station. One of the primary concerns was the performance of the multi-spectral scanner, an essential Earth-resource instrument. The scanner's focus adjustments often disrupted its alignment, causing significant operational issues. Astronaut Weitz noted that these adjustments sometimes resulted in the scanner losing alignment completely, a problem that led to three out of twelve data passes being of marginal quality.

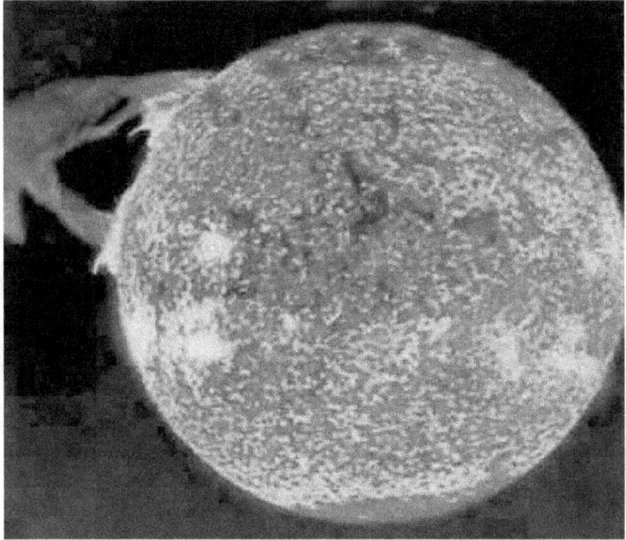

Skylab's cameras recorded thousands of observations of the Sun. Such detailed observations cannot be made from Earth because of atmospheric interference. (Naval Research Laboratory)

Calibration of the body-mass measuring device also proved troublesome. This device, crucial for tracking the astronauts' weight by measuring the oscillatory frequency of a spring-mounted chair, required frequent recalibration. The calibration process involved using objects of known mass, such as heavy batteries, which were challenging to keep in place. Despite these difficulties, the hardware's overall performance on Skylab exceeded the crew's expectations. The support team received numerous compliments for their efforts, particularly for the high-fidelity training simulators. Weitz praised the quality of the trainers, and Kerwin appreciated that training personnel had incorporated potential failures and challenging scenarios into the training.

Most operational checklists proved effective, though some issues arose with inflight changes. One notable problem was cataloging teleprinter messages for later reference. Conrad suggested that future crews use notebooks to record these messages permanently. Additionally, Channel B communications, used for secondary transmission, experienced delays that sometimes caused responses to queries to be lost for days.

The astronauts' living conditions on Skylab were generally satisfactory, but the food left much to be desired. The unique space environment affected their taste buds, making previously enjoyable foods taste markedly different. Bread well-received on Earth was described as much less appealing in space. The astronauts 'food preferences remained consistent, though their taste for certain items changed. Many crew members, including Weitz, reported an increased craving for spicier foods, likely due to a reduced sense of taste and smell, exacerbated by head congestion from blood pooling in the upper body.

Among the food options, German potato salad, with its strong vinegar and onion flavor, was notably popular, with the crew consuming all four cans on board. Frozen foods were generally preferred over reconstituted items, which were often found less satisfactory. Conrad advised that future missions include spicier foods and allow more time for reconstituting dehydrated items. He found that pre-reconstituted foods, when reheated, were easier to consume than those that were reconstituted just before eating.

A rigid menu with limited variety constrained the astronauts 'diet. Kerwin noted that foods selected for the mission became less palatable over time due to their repetitive nature. He recommended implementing a more flexible food plan, similar to one initially considered for Skylab, which would allow for interchangeable food items within defined categories.

Despite these challenges, the crew thoroughly enjoyed their Skylab experience. The unique environment offered opportunities for activities and experiences unavailable on Earth, making the mission successful. The medical experiments conducted in space continued to provide valuable insights, extending the mission's contributions beyond the immediate realm of spaceflight.

The Skylab missions were marked by significant achievements and challenges, with the second mission particularly notable for its dramatic developments. Despite facing various personal and mechanical issues, the mission was a considerable success, with both the ground and flight crews demonstrating remarkable perseverance.

One of the mission's defining moments came during the last extravehicular activity. The primary objective of this spacewalk was to replace the Advanced Technology Module (ATM) film. A secondary goal, however, involved reactivating a malfunctioning power module in the station's electrical system. Engineers at the Huntsville control center had diagnosed the problem as a stuck relay and suggested that a forceful blow to the battery housing might free it. Following precise instructions relayed via teleprinter, Commander Conrad used a hammer to strike the housing. The power module was reactivated within minutes, restoring electricity and resolving a critical issue.

Astronaut Weitz on solar flare watch. All of the solar experiments could be controlled or monitored from this center.

Chapter 5 - Skylab 3

Second Crewed Mission

The second mission, launched on July 28, 1973, encountered its set of challenges. Early in the mission, failures of batteries and gyroscopes raised serious doubts about Skylab's ability to function without a crew for an extended period. To address these concerns, the launch of the second crew was expedited by three weeks. Although the launch proceeded smoothly, the crew faced significant problems once in space.

June 1973) -The three members of the prime crew of the second manned Skylab mission (Skylab 3) left to right, astronaut Alan L. Bean, commander; scientist-astronaut Owen K. Garriott, science pilot; and astronaut Jack R. Lousma, pilot. Photo credit: NASA

All three astronauts experienced motion sickness so much that they fell behind schedule. Mechanical difficulties also threatened to cut the mission short, but these were eventually resolved.

The second crew, which arrived aboard Skylab in 1973, was led by Alan L. Bean, a U.S. Navy captain and astronaut since 1963. Bean had previously served as the lunar module pilot for Apollo 12 in 1969. He was joined by Pilot Jack R. Lousma, a major in the U.S. Marine Corps and an astronaut since 1966, and Scientist Pilot Owen K. Garriott, a civilian with a Ph.D. in electrical engineering who had been selected as an astronaut in 1965. Lousma and Garriott had yet to fly in space before their Skylab mission.

(28 July 1973) - A close-up view of the Skylab Space Station photographed against an Earth background from the Skylab 3 Command and Service Modules (CSM) during station-keeping maneuvers before docking. Aboard the Command Module (CM) were astronauts Alan L. Bean, Owen K. Garriott and Jack R. Lousma, who remained with the Skylab Space Station in Earth orbit for 59 days. This picture was taken with a hand-held 70mm Hasselblad camera using a 100mm lens and SO-368 medium speed Ektachrome film. Photo credit: NASA

The issue of motion sickness was a recurring challenge for astronauts. While the first Skylab crew had been fortunate enough to avoid this problem, subsequent missions revealed that the larger spacecraft and increased freedom of movement could contribute to motion sickness. This condition had affected nearly half of the Soviet cosmonauts in the Vostok and Voskhod missions. It was also seen in the Apollo program, where nine out of 29 astronauts experienced nausea and vomiting. The second Skylab crew was not exempt from this issue. Less than an hour after launch, pilot Jack Lousma reported nausea, underscoring Conrad's earlier caution about the potential for motion disturbances in space.

Prime crewmen for the Skylab 3 mission. Pictured in the One-G trainer Multiple Docking Adapter (MDA) at JSC are, left to right, Scientist-Astronaut Owen F. Garriott, science pilot; and Astronauts Jack R. Lousma and Alan L. Bean, pilot and commander, respectively.

On July 29, 1973, the crew of Skylab's second mission faced a particularly challenging day. Pilot Jack Lousma, who had previously experienced nausea, found some relief from a medication containing scopolamine and dextroamphetamine, which helped block nerve endings in the stomach. This allowed him to eat lunch, but the respite was short-lived. As the crew began activating the workshop that afternoon, their symptoms worsened, and by 6:00 p.m., all three astronauts were grappling with severe motion sickness, with Lousma suffering the most.

The following day brought little improvement. Breakfast went largely uneaten, and by 8:30 a.m., Bean reported that their efficiency was compromised while the crew was moving around and working. The astronauts requested a break to rest and recover, and Houston granted a midafternoon rest period. However, this break was largely consumed by efforts to address an electrical problem in the spacecraft. By the evening, the crew had fallen nearly a full day behind schedule, prompting NASA officials to postpone a planned extravehicular activity (EVA) for at least one day.

Seated at the control and display console in the docking adapter, the astronaut manually controlled many important operations and monitored scientific observations. In this photograph, Jack Lousma of the second crew observes the Sun, watching for the appearance of solar flares. The astronaut's patience was rewarded late in the mission.

Back on Earth, the crew's condition sparked a debate over the most effective treatment for motion sickness. Dr. Ashton Graybiel, the principal investigator for experiment M131, suggested that rapid head movements could help astronauts adjust more quickly to weightlessness. Graybiel proposed that the crew perform head movements—30 to 40 per minute for 10 minutes, three times a day—as a remedy, arguing that resting alone would not alleviate their symptoms. Although Dr. Berry supported Graybiel's approach, many Houston officials, particularly those outside the medical office, were skeptical.

Despite the controversy surrounding their treatment, Houston instructed the Skylab crew to persist with their activation tasks at their pace while also attempting the recommended head movements. The astronauts, however, approached these exercises with reluctance, as they found that movement exacerbated their nausea. By July 30, astronaut Garriott reported that although the crew was still struggling, their persistence in managing their condition and fulfilling their duties highlighted their resilience in the face of adversity.

Skylab's floors were made up of triangular grids. Triangular shoe cleats fitted into the grid cavities. With a twist of his foot, a crewman could position himself wherever he chose. A number of other types of restraints were also provided.

Unsecured objects floated about the workshop and were collected on the screens, such as this one being vacuumed by Jack Lousma. Crewmen began their search for missing items there.

The crew performed the head movement exercises twice, with Bean completing them once, while Lousma avoided them. Although the worst of the illness subsided by the third day, the crew faced significant delays due to various activation problems. Bean attributed much of their difficulty to an influx of unscheduled tasks. He noted, "We seem to have about as many new chores as old ones. We're having difficulty progressing because we're doing other work." The crew spent considerable time troubleshooting the workshop's dehumidifier, repairing the urine separator, and searching for misplaced personal items. Bean remarked on the constant challenge of finding necessary equipment, such as shaver heads, which disrupted their routine.

Reflecting on their experience, Bean attributed much of their initial sickness to the hectic pace of the first week. He observed that the activation demands were overwhelming and recommended prioritizing meals and rest over activation requirements for future missions. He suggested that, if necessary, taking an extra day or two for activation would be beneficial to ensure that astronauts could maintain a regular eating and sleeping schedule.

By August 1, the crew's condition had significantly improved. A telecast to Mission Control showed them in high spirits during lunch, with Bean demonstrating his newfound proficiency at eating while hanging upside down and Lousma noting that "the food tasted a lot better." The crew now followed a new regimen of six smaller meals a day, a change recommended by Houston doctors after it became clear that full stomachs had contributed to their earlier discomfort. That afternoon, Lousma experienced only mild dizziness while experimenting with M131, and Garriott completed the first run on the ergometer and lower-body negative-pressure device.

By the evening of August 2, the medical office had cleared the Skylab crew for an extravehicular activity (EVA) scheduled for August 4. Despite ongoing concerns about motion sickness, which had previously affected individual astronauts but never an entire crew, the team demonstrated remarkable adaptability and resilience. Their ability to manage these unexpected challenges was crucial, as motion sickness posed a significant concern for future missions. The astronauts' response to the illness did not align with past performance; notably, Commander Alan Bean had previously fl to the Moon without exhibiting any symptoms. Science Pilot Jack Lousma had displayed considerable resistance to motion sickness during ground tests. The exact cause of the illness remained uncertain, but the potential consequences were evident. George Low, then Associate Administrator for the Office of Manned Space Flight, expressed his concern: "Were we to lose three or four days out of each seven-day Space Shuttle flight because of motion sickness, the entire Shuttle effort would be in jeopardy." Consequently, preventing motion sickness became a top priority following the second mission.

The crew's first EVA faced further delays on August 2 due to a malfunctioning steering rocket, which threatened the entire mission at one point. The Apollo spacecraft's reaction control system, essential for stabilizing the spacecraft in orbit or altering its velocity, comprised four sets of rockets, each equipped with four thrusters. These thrusters could also facilitate a return to Earth in the event of a main engine failure. The malfunction was unexpected, given the reliable history of these reaction control rockets.

The issue arose when quad B developed a leak on launch day. Skylab procedures allowed spacecraft operation with one quad out of commission, but the situation worsened when, six days later, temperatures in quad D dropped below normal limits. This triggered a master alarm, alerting Mission Control and awakening the crew. Initially, the malfunction seemed minor, and the problem was not immediately linked to the earlier leak. The astronauts activated the reaction control system heaters and continued their duties. However, within the next hour, Mission Control received alarming reports of a second leak. The temperature and pressure in quad D plummeted, and the crew observed a stream of sparks outside their window, similar to what they had seen previously.

Engineers at Johnson Space Center (JSC) feared the worst—that the two leaks indicated a broader issue with the oxidizer system, potentially due to contamination of the nitrogen tetroxide. Such a problem could lead to the failure of the remaining rockets and potentially damage electrical circuits within the service module. Although quad D had lost less than 10% of its oxidizer, the possibility of the leak expanding posed a severe risk. While the astronauts could maneuver the spacecraft with two or even one functioning quad, this situation was best avoided.

By mid-morning, the press was briefed on the severity of the situation, and Skylab's rescue capability, introduced three years earlier, was suddenly perceived as a crucial asset. The spacecraft manager, Glynn Lunney remarked that without this rescue capability, the team would have been scrambling to bring the spacecraft as quickly as possible. The news electrified Kennedy Space Center, and within three hours, preparations for a potential rescue mission were underway. The team aimed to ensure a swift and effective response to the unfolding crisis by expediting the spacecraft's mated assembly with its Saturn launch vehicle.

Astronaut Jack R. Lousma, Skylab 3 pilot, participates in the August 6, 1973, extravehicular activity (EVA) during which he and Astronauts Owen K. Garriott, science pilot, deployed the twin pole solar shield to help shade the Orbital Workshop (OWS). Note the reflection of the Apollo Telescope Mount and the Earth in Lousma's helmet visor.

Rescue Mission Considered

The Skylab mission had always included contingency plans for rescuing the crew if the Apollo module lost its capability to return safely to Earth. With the loss of two of the four thruster assemblies, the possibility of completing the mission became uncertain, making rescue operations a significant consideration.

CSM 119 on display at the Apollo/Saturn V Center

The designated rescue vehicle was a modified Apollo command and service module. This adaptation involved removing the storage lockers from the aft bulkhead and installing two additional couches. The modifications also included enhanced life support and communications umbilicals to accommodate up to five crew members. The plan was for two crew members to occupy the rescue vehicle during its ascent into orbit. Upon docking with Skylab, these two astronauts would join the Skylab crew, who would then return to Earth aboard the rescue vehicle, leaving the center couch vacant.

AS 209 on display at the Kennedy Space Center Visitor Complex

Following a thorough examination of records at Kennedy Space Center, the possibility of contaminated nitrogen tetroxide was ruled out. Johnson Space Center (JSC) officials were reassured that the remaining reaction control quads were still functional. If they were not, simulations indicated that the spacecraft could safely return to Earth with the remaining operational quads.

With this in mind, Christopher Kraft, Director of Flight Operations, informed the crew that their extravehicular activity (EVA) would be delayed again. The delay was necessary to allow Mission Control to develop procedures for a safe reentry with only two operational quads. Kraft emphasized that while rescue operations were being undertaken as a precaution, the mission was to proceed as if everything were proceeding normally.

The situation with the leaking thrusters highlighted the Skylab operation's strengths and vulnerabilities. An investigation revealed that the failure in quad D was due to loose fittings in the oxidizer lines, a problem that had gone unnoticed during two years of rigorous testing. Although NASA officials were initially uncertain whether the crew could deorbit the spacecraft with only one or two functioning quads, Skylab's rescue capability alleviated the immediate pressure to decide. Within a few hours, the spacecraft's condition was accurately assessed, and the mission continued with renewed focus and optimism.

Deploying The Twin-Pole Sunshade

The Skylab astronauts faced two crucial tasks during their mission: replacing the film in the Apollo Telescope Mount (ATM) before conducting any further solar observations and deploying Marshall's twin-pole sunshade. The latter was particularly pressing, as the original parasol shade was at risk of disintegration under the harsh ultraviolet radiation of space. The timing of this replacement had been a matter of debate among mission planners. Bill Schneider and Rocco Petrone had advocated for deploying the new sunshade before the first crew departed the workshop. However, Chris Kraft, concerned about the strain of another major extravehicular activity (EVA) on Pete Conrad's crew, opted to delay the deployment until the second mission. Ultimately, medical considerations prevailed, and the decision was made to wait.

Marshall's twin-pole sunshade design was selected over an improved parasol, as it could be deployed without uncovering the workshop, even for a moment. This was a crucial factor, as exposing the workshop to the unfiltered sunlight could have had disastrous consequences. The engineers at Marshall were confident in their deployment procedure. Unlike the previous EVA, where Dr. Joe Kerwin had to free a jammed solar array while in a precarious position, this operation would allow the astronauts to work with firm footing, reducing the risks involved.

The deployment plan was meticulously detailed. Astronaut Owen Garriott would begin the operation, positioning himself at the workstation just outside the airlock hatch. His task was to connect the eleven sections of the pole to support the sunshade. Meanwhile, Jack Lousma, stationed at the center of the mount, would secure the foot restraints and attach the shade's base plate to the ATM truss. Once the two 17.5-meter poles were fully assembled, Lousma would fasten them to the base plate, forming a V-shaped structure. The sail would then be attached to a rope running the length of the poles, and Lousma would carefully hoist the shade into place. Astronaut Alan Bean, positioned inside the docking adapter, would monitor the entire operation.

Astronaut Jack Lousma begins the extravehicular activity to erect the twin-pole shield to shade the workshop.

The crew had undergone extensive preparation for this task. With more than 100 hours of EVA training under their belts, including the sail deployment in Huntsville's underwater training tank, they were well-prepared for the challenges ahead. Given the complexity of the operation, ample time was allocated, as a single mistake during EVA could lead to catastrophic consequences.

Preparations began on July 5th, with the crew thoroughly reviewing procedures and inventorying the necessary hardware. The morning of July 6th was spent donning their cumbersome spacesuits and testing support systems. Shortly after noon, they depressurized the airlock and opened the hatch, stepping out into the vacuum of space to begin their work.

The operation did not proceed as smoothly as anticipated. The first setback occurred when a rubber grommet, designed to fit over the locking nut of each pole section, caught on the storage rack. This seemingly minor issue led to a significant delay, as it took nearly 20 minutes to remove and connect the first three sections of the pole—a pace that, if continued, would have extended the deployment well beyond the scheduled time. However, Garriott adjusted his position, allowing him to remove the rods from a different angle, speeding up the process.

This photograph, taken in the neutral buoyancy simulator, shows the base plate fabricated to hold the twin poles in place. Also shown is the bag holding the fabric sail and the lines by which it was drawn into place. Deployment procedures were carefully worked out in the simulator.

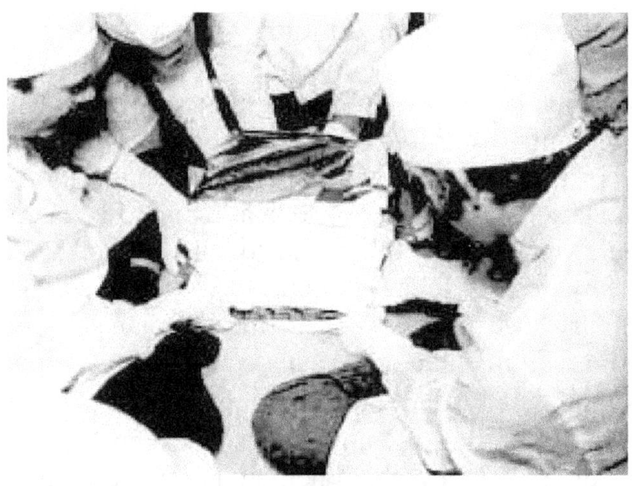

Highly experienced Navy Seal Team members, professional parachute riggers, used a unique accordion fold to pack the twin-pole shield before it was stowed in the command module.

Concern over the possibility that materials used for the parasol would deteriorate with prolonged exposure to the Sun's rays and that more complete shielding was required to control workshop temperatures better prompted the installation of a second Sun shield during the second manned mission. This time the crew exited the space station and installed a twin-pole device to position the shield over the parasol.

As with many EVAs, the astronauts encountered additional challenges. They spent valuable time untwisting a rope that had become tangled, eventually discovering they could separate the pole sections, pass the line through, and rejoin the poles. Lousma faced further difficulties when hoisting the shade along the poles; the material initially refused to unfurl properly. Yet, with the sun's help, the folds in the fabric gradually straightened, and the sail opened successfully.

Though fraught with minor frustrations, the deployment ultimately took nearly four hours. Despite the challenges, the astronauts remained in good spirits, seemingly enjoying the task. They concluded the EVA by exchanging the ATM film, retrieving experiment samples, and inspecting the workshop for evidence of several malfunctions, including issues with the Apollo quads. When Garriott and Lousma finally reentered the airlock, they had spent an impressive six and a half hours outside, marking the longest spacewalk up to that time.

The twin-pole shield was successfully erected over the parasol. A rectangular fabric shield, which was thermally treated, it remained in place for the remainder of the mission.

The effects of their labor were immediate. The twin-pole sunshade deployed over the parasol, and the temperature inside the workshop dropped noticeably, a testament to the effectiveness of the new shield. Garriott, standing on the telescope mount, had also successfully attached Experiment S174, further contributing to the mission's scientific objectives. The crew's efforts not only ensured the continued functionality of Skylab but also set a new standard for what could be achieved during an EVA, demonstrating the resilience and adaptability of both the astronauts and the technology supporting them in the unforgiving space environment.

The experiment was designed to collect dust particles in space and study their impact phenomena, addressing immediate scientific needs. However, its uneven deployment had resulted in certain hot spots within Skylab. During periods of maximum sunlight, particularly in the last week of Conrad's crew's mission, temperatures inside the workshop reached as high as 28°C. While this temperature was tolerable during the workday, it became uncomfortable for sleeping. Installing the second sunshade brought the internal temperatures closer to the levels intended initially by the thermal engineers, ensuring a more habitable environment for the astronauts.

More importantly, the successful deployment of the second shade significantly bolstered confidence in extravehicular activities (EVA). It demonstrated that with thorough preparation, astronauts could accomplish a wide array of tasks in space, overcoming the inherent challenges of working in a microgravity environment. This achievement laid a strong foundation for future missions, where EVA would become a routine and essential component of space operations.

Solar Viewing

With the thermal situation under control, the crew wasted little time focusing on solar observations. On August 7th, astronaut Owen Garriott began a three-hour observation of the sun's outer atmosphere, known as the corona. Although the sun showed no significant activity during this period, Garriott engaged the ground team with a steady stream of questions, eager to maximize the scientific output of the session.

On August 9th, the sun became considerably more active two days later. Garriott captured images of a medium-sized solar flare, marking the first significant solar event of the mission. The following day, astronomers at the Canary Island Observatory detected an even more significant solar event. This information was quickly relayed to the Skylab crew amid a scheduled half-day of rest. Despite being behind schedule, Garriott and Alan Bean quickly crewed the solar telescopes to capture the event. Over the next hour, they filmed a massive eruption of solar radiation. Dr. Ernest Hindler of the High Altitude Observatory later described the coronal transient as "a magnificent specimen of this type," noting that such an event might only occur two or three times a year.

In the following ten days, solar observations intensified, with the crew dedicating up to 14 man-hours on August 20th alone to this crucial task. The hydrogen-alpha (H-alpha) telescopes were the primary instruments used to locate solar activity and recognize the early stages of solar flares. Skylab's x-ray and ultraviolet instruments were aligned with these H-alpha telescopes, enabling a coordinated approach. When an astronaut aligned the crosshairs of the H-alpha monitor on a specific solar activity, the other instruments automatically focused on the same target. The H-alpha telescopes provided photographs and television images, with a zoom capability allowing the crew to vary the field of view as needed.

A second monitor on the Apollo Telescope Mount (ATM) panel displayed images from the extreme ultraviolet spectroheliograph. In these wavelengths—approximately 20 times shorter than those visible to the unaided eye—the sun appeared blotchy, with many bright points indicating active regions. This allowed a detailed examination of solar activity that was impossible from Earth.

The white-light coronagraph, developed at the High Altitude Observatory in Colorado, was another critical tool for studying the solar corona. This instrument featured four coaxial disks at the front of the telescope that blocked out the sun's bright light, allowing only the faint corona to be observed. While the wavelengths captured by the coronagraph were visible to the naked eye, the instrument provided a view rarely seen on Earth, effectively giving the crew the experience of a solar eclipse every hour of the day. The images from the coronagraph were recorded on 35-mm film and could be displayed on a console monitor or transmitted to Earth via television.

The crew's work with the solar telescopes not only advanced our understanding of the sun but also demonstrated Skylab's capabilities as a platform for solar research. Their observations, captured with state-of-the-art instruments, offered scientists on Earth unprecedented insights into solar phenomena, reinforcing the importance of

Skylab's mission in the broader context of space exploration.

During Skylab's mission, the astronauts engaged in a profound two-week period of solar observation, culminating in a remarkable discovery on the 21st of the month. The crew's attention was drawn to a massive solar prominence on the sun's eastern edge, first identified by an astronomer working with the National Oceanic and Atmospheric Administration (NOAA) at their site in the Canary Islands. NASA was promptly alerted, and investigators at the Johnson Space Center (JSC) swiftly prepared an observing program tailored for the Skylab crew.

Astronaut Alan Bean was the first to visually identify the solar prominence, describing it as a "big bubble" on the sun's edge. Over the next several hours, solar scientists eagerly watched as the prominence, nearly three-quarters the size of the sun itself, arched outward through its corona, forming a colossal loop structure. This event was the most significant solar phenomenon observed since Skylab's launch. The crew's judicious use of limited film in the white light coronagraph earned them considerable praise, particularly for Bean's role in capturing this extraordinary event.

The success of this solar observation highlighted the contrast between Skylab's first and second missions. At a press conference on August 10th, NASA officials acknowledged the marked improvement in operations. They attributed this progress to technical adjustments and the more open communication established by the second crew, led by Commander Pete Conrad. The crew's constructive feedback had facilitated better collaboration with the scientists on the ground, resulting in a more seamless exchange of information. Unlike the previous crew, the second crew engaged more directly with investigators, often through the capsule communicator, enhancing the rapport between the astronauts and the scientific community. Owen Garriott, in particular, was noted for his inquisitive nature, frequently asking questions that led to a deeper exchange of ideas and knowledge.

In the days following a critical extravehicular activity (EVA), the Skylab crew was heavily involved in earth-resource observations. Flight planners had strategically scheduled 26 earth-resource passes during the 58-day mission, concentrating them at the beginning and end of the mission due to poor lighting conditions in the northern hemisphere during the middle three weeks. Typically, Houston scheduled one earth-resource pass daily, each lasting approximately 35 minutes. However, an additional two hours were often required for the crew to adjust camera settings, replace film, and load maneuver parameters into Skylab's computer. As the mission progressed, the crew became more efficient, eventually halving the preparation time for these tasks.

Earth-resource observations were a collaborative effort, unlike solar observations, often carried out by a single astronaut. Garriott usually operated the S190B earth-terrain camera through Skylab's anti-solar scientific airlock, opposite the parasol. Meanwhile, Alan Bean and Jack Lousma alternated between handling the viewfinder tracking system for the S191 spectrometer and managing the controls at the main display console.

The crew conducted nine earth-resource passes before August 13th, with varying degrees of success. While the sensors performed well, heavy cloud cover often impeded site verification on several runs. One of the most typical and notable earth-resource passes occurred on August 8th. The crew initiated this pass off the coast of Oregon, operating the earth-resource cameras for 35 minutes as they documented a 13,500-kilometer stretch of land and sea, extending from the Pacific Northwest down to a point south of São Paulo, Brazil. This meticulous documentation contributed valuable data to earth-resource studies despite the challenges posed by weather conditions.

These detailed observations and the crew's ability to adapt to the evolving demands of the mission underscored Skylab's significant contributions to solar science and earth-resource management. The success of these missions advanced our understanding of the sun's behavior and its impact on Earth and set a precedent for future space-based observations and collaborations between astronauts and scientists.

As Skylab continued its mission, the crew embarked on various earth-resource observations with objectives that spanned diverse geographical regions. The data collection efforts focused on critical areas such as soil moisture levels in Oklahoma, mineral formations in Utah, the rapid urban expansion of Houston, and the vast natural resources of the Amazon rainforest. These observations were intended to provide valuable insights into these regions' environmental and geological conditions.

The Earth resources experiment program used a variety of scientific equipment to record data concerning Earth. These instruments, located in the docking adapter, surveyed Earth features and obtained photographs which provided valuable data to foresters, geologists, oceanographers, meteorologists, and others.

During a briefing on August 15th, Richard Wilmarth, the coordinator of the earth-resource program, expressed satisfaction with the volume of data collected thus far. However, newsmen at the briefing raised concerns about the three-week interruption in operations and the limited number of observation sites in the southern hemisphere. In response, Wilmarth indicated that NASA considered conducting additional observation runs to address these gaps.

The pause in earth-resource activities allowed the Skylab crew to engage in some of the 22 corollary experiments, which encompassed a wide range of scientific disciplines, including space technology, space physics, and stellar astronomy. These experiments, collectively called corollary experiments, offered the crew a chance to contribute to a broader understanding of space and its phenomena.

Eight highly sensitive instruments, mounted on the spar in the solar observatory canister, studied the Sun in great detail.

Astronauts Alan Bean and Jack Lousma were primarily responsible for carrying out the corollary experiments, allowing Owen Garriott to remain focused on the Apollo Telescope Mount (ATM) console. On the morning of August 13th, Bean and Lousma tested the M509 maneuvering device, a large backpack that NASA was developing for use during extravehicular activities (EVA).

The M509 was designed to enhance the astronaut's mobility in space, but Bean found the device lacking in speed and overly precise in attitude control. He emphasized the need for a device that "flies like a spacecraft," ensuring that an astronaut's intuitive responses would be effective. In subsequent testing sessions, the two astronauts experimented with a hand-held gas pistol unit and a foot-controlled unit, both intended to improve maneuverability. However, Bean found these units unsatisfactory, citing their unnatural feel and the extensive training required to master them. Despite the challenges, the crew devoted over 75 man-hours to testing and photographing these units in action.

The remainder of the time allocated for corollary experiments was primarily dedicated to Karl Henize's stellar astronomy experiment (S019) and Dr. Donald Packer's airglow photography experiment (S063). Henize, a former Northwestern University professor and scientist-astronaut, utilized a reflecting telescope and prism with a 35-mm camera positioned in Skylab's anti-solar airlock. The experiment required precise operations: a crew member would extend a rotating mirror through the airlock, focus the telescope on the desired star field, and take two or three photographs with exposure times ranging from 30 to 270 seconds. Although the process typically took less than an hour, it necessitated careful scheduling during Skylab's night phase to ensure optimal results.

A wide variety of terrain features was recorded by Skylab's cameras: (1) The colorful plateaus of southeastern Utah, (2) lofty Mount Rainier in Washington, (3) the wind-blown dunes of the Spanish Sahara Desert, and (4) volcanoes of the Hawaiian Isles were all studied by Skylab.

Dr. Packer's airglow photography experiment, developed at the Naval Research Laboratory in Washington, D.C., involved capturing images of the Earth's ozone layers and the horizon's airglow. The photography was conducted from both the wardroom window and the scientific airlock, using reflected light as well as in total darkness. This experiment aimed to provide new insights into the composition and behavior of the Earth's atmosphere, particularly the ozone layers that play a critical role in protecting the planet from harmful ultraviolet radiation.

Astronaut Gerald P. Carr flies the Astronaut Maneuvering Equipment in the OWS Description: Astronaut Gerald P. Carr, Skylab 4 commander, flies the M509 Astronaut Maneuvering Equipment. Carr was strapped into the back-mounted, hand-controlled Automatically stabilized Maneuvering Unit (ASMU). The M509 exercise was in the forward dome area of the OWS. THe dome area was about 22 feet in diameter and 19 feet form top to bottom.

In late August, the schedule of observations was expanded to include the drought-stricken regions of Mali and Mauritania. The crew focused on capturing data that could assist in understanding and mitigating the impacts of severe droughts in these areas. This addition underscored Skylab's commitment to addressing pressing environmental challenges on Earth, even as the station's primary focus remained on advancing space exploration and scientific research.

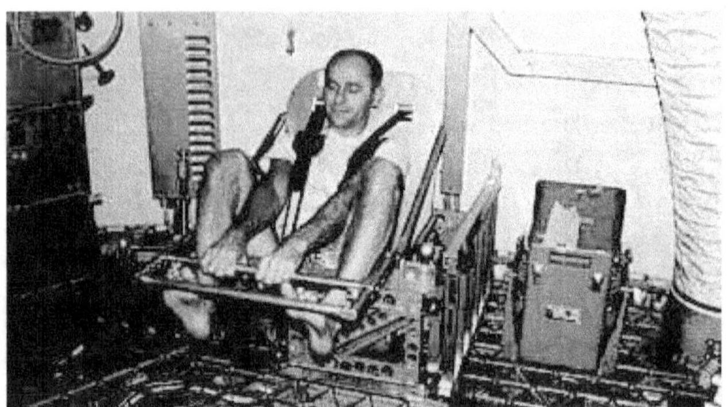

Here, Comdr. Alan Bean conducts tests to measure body mass in zero gravity.

An astronaut maneuvering unit, shown here, was flown in the workshop to test it under weightless conditions for possible future application.

Astronaut Alan L. Bean, Skylab 3 commander, flies the M509 Astronaut Maneuvering Equipment in the forward dome area of the Orbital Workshop (OWS) on the space station cluster in Earth orbit. Bean was strapped in to the back-mounted, hand-controlled Automatically Stabilized Maneuvering Unit (ASMU). This ASMU exerperiment was being done in shirt sleeves. The dome area where the experiment was conducted was about 22 feet in diameter and 19 feet from top to bottom.

Through these diverse activities, the Skylab mission demonstrated the versatility and value of space-based observations, contributing to our understanding of the Earth and our ability to operate effectively in space. The crew's efforts in conducting these experiments and observations advanced scientific knowledge and laid the groundwork for future space missions and technological developments.

By the second week of the mission, mechanical malfunctions had become an unfortunate and persistent challenge for NASA engineers. Skylab, the pioneering space station, seemed to be aging more rapidly than anticipated, with various issues that demanded constant attention. One of the most persistent problems was a leak in the dehumidifier, which, though not posing a significant danger, required daily maintenance. This nuisance became a regular part of the astronauts' routine, a reminder of the complex systems that needed to function flawlessly in the harsh space environment.

On August 20, astronaut Alan Bean dedicated an entire day to inspecting the dehumidifier system. He added nitrogen to the system with meticulous care and checked each connection using a stethoscope and a soap solution, much like one would test a bicycle tire for leaks. By the end of the day, engineers concluded that all the pipe connections were intact and functioning properly. However, suspicions then turned to the separator plates within the heat-exchange unit, where a more elusive issue might have been lurking.

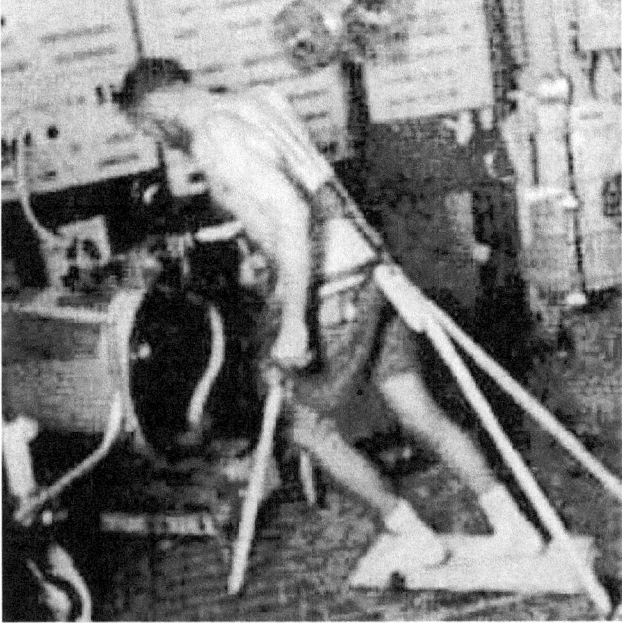

Astronaut Carr keeps in shape using the exerciser on Skylab 4. The second crew assembled the treadmill in space and used it throughout the rest of the mission.

Mechanical failures seemed to crop up almost daily. On the same day that Bean inspected the dehumidifier, another issue arose with the mechanism used to extend the mirror for scientist-astronaut Karl Henize's experiment. The mirror jammed midway out of the airlock, and despite numerous attempts, it could neither be retracted nor fully extended until the following morning. Such issues, though not catastrophic, added to the growing list of concerns for both the crew and ground control.

More worrisome were the leaks in the coolant loops, which were critical for cooling Skylab's various electronic systems, including the controls for the Apollo Telescope Mount (ATM) and the earth-resources package. On August 5, Huntsville telemetry indicated a pressure loss in the primary cooling loop. This news cast a shadow over the engineers working in George Hardy's office, where there were already fears of a leak in the secondary loop. Contingency plans were quickly formulated in case of a total failure of the cooling system. However, by the next day, further data provided some relief. The primary system was expected to run for at least another three weeks, and the secondary loop appeared likely to last the entire mission. Huntsville engineers hoped to devise a method to replenish the coolant before the final flight.

Astronaut William Pogue lifting the work chamber hatch at the materials processing facility.

Of all Skylab's mechanical issues, the erratic behavior of the gyroscopes was perhaps the most troublesome. Since the station's launch three months earlier, Huntsville engineers had grappled with faulty readings from the nine rate gyroscopes. A detailed investigation linked the gyroscopes' high drift rates to gas bubbles forming in their float chambers, likely caused by exposure to the vacuum of space. Huntsville engineers redesigned the gyroscopes to address this and prepared a backup package of six rate gyros, colloquially known as the "six-pack." The second crew carried up this package was ready to be installed on an experiment rack in the docking adapter if necessary. This location, close to Skylab's center of gravity, allowed proper alignment and provided an easy tie-in with the existing system.

The decision to install the six-pack was a difficult one. Although most of Skylab's original nine gyroscopes exhibited some instability, Mission Control maintained at least one good gyroscope on each axis and a generally serviceable backup. Installing the new gyroscope package would require work outside the workshop, and failure during the installation could potentially end the mission prematurely. However, it was agreed that the decision could not be delayed beyond the second extravehicular activity (EVA). Installing the gyroscopes on the final EVA, just one day before the mission's end, would not leave the crew enough time to make necessary adjustments.

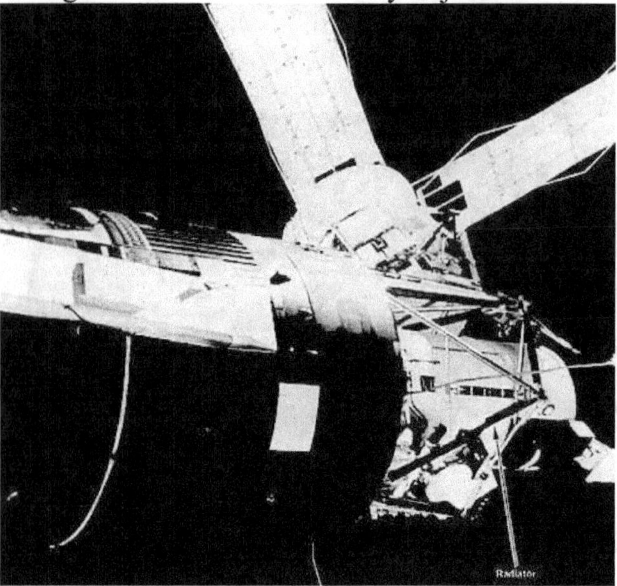

Radiating heat to space was necessary to cool Skylab and its systems. Panels were attached to the cylindrical portion of the airlock and docking adapter, and tubes, through which liquid coolant was circulated, were welded to the interior side of the skin. These radiating devices were effective.

On August 21, NASA management made the call to proceed with the six-pack installation. The original gyroscopes showed continued deterioration, and Houston did not want to risk an uncrewed period with only one functioning gyroscope per axis. The astronauts were scheduled to install the new gyroscopes on August 24, coinciding with the replenishment of the ATM film magazines. The EVA proceeded flawlessly, and when power was restored, Skylab had nine fully operational rate gyroscopes—comprising the newly installed six-pack and three from the original set. Skylab engineers could implement the redundancy management procedures planned from the mission's outset for the first time in nearly three months.

Despite the challenges posed by illness and mechanical failures, which disrupted the flight schedule for ten days, the crew settled into a routine after the first EVA. Their day began with a 6:00 a.m. CST reveille, a loud buzzer rousing the astronauts from sleep. In the hour before breakfast, they dressed, shaved, and prepared for another day of work aboard the aging but resilient Skylab.

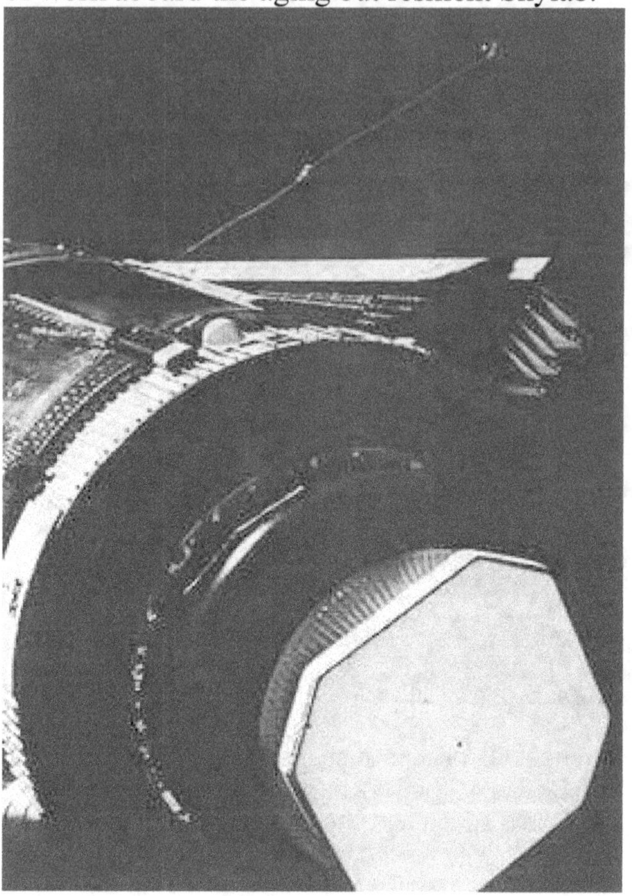

A large octagonal radiator on the aft end of the workshop also radiated heat to space.

The routine of selecting clothing aboard Skylab was straightforward, as the astronauts had a single standard uniform. This consisted of black trousers and turtleneck T-shirts designed for practicality in the confined environment of the space station. The trousers were cleverly designed with zippers, allowing the astronauts to convert them into shorts if they became too warm. The astronauts often stripped down to their undershorts for more physically demanding activities, such as using the onboard exercise bicycle. The uniform also included a jacket, essential for the cooler temperatures found in the airlock and docking adapter. Without laundering their clothes, the astronauts would wear each outfit for a few days before discarding it.

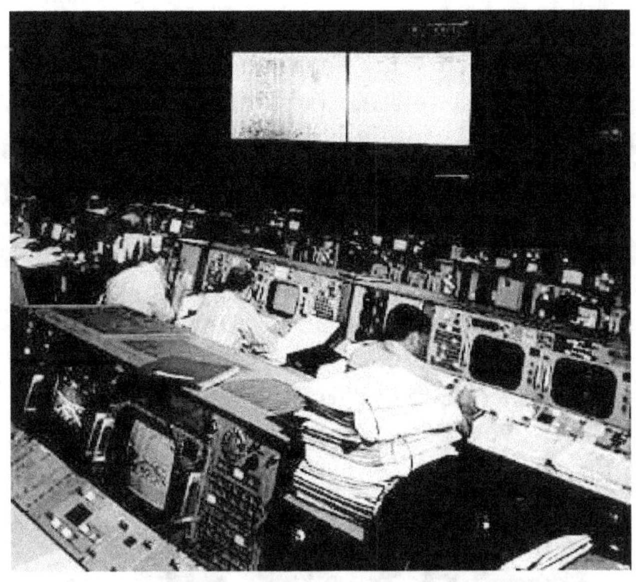

As Skylab flight crews worked in orbit, engineers and technicians at Johnson's Mission Control Center and the Marshall Space Flight Center in Huntsville, monitored the consoles. They analyzed problems as they arose, performed tests as necessary. and relayed instructions to the flight crews for corrective action.

Dressing in space presented unique challenges, especially when putting on socks or tying shoelaces. In the microgravity environment, astronauts found themselves straining their stomach muscles as they bent over to attend to their feet. Despite these difficulties, the clothes were generally well-received, though some astronauts did express concerns about the limited supply of socks.

Living in Skylab required adaptation to several unconventional aspects of daily life. The waste-management compartment, for example, was reminiscent of a bathroom on a commercial jetliner, both in its size and metallic, functional appearance, complete with gurgling noises. The compartment was a tight space that took some getting used to, mainly because it lacked the triangular gridwork on the floors of the rest of the workshop. Instead, engineers had designed a smooth floor for easier cleaning, but this design made it difficult for the astronauts to gain traction, leading one member of the third crew to complain that "you just ricochet off the wall like a BB in a tin can." Keeping control of personal items in this environment was another challenge, as they would float away unless securely anchored. To counter this, Alan Bean ingeniously used Velcro to attach his toiletries to the cabinet, preventing them from drifting off.

Without gravity, conventional washing methods were impractical. The lack of a regular sink meant that astronauts had to wash their hands from a valve recessed into the wall, and bathing was primarily done using wet washcloths. Though a shower was available, it was time-consuming, taking about an hour to complete. The first crew, committed to maintaining their hygiene, showered once a week and did not mind vacuuming the excess water afterward. Later crews, however, opted for a daily scrubbing with washcloths, finding it more convenient.

The bathroom's compact size meant that only one person could use it at a time, leading to some initial scheduling difficulties, particularly in the first hour of the day. To alleviate this, astronaut Paul Weitz shaved at night, while members of the third crew, Gerald Carr and William Pogue, eventually stopped shaving altogether. Alan Bean's crew managed to find the needed time by extending their morning preparations into breakfast.

At 7:00 a.m., the crew would gather around the wardroom table for breakfast, a meal that was both a functional and social occasion. The table was equipped with parallel bars under the food trays that served as makeshift chairs, though many astronauts preferred to stand, as sitting required an uncomfortable bending at the waist that strained their stomach muscles. A typical breakfast consisted of familiar items like bacon and eggs, bread, coffee, and orange juice. While the meals represented a definite improvement over the food provided during the Apollo missions, the astronauts still found the fare to be somewhat bland and the menu too repetitive.

Eating in space presented its challenges, chief among them the difficulty of keeping everything in place. Opening the lid on a warming tray would often cause a can or two to float away, and silverware and food particles similarly tended to drift. All three Skylab crews voiced their dissatisfaction with the size of the utensils provided, with Alan Bean, typically the least critical of the group, finding their small size "ridiculous."

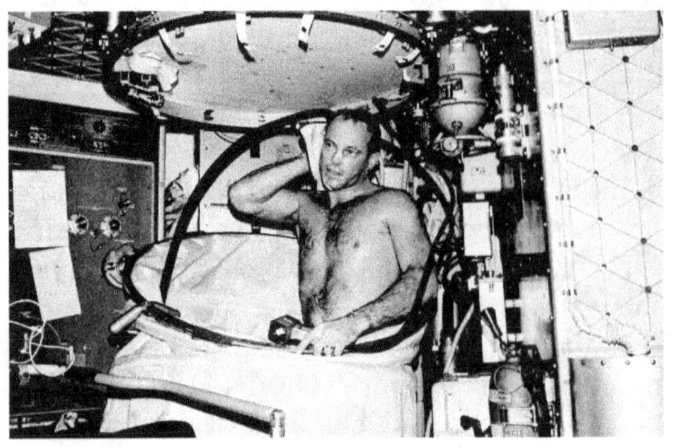

Astronaut Jack R. Lousma, Skylab 3 pilot, takes a hot bath in the crew quarters of the Orbital Workshop (OWS) of the Skylab space station cluster in Earth orbit. In deploying the shower facility the shower curtain was pulled up from the floor and attached to the celling. The water comes through a push-button shower head attached to a flexible hose. Water was drawn off by a vacuum system. NASA

Another minor but persistent issue was the presence of gas bubbles in the water supply. The air used to pressurize the water tanks could not float to the surface in the weightless environment, leading to bubbles that made drinking water slightly more cumbersome. Despite these inconveniences, the crew adapted to their surroundings, making the best of life aboard Skylab as they continued their groundbreaking work in space.

In the microgravity environment of Skylab, daily life brought both challenges and moments of camaraderie for the astronauts. The process of rehydrating their food, a routine task, often became a source of frustration. Occasionally, the bubbly water used for rehydration would burst the plastic bags, sending food splattering around the wardroom. This issue, coupled with the gas produced by the rehydrated food, contributed to frequent flatulence, a discomforting experience that one crew member humorously noted involved "farting about 500 times a day," establishing far from an ideal way to spend time in space.

Despite these minor annoyances, mealtimes were among the more pleasant hours spent aboard Skylab. They offered the crew a much-needed break from their busy schedules and provided opportunities to relax, gaze out the wardroom window at the Earth below, or simply enjoy the company of their fellow astronauts. However, the demanding schedule of space life often curtailed any hope of a leisurely breakfast. Mornings were filled with tasks that had to be completed before the workday began, including setting up for the noon meal, checking spacecraft systems, loading film, and attending to various medical and hygiene tasks like collecting and processing urine, weighing fecal samples, and measuring leftover food. Often, the crew found themselves behind schedule even before the official workday had begun.

The astronauts used both spring-wound and safety razors for shaving. Pilot Lousma was shown here.

Spaghetti and meat sauce came premixed and ready to be heated.

A typical day on Skylab was meticulously planned. By 8:00 a.m., astronaut Owen Garriott would be stationed at the Apollo Telescope Mount (ATM) console, ready to conduct solar observations. Meanwhile, his crewmates, Alan Bean and Jack Lousma, would either participate in a medical experiment or test maneuvering units designed for spacewalks. The crew's tasks were diverse and shifted throughout the day. By mid-morning, the roles would change—Lousma might move to the solar telescopes while Garriott returned to the workshop's lower level for his daily exercise session on the ergometer. This routine had gained importance after the first two weeks in space. Recognizing the need for physical fitness, flight planners had started allocating 90 minutes each day for exercise and hygiene beginning on August 10.

The crew's afternoon was typically dedicated to conducting various experiments. If there were no major experiments or repair work scheduled, Bean could perform a "corollary" experiment—secondary tasks that provided an excellent way to fill out the workday since most could be completed within an hour or two. Solar viewing continued through lunchtime, with the crew eating in shifts to ensure that the scientific work continued uninterrupted.

The evenings brought more experiments and the necessary but sometimes tedious household chores. After dinner, usually served at 6:00 p.m., the astronauts would review the next day's schedule, a task that could be time-consuming, especially when it involved adjustments to the ATM experiments. While the teletype machine on Skylab was an improvement over the handwritten schedules used during Apollo missions, the daily instructions often required two meters of teletype paper, making it difficult for the crew to secure the printout to the wardroom table.

(1 March 1973) — The three members of the prime crew of the first manned Skylab mission dine on specially prepared Skylab space food in the wardroom of the crew quarters of the Skylab Orbital Workshop (OWS) trainer during Skylab training at the Johnson Space Center. They are, left to right, scientist-astronaut Joseph P. Kerwin, science pilot; astronaut Paul J. Weitz, pilot; and astronaut Charles Conrad Jr., commander. Photo credit: NASA

As part of their evening routine, the crew transmitted a status report to ground control, which included detailed medical data on their eating, sleeping, and exercise habits. On August 30, Bean's report was typical of the information shared: totals on water consumption, urine output, and ergometer exercise were provided, along with insights into the crew's sleep patterns and dietary adjustments. Bean had added 15 salt tablets to his prescribed menu, Garriott opted for additional salts, peach ambrosia, and jam, while Lousma, the biggest eater among them, had made his dietary adjustments. Each evening, the crew also held a private medical conference with the flight surgeon, who confirmed that, despite initial space sickness, the second crew was adapting well to life in orbit.

Scientist-Astronaut Owen K. Garriott, science pilot of the Skylab 3 mission, was stationed at the Apollo Telescope Mount (ATM) console in the Multiple Docking Adapter of the Skylab space station in Earth orbit. From this console the astronauts actively control the ATM solar physics telescope.

Flight planners hoped the crew would complete their evening chores by 8:00 p.m., leaving two hours for relaxation. However, this seldom happened. Unlike the first Skylab crew, who made a point of eating dinner together, Bean, Garriott, and Lousma often worked through mealtime. One astronaut would remain at the ATM console, another would review the next day's instructions, and the third would grab a quick bite. Dinner often turned into a late-night snack, consumed just 30 minutes before bedtime. When the day's activities left little time, physical exercise was postponed until the evening, with sleep sometimes delayed by an hour or more.

Even amid their demanding schedules, the astronauts found ways to inject humor into their routines. Jack Lousma, known for his lighthearted banter, often entertained his "space fans" on the channel B tape, offering commentary on everything from the cramped conditions of the airlock to Garriott's haircuts. He also took every opportunity to promote the Marine Corps, his military branch. Lousma's playful spirit was captured on film in one memorable routine with makeshift "barbells." After pretending to strain mightily to lift the weights from the floor, he would suddenly soar into the air, the bells floating above his head in the weightlessness of space. Lousma's antics and his skill behind the camera provided an excellent and entertaining glimpse into life aboard Skylab. His tours of the workshop, filmed in early September, remain valuable records of the day-to-day existence in space, showcasing both the challenges and the lighter moments that characterized the mission.

Garriott displays the lightweight portable television camera which the astronauts used to televise their activities.

During their mission aboard Skylab, astronaut Owen Garriott found solace in the simplicity and curiosity of science demonstrations, providing a necessary diversion from the relentless demands of spaceflight. Despite the rigorous schedule, Garriott had envisioned conducting various scientific demonstrations during his designated weekly holidays. Although these holidays were frequently skipped due to the crew's workload, Garriott still managed to carve out time for these activities. His demonstrations focused on the intriguing effects of weightlessness on various objects, including water drops, magnets, and spinning objects. These simple yet profound experiments were not just a respite but also a means to share the wonders of space with those on Earth.

In the spirit of scientific inquiry, one of Garriott's most successful demonstrations was born out of spontaneous inspiration. While working on a student experiment involving a nut and bolt, he decided to spin the nut in a weightless environment and then attract it with a magnet. The result was a captivating display where the spinning nut was influenced by magnetic torque, showcasing the unique behavior of objects in microgravity. Although less critical than other mission objectives such as solar observations or Earth resource studies, these demonstrations resonated with the public. They were easy for laypeople to understand, which brought Skylab valuable public attention and enthusiasm.

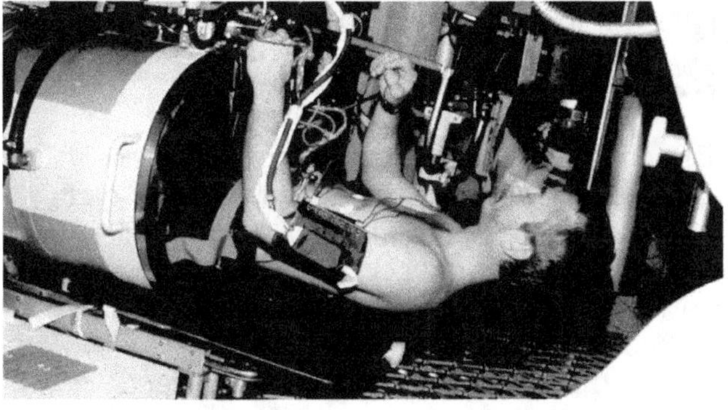

Scientist-astronaut Owen K. Garriott, science pilot of the Skylab 3 mission, lies in the Lower Body Negative Pressure Device in the work and experiments area of the Orbital Workshop (OWS) crew quarters of the Skylab space station cluster in Earth orbit. The LBNPD (M092) experiment provided information concerning the time course of cardiovascular adaptation during flight and provided in-flight data to predict the degree of orthostatic intolerance and impairment of physical capacity to be expected upon return to Earth environment. NASA

In contrast, the mission commander Alan Bean exhibited little need for such diversions. known as the most industrious crew member, Bean was deeply focused on maximizing the mission's scientific output. He rarely took time to gaze out of the spacecraft's windows, instead dedicating himself to extending the hours spent on experiments. During the mission's early days, the crew fell behind schedule. However, after a crucial extravehicular activity (EVA) on August 6, Bean's determination to catch up intensified. When he inquired with Mission Control about the extent of their delays, the response only fueled the crew's resolve to achieve, and even exceed, their mission objectives.

The crew's turnaround over the next two weeks was remarkable. Initially, the workload provided by Houston had been overwhelming, but soon, the astronauts caught up and even began to outpace the planned activities. By August 12, Bean requested additional tasks from Mission Control, noting that the crew was no longer working as strenuously as they had before the flight. He encouraged the ground team to push them harder, emphasizing that they had the energy, time, and capability to do more.

Mission Control responded to Bean's challenge by increasing the crew's daily workload. During the mission's third week, the time spent on experiments and repair activities was extended from 8 to 12 hours per astronaut. This adjustment paid off handsomely; by the end of the mission, the crew had surpassed their experiment goals by an impressive 50%.

View of Scientist-Astronaut Owen K. Garriott, Skylab 3 science pilot, in his sleep restraints in the crew quarters of the Orbital Workshop (OWS).

As the mission progressed into early September, Bean sought to extend their stay beyond the planned 59-day duration. However, this request was denied by Houston's medical office, which required more data before committing to longer missions. The decision was also influenced by the dwindling supplies of food and film aboard Skylab. By mid-September, the flight controllers had reduced the time allocated to the Apollo Telescope Mount (ATM) observations to eight hours daily. While the resumption of Earth-resource passes helped fill some of the time,

Mission Control struggled to keep the crew fully occupied.

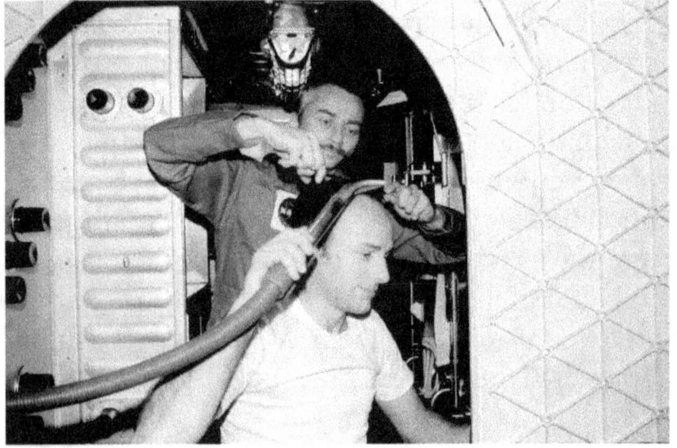

Scientist-Astronaut Owen K. Garriott, Skylab 3 science pilot, trims the hair of Astronaut Alan L. Bean, commander, in this on-board photograph from the Skylab Orbital Workshop (OWS). Bean holds a vacuum hose to gather in loose hair.

In a post-flight briefing, Bean expressed his frustration with the lack of meaningful work during the latter part of the mission. Although the scientific experiments were valuable, he believed they could not fill the available time. Instead of reducing the demanding 70-hour workweek, Bean proposed that future missions include additional experiments better to utilize the crew's time and Skylab's capabilities. Garriott supported this idea, advocating for the next crew to take more ATM film.

The second crew's recommendations and ability to handle a 12-hour workday convinced flight controllers to adopt this standard for the final mission. As a result, the flight plan for Skylab's third and last crew was expanded to include new experiments, ensuring that they would rise to the challenge set by their predecessors.

The Final Mission

While the second crew aboard Skylab had set unprecedented records for productivity in orbit, the third crew was immersed in the intensive preparation required for their mission. Commanded by Gerald Carr, the crew spent long hours in Houston's simulators, honing their skills and mastering the various tasks they would need to perform in space. Carr and his colleagues had limited access to the crucial training facilities for much of their training period, as the second crew

had priority until they departed for the Cape. It wasn't until late July, with just over three months remaining before their scheduled launch, that Carr's crew finally had uninterrupted use of the simulators.

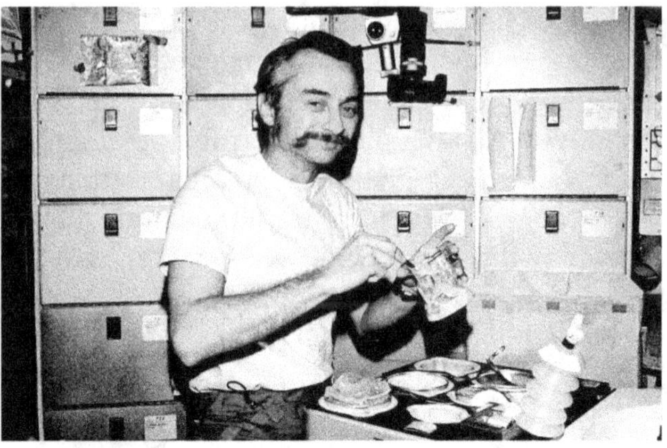

This photograph was taken during the Skylab-3 mission (2nd manned mission), showing Astronaut Owen Garriott enjoying his meal in the Orbital Workshop crew wardroom. The tray contained heating elements for preparing the individual food packets. The food on Skylab was a great improvement over that on earlier spaceflights. It was no longer necessary to squeeze liquified food from plastic tubes. Skylab's kitchen was so equipped that each crewman could select his own menu and prepare it to his own taste.

Their training regimen was rigorous and multifaceted. The crew dedicated significant time to practicing essential procedures such as rendezvous, docking, and reentry. In Huntsville, they rehearsed extravehicular activities in a large water tank designed to simulate the weightlessness of space. Meanwhile, back in Houston, they spent countless hours running simulations at the Apollo Telescope Mount (ATM) console and familiarizing themselves with the more than fifty pieces of experiment hardware they would need to operate aboard Skylab.

As the mission neared, the workload increased. Driven by Skylab's early successes, mission planners and experimenters introduced new tasks for the final crew. Inspired by the results of earlier solar observations, astronomers requested additional studies. Medical experts also sought extra measurements and photographs to expand their research. However, planning these new experiments was not always seamless; Carr's crew would later express concerns that the training for some of these tasks had been inadequate, leaving them underprepared for certain activities.

Late in the summer of 1973, a significant change was introduced to the mission. NASA officials decided to use Skylab to observe Comet Kohoutek, a celestial event discovered by Czech astronomer Lubos Kohoutek in March of that year. The comet's early discovery, nine months before reaching its closest point to the sun, offered astronomers an unusual opportunity to prepare detailed observation plans. With its vantage point above Earth's atmosphere, Skylab was ideally positioned to capture this once-in-a-lifetime event as the comet made its solar passage in late December. However, this new objective required the addition of complex maneuvering procedures to the crew's already packed training schedule, further intensifying their preparations.

The addition of these new experiments significantly increased the third crew's workload. Yet, beyond the scientific tasks, another challenge loomed over them: the high expectations set by the second crew's remarkable performance. In an October 2, 1973 press conference, Kenneth S. Kleinknecht, Skylab manager at the Johnson Space Center (JSC), praised the second crew's accomplishments, noting that their success demonstrated that humans could achieve more in space than initially thought. This success led management to consider extending the final mission to 70 days—a decision to incur additional costs of approximately half a million dollars per day. To justify this potential extension, flight planners had to ensure that the crew would have enough work to occupy their time fully.

The manager of the Orbital Assembly Project Office in Houston also acknowledged the implications of the second crew's productivity, suggesting it was indicative of future expectations. Plans for the final mission were adjusted accordingly. The crew's daily schedule was expanded to include 28 man-hours dedicated to experiment work alongside 12 new Joint Observing Programs for the ATM. Additionally, 10 to 14 new Earth-resource passes were added to the 20 already planned, and the crew was tasked with collecting extra medical data. Continuing the handyman tradition of previous crews, they were also assigned technical maintenance tasks, including recharging the coolant in a refrigeration system and troubleshooting a failed Earth-resources microwave antenna.

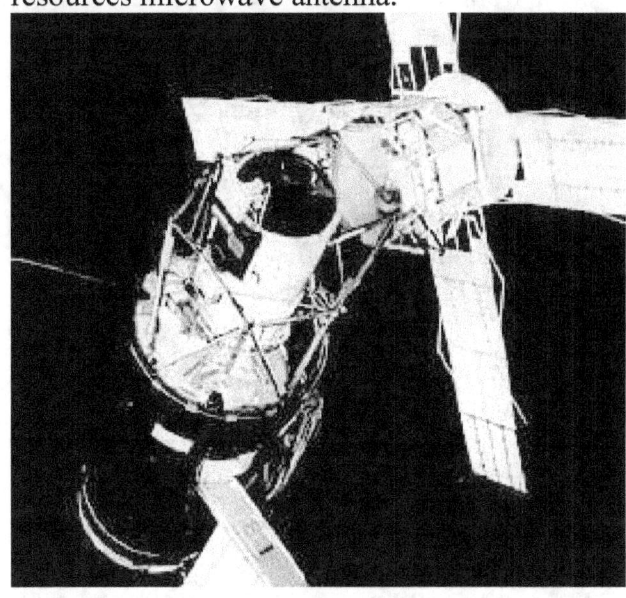

Some areas on the Skylab exterior became discolored. Engineers attributed this to an interaction between contaminants and solar ultraviolet radiation.

Despite the increased workload and the high stakes, the crew's training and launch preparations progressed smoothly throughout October, setting the stage for a planned liftoff on November 11. However, just five days before the scheduled launch, an inspection of the Saturn IB launch vehicle revealed a critical issue: cracks in each of its eight stabilizing fins. These cracks, likely caused by stress corrosion, posed a serious risk, as the fins could have potentially ripped off during the rocket's passage through maximum aerodynamic pressure early in flight. The situation demanded immediate action. Replacement fins were quickly fl in from NASA's Michoud Assembly Facility in Louisiana and installed while the Saturn IB remained on its launch platform, known as the "milkstool." Special work platforms, resembling painters' scaffolds, were suspended from the mobile launcher to allow technicians to reach the rocket base and complete the necessary repairs.

The delay caused by these repairs pushed the launch date back by five days, but it was a small price to pay to ensure the safety and success of the mission. As the crew prepared for their journey, they carried with them the weight of expectations, not only from the remarkable achievements of their predecessors but also from the ambitious goals set for their mission. The final Skylab crew was ready to take on the challenges that awaited them, determined to build upon the legacy of those who had gone before.

The repair crews faced grueling 12-hour shifts, but their dedication paid off when they completed the necessary repairs on November 12, 1973. With the repairs finalized, the launch of Skylab's third mission was rescheduled for November 16. However, a significant concern loomed over the mission: the possibility of motion sickness among the astronauts during the initial days in orbit. The unfortunate experience of the second Skylab crew had heightened the anxiety of NASA's top managers. In response, a group of NASA and external medical experts convened in late October to assess the data on space sickness. Their recommendation was to administer medication upon reaching orbit to mitigate the symptoms of motion sickness.

The mission commander, Lieutenant Colonel Gerald P. Carr, and his crewmates, expressed reservations about this recommendation. The astronauts were concerned about the undesirable side effects of the proposed medications. A compromise was reached: Carr would delay taking his medicine until after completing the critical task of rendezvous. On the second and third days, all three astronauts were instructed to take the medication routinely, and after that only if symptoms appeared. To further reduce the risk, they were advised to minimize head movements and spend the first night in the confined space of the command module, as moving around in the larger workshop was believed to exacerbate the onset of motion sickness. Despite their agreement, the astronauts remained skeptical, suspecting that even the medical experts did not fully comprehend the nature of space sickness.

Skylab 2 CM

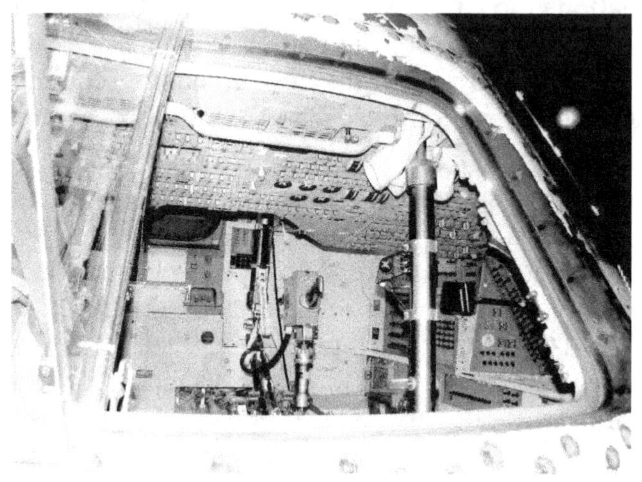

Skylab 2 CM interior

The S-IB first stage for the Skylab 4 mission's SA-208 Saturn IB rocket arrives at the Vehicle Assembly Building (VAB) at NASA's Kennedy Space Center in Florida.

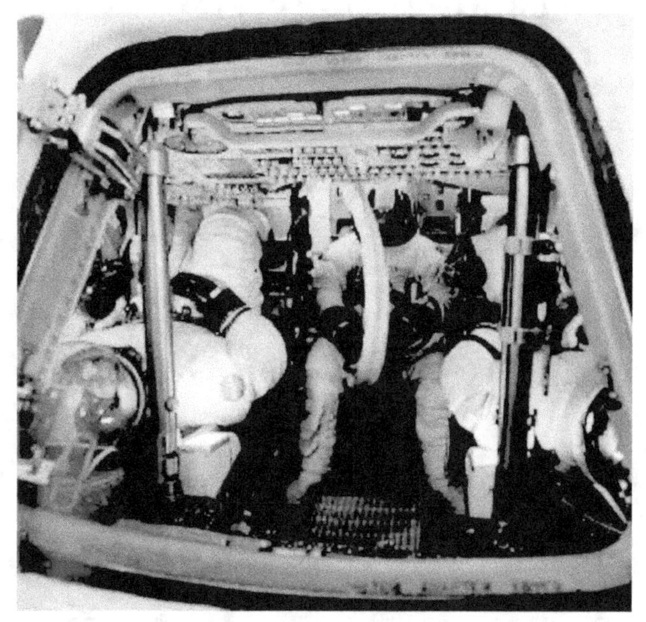

Chapter 6 - Skylab 4

On the morning of November 16, 1973, at precisely 9:01 a.m. EST, Skylab's last mission roared into the clear Florida sky. The launch proceeded smoothly, marking the beginning of the flight for the all-rookie crew.

These three men were the prime crewmen for the Skylab 4 mission. Pictured in their flight suits with a globe and a model of the Skylab space station are, left to right, Astronaut Gerald P. Carr, commander; Scientist-Astronaut Edward G. Gibson, science pilot; and Astronaut William R. Pogue, pilot.

The third and final Skylab crew, launched in 1974, comprised Commander Gerald P. Carr, Pilot William R. Pogue, and Scientist Pilot Edward G. Gibson. Gerald Carr, a lieutenant colonel in the U.S. Marine Corps, had been appointed as an astronaut in 1966. William Pogue, a lieutenant colonel in the U.S. Air Force, also became an astronaut in 1966. Edward Gibson, a civilian with a Ph.D. in engineering and physics, had been selected in 1965. Like their predecessors, this crew had yet to fly in space before their mission aboard Skylab.

Gibson in the crew wardroom. Space meals were ready to eat. The third crew ate Thanksgiving dinner at this food table.

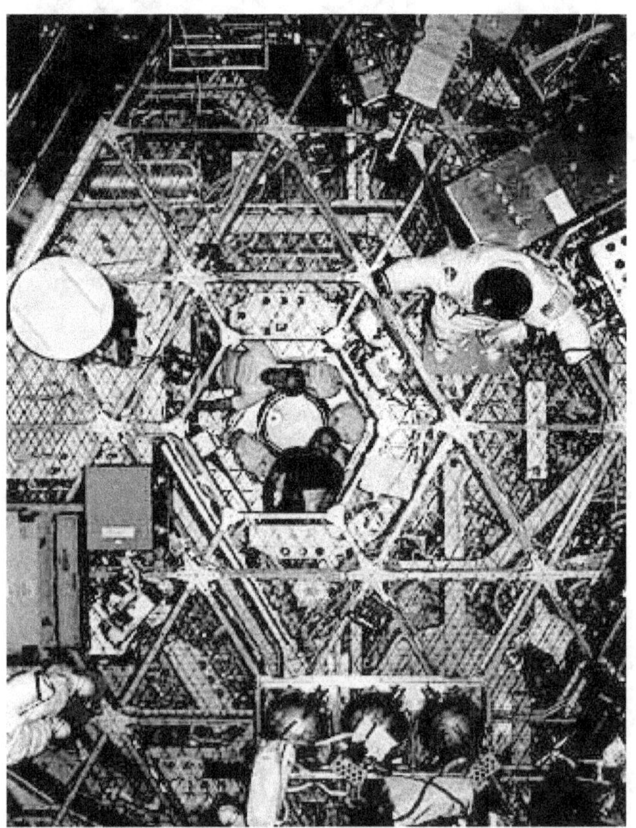

Here, Gibson and Carr peer through the octagonal opening which separated the workshop's two levels. Behind them was the trash airlock.

As they orbited the Earth for the first time, the crew members were visibly excited, eagerly peering out of the spacecraft windows, so much so that Carr reported to Mission Control that the windows were smudged from their enthusiastic viewing

Skylab 4 Command and Service Module, showing the location of two of the four quads of attitude control thrusters on the Service Module

By the fifth orbit, as the spacecraft soared between Australia and Guam, Carr sighted the Skylab workshop. Within ten minutes, he had maneuvered the Apollo spacecraft to within 30 meters of the workshop. Demonstrating remarkable precision, Carr guided the spacecraft for docking. However, the docking gear, which had previously caused issues, proved troublesome once again. After two failed attempts, Carr hard-docked the command module to the multiple docking adapter, accomplishing the task almost eight hours after launch.

With docking complete, the crew found themselves out of communication with Houston for 41 minutes during their journey from Bermuda to Carnarvon, Australia. Seizing the opportunity, they began organizing the command module, stowing away gear used during the rendezvous and docking operations. Before doing so, Carr and scientist-pilot Edward G. Gibson took their anti-nausea pills. Pilot William R. Pogue, however, had delayed taking his medication, and the effects of weightlessness soon took their toll. Just before re-establishing ground contact, Pogue requested a vomit bag from Gibson, acknowledging that he felt nauseated. Despite his efforts to go slow and mitigate the symptoms, Pogue eventually vomited, albeit only slightly, but the nausea persisted.

As Houston reconnected with the crew shortly before 6 p.m., the flight controllers reiterated the physicians' caution about entering the Skylab workshop too soon. Carr had earlier requested permission to begin activating the workshop that evening, but mission planners saw no advantage and advised against it. Carr agreed to wait until the following day.

Once Skylab went out of radio range again, Carr and Gibson debated how to report Pogue's illness in the evening status update. Carr was inclined to withhold the full details, opting to say that Pogue was simply not hungry to explain the food he had not eaten. As they prepared their second meal, the irony of the situation weighed on Carr and Gibson. Pogue, who had earned the nickname "Iron Belly" for his reputed resistance to motion sickness, was now the one suffering. Known for his steadfast inner ear, likened to having "cement" in it, Pogue found himself miserable. His crewmates helped him move to the docking tunnel, hoping a cabin fan's airflow might relieve him. Unfortunately, Pogue's condition showed no signs of improvement.

When Houston reestablished communication, Carr requested a postponement of the status report, citing that the crew had not yet started eating. Houston agreed, granting Carr and his crew two more hours to deliberate on handling the situation. During this time, they failed to remember that an onboard tape recorder had been running continuously, capturing every conversation. Believing that only the three knew what had transpired, Carr and Gibson decided to play down the severity of Pogue's illness, aiming to avoid unnecessary repercussions. Pogue had vomited only a small amount; it was not a violent episode, and they were hopeful that he would recover fully before the crew was scheduled to move into the workshop the following day.

Gibson expressed concern that the flight surgeons might overreact if informed of the vomiting. Carr wavered, considering the idea of reporting Pogue's illness but omitting the fact that he had vomited. "I'd just say he doesn't feel like eating," Carr mused. However, just minutes before the scheduled medical conference, Carr turned to Pogue and said, "I think we better tell the truth tonight. We're going to have to turn in a fecal/vomitus bag, although we could just throw it down the trash airlock and forget the whole thing." Gibson was tempted by the idea, remarking, "I think all the managers would be happy." The prospect of disposing of the evidence and reporting only nausea seemed appealing, as the flight surgeons more seriously viewed vomiting.

Ultimately, they decided to keep the incident quiet. During the medical conference, Pogue's nausea was mentioned, but his vomiting was not disclosed. Later that evening, Carr delivered the status report, noting only that "the pilot had no strawberries for lunch and has not eaten meal C." The crew retired for the night, hoping to leave the incident behind them.

The following day, after a restful night's sleep, all three astronauts felt considerably better. Pogue, although still recovering, decided to take it easy for the day. Meanwhile, Carr and Gibson prepared breakfast while enjoying the breathtaking view of the Alps and southeastern Europe. By 8:45 a.m., they were ready to enter the Skylab workshop. It took them about 30 minutes to pressurize the multiple docking adapter, remove its hatch, and stow both the hatch and the docking probe in the command module. At 9:16 a.m., Carr turned on the lights, and the crew began their work, connecting communications, initiating the environmental control system, and powering up the workshop.

(8 Nov. 1973) - The three members of the Skylab 4 crew were photographed standing near Pad B, Launch Complex 39, Kennedy Space Center, Florida, during preflight activity. They are, left to right, scientist-astronaut Edward G. Gibson, science pilot; astronaut Gerald P. Carr, commander; and astronaut William R. Pogue, pilot. The Skylab 4/Saturn 1B space vehicle was on the pad in the background. Skylab 4, the third and last manned visit to the Skylab space station in Earth orbit, will return additional information on the Earth and sun, as well as provide a favorable location from which to observe the recently discovered Comet Kohoutek. NASA

Unbeknownst to the crew, the tapes from the onboard recorder were being routinely transcribed in Houston, and the candid discussions between Carr and Gibson regarding Pogue's illness were soon uncovered. The response from Mission Control was swift. A medical conference was convened in the afternoon. By the end of the day, Alan Shepard, the chief of the Astronaut Office, personally took the microphone in Mission Control to deliver a public and official reprimand. "I just wanted to tell you," Shepard said, "that on the matter of your status reports, we think you made a fairly serious error in judgment here in the report of your condition." Carr, accepting the rebuke, responded, "Okay, Al. I agree with you. It was a dumb decision." And with that, the matter was ostensibly settled.

However, the incident raised concerns among reporters at the evening's change-of-shift press briefing. They questioned whether this event signaled a breakdown in the open and honest communication between the crew and flight controllers. Flight Director Neil Hutchinson assured them that he did not believe so, but he warned that if there were any further signs of a lack of candor, immediate steps would be taken to correct the situation.

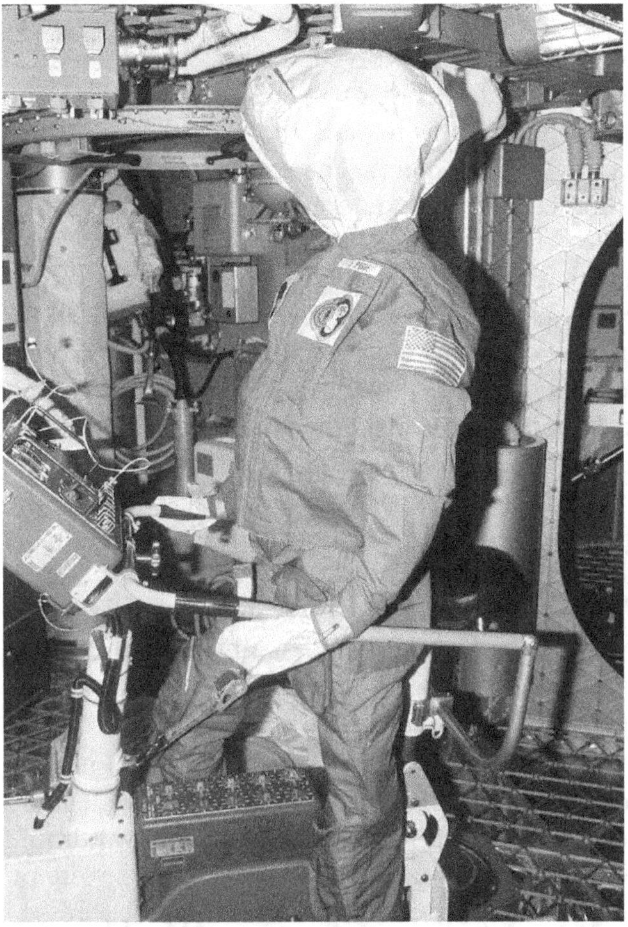

(July-September 1973) --- This photograph was an illustration of the humorous side of the Skylab 3 crew. This dummy was left behind in the Skylab space station by the Skylab 3 crew to be found by the Skylab 4 crew. The dummy was dressed in a flight suit and propped upon the bicycle ergometer. The name tag indicated that it represents William R. Pogue, Skylab pilot. The dummy for Gerald P. Carr, Skylab 4 commander, was placed in the Lower Body Negative Pressure Device. The dummy representing Edward G. Gibson was left in the waste compartment. Astronauts Alan L. Bean, Owen K. Garriott and Jack R. Lousma were the Skylab 3 crewmen. Gibson was the Skylab 4 science pilot. Photo credit: NASA

The long-term impact of this incident on the crew's relationship with Mission Control was difficult to measure. Managers believed—and the tape-recorded evidence supported their view—that the astronauts meticulously reported every mistake on channel B after the incident. Yet, they remained reluctant to discuss their problems on the public air-to-ground channel. Carr later noted that the crew could hardly relish the prospect of having their shortcomings broadcasted across the country. The private line, previously used by Pete Conrad for operational purposes, was no longer an option except in true emergencies, leaving them only channel B. However, with the built-in time lag of nearly 24 hours before Mission Control could review the transcripts and knowing that these transcripts would eventually be made public, the crew hesitated to be entirely frank. This hesitancy was exacerbated by the lack of rapport between the crew and their flight controllers. This disconnect had begun during their training and contributed to growing frustrations over the subsequent six weeks.

Despite these tensions, the flight control teams were eager to return to their routine, grateful to have men back in the Skylab workshop after several weeks of uncrewed operations. They swung into their tasks with renewed energy, determined to keep the mission on track.

Activating Skylab's workshop was the first critical task upon the crew's arrival, a task that proved far more complex than anticipated. Although one flight controller casually compared it to "only a little more complicated than when you come back from vacation," Commander Gerald Carr and his crew soon discovered it was anything but routine. The activation process demanded more time and effort than expected, as each job encountered unforeseen challenges. Every task took longer than planned, compounded by inevitable mistakes and frequent interruptions from mission control in Houston, which required constant attention.

One such setback occurred when Astronaut William Pogue was flushing the potable water system with an iodine solution to prepare for the connection of a new water tank. Inadvertently, Pogue left a valve in the wrong position, causing the disinfectant to flow into the waste tank, costing the crew an hour of precious time. By the end of their first day, they were already two hours behind schedule—a deficit they would not recover the following day. Despite these delays, mission planners pressed forward, setting up a regular flight plan for Monday.

Two Skylab 4 crewmen were seen passing trash bags through the trash airlock of the Orbital Workshop OWS of the Skylab space station in Earth orbit. The trash airlock leads to the OWS waste disposal tank. Astronaut William R. Pogue, pilot, holds onto the OWS crew quarters ceiling as he prepares to jump onto the OWS airlock hatch cover to force another trash bag further down into the airlock. Astronaut Gerald P. Carr, commander, assisted by holding onto the trash bags. A third trash bag was floating in the zero-gravity environment near Pogue's right leg. The wardroom can be seen behind Pogue.

The primary task for Monday was recharging the primary coolant loop, essential for cooling the spacesuits and airlock batteries. The successful completion of this task was crucial, as it would allow the crew to proceed with their first extravehicular activity (EVA) as scheduled. Without proper cooling, this critical operation would likely require two separate excursions. Pogue, using equipment identical to that employed for ground-based refrigeration systems, completed the recharging without any issues.

However, this repair was only one of the many extra tasks the third crew had to tackle during their first week in space. One particularly time-consuming task involved a new set of medical measurements designed to study how blood and body fluids shift toward the head in the zero-gravity environment. The crew needed to take girth measurements at more than 50 points on their bodies and capture photographs on infrared-sensitive film. The entire process consumed approximately four man-hours. The tapes used for the measurements were cumbersome, and the crew had not practiced with them before the flight. Compounding the challenge, the photography was difficult due to inadequate restraints for the photographer. Pogue, attempting this task for the first time on the fifth day, struggled to maintain his position in the microgravity environment. As he tried to steady himself by wedging a shoe between two water tanks, he accidentally turned a valve, which later led to a loss of pressure—an issue discovered that night.

Tuesday, the crew's fourth day in the workshop, was another tightly packed day, leaving them with no time to gaze out the wardroom window, even though visual observations were on their list of optional activities. Carr later expressed the crew's frustration to Houston, stating, "If we're ever going to get caught up, we're going to have to whack something out [of the flight plan] tomorrow. We haven't had time to stow everything properly, and this place was really getting to be a mess."

The first week's major event was the planned EVA on November 22, Thanksgiving Day, during which Pogue and Edward Gibson were to reload the Apollo Telescope Mount (ATM) cameras and troubleshoot an inoperative antenna on the microwave sensor. This task presented a significant challenge as there were no restraints on the underside of the multiple docking adapter where the antenna was located. However, the crew's confidence was bolstered by the procedures they had rehearsed in Huntsville's large water tank. Just before noon on Thursday, Gibson and Pogue suited up in the workshop's forward dome and squeezed into the airlock. After meticulously checking their equipment—knowing that stepping out into the vacuum of space left no room for error—they depressurized the airlock and opened the hatch.

Pogue's first task was to photograph the contamination surrounding the workshop, but after just a few exposures, the camera malfunctioned. The shutter speed knob spun uselessly in his gloved hand. Undeterred, Pogue then assisted Gibson in reloading the ATM cameras. Once that was completed, they maneuvered around the airlock to reach the malfunctioning antenna on the Earth-facing side of the cluster. Despite initial difficulties, they discovered that Gibson, who had better control of his movements, was better suited to restrain Pogue while he worked on the electronics module.

Scientists on the ground suspected that the problem with the antenna lay in one or both of the potentiometers that controlled its oscillations. Pogue carefully opened the module and cleaned the potentiometers, but when Carr applied power to the antenna, it remained unresponsive. Further tests revealed the issue was within the pitch circuit, which controlled the fore-and-aft oscillations, and it could not be corrected. Consequently, Pogue installed a pin to lock the pitch gimbal and a jumper to bypass the faulty circuit. When Carr activated the antenna again, it functioned, though only in a limited capacity, scanning side-to-side across the spacecraft's ground track. This partial restoration of functionality was a significant achievement, and the experimenters on Earth were delighted.

Pogue and Gibson returned to the airlock after a successful 6½-hour EVA, marking another milestone in the evolution of spacewalks since the Gemini missions. Despite the physical demands, neither astronaut felt overly exerted, a testament to how far EVA procedures had advanced. After such a strenuous day, Carr decided there was no need to push the crew to complete the post-EVA checklist that evening. Exhausted, the crew agreed to leave it for the following day.

However, by the next day, the astronauts were still behind schedule. Mission Controller Neil Hutchinson informed reporters that the crew might take Saturday off instead of Monday, as their first scheduled day off on November 19 had already been canceled before launch. Despite these setbacks, the crew's perseverance and problem-solving skills during this challenging first week laid the groundwork for the successful continuation of their mission aboard Skylab.

The mission schedule allowed some flexibility, and Hutchinson recognized that the crew needed a break. He acknowledged that the flight planners had miscalculated the time required for various tasks, overloading the crew with too much work. This issue had surfaced during previous missions, but the highly driven second crew left a lasting impression on the planners, who now aimed to bring Commander Gerald Carr, William Pogue, and Edward Gibson up to the same level of efficiency that Alan Bean, Jack Lousma, and Owen Garriott had achieved.

During the first Skylab mission, Pete Conrad had been quick to push back when Mission Control pressed him too hard, but Carr, a different kind of leader, hesitated. His reluctance was likely exacerbated by his earlier misjudgment regarding Pogue's illness, which may have made him even more cautious about admitting that his crew was struggling to keep up with the demanding flight plan—especially over the open communications loop.

A well-needed break on the 24th provided the crew with some much-needed breathing space, as Mission Control deliberately refrained from any communication that might seem intrusive or demanding. That evening, Carr took the opportunity to review the first week of the mission with the flight controllers. Speaking into the channel B tape recorder, he described the experience as "frantic." He explained that simply learning to move around in the microgravity environment took a great deal of time and effort, and the crew's voices likely betrayed their mounting frustration. No matter how hard they tried, they couldn't seem to catch up with the flight plan, leading to a sense of demoralization.

Despite these challenges, Carr remained cautiously optimistic. They had managed to complete all the work scheduled up to that point, but the risk of falling behind again loomed large. He urged the flight planners to create more realistic schedules that the crew could reasonably keep up with. Sunday marked a return to the demanding schedule, with tasks such as conducting a cardiovascular assessment on Gibson, replacing a video display tube, installing a new automatic timer on the Apollo Telescope Mount (ATM) console, and checking out the Earth-resource sensors.

That evening, Flight Director Donald Puddy offered some encouragement, commenting positively on the crew's accomplishments. The day off had clearly lifted their spirits, and Puddy expressed confidence that within a few days, concerns about lagging behind the flight plan would likely disappear. The first Earth-resource pass was scheduled for Monday, weather permitting, with ATM observations set to commence on Tuesday.

However, the mission encountered an unexpected complication on the night of the 23rd. Without warning, one of the control moment gyros—a crucial component of Skylab's attitude control system—began to overheat and slowdown alarmingly. The data suggested that an inadequately lubricated bearing had seized up. In response, flight controllers shutdown the malfunctioning gyro and switched the workshop computer to operate on two gyros instead of three. This sudden development raised serious concerns about the mission's future.

Under normal circumstances, losing one control moment gyro might have been a minor issue, but the situation was made more critical by the already depleted supply of gas for the attitude-control thrusters. In the first few days following Skylab's launch, the workshop had consumed attitude-control fuel at an alarming rate. By the time Carr's crew reached Skylab, only about one-third of the original fuel supply remained. This combination of factors turned what might have been a manageable problem into a significant challenge, threatening the mission's success and requiring the team to reconsider their approach to completing their objectives.

With many Earth-resource passes yet to be conducted and the maneuvers required to observe Comet Kohoutek demanding significant fuel, the situation grew increasingly critical. The mission's success hinged on the ability to gauge the propellant needed for each maneuver precisely. Consequently, experts in Huntsville and Houston immediately began working on more accurate methods to assess Skylab's momentum state and develop new computer programs to optimize fuel usage. Every experiment that required maneuvering the spacecraft now became substantially more complicated.

On Monday, the scheduled Earth resource pass had to be canceled due to cloud cover over the observation site. This change in plans allowed the crew to focus on other tasks, including a cardiovascular experiment on Commander Carr, stellar spectroscopy, and an observation of Comet Kohoutek. Scientist-astronaut Edward Gibson also prepared the solar instruments for the first observing period scheduled for Tuesday. Despite the shift in focus, it was another demanding day. Carr and Pogue both expressed frustrations, noting they made errors and felt rushed.

Tuesday's schedule set the tone for the next two weeks of operations. The astronauts began their early-morning chores by 6:30 a.m., and by 8:22 a.m., Carr had initiated operations on the Apollo Telescope Mount (ATM). Shortly afterward, they received news that the observing schedule would be more intense than initially planned due to significant solar activity, which presented a good opportunity to observe a solar flare. Carr, primarily responsible for ATM duties that day, dedicated much of his time to monitoring the sun's activity.

While Carr was engaged with solar observations, Pogue and Gibson were occupied with various tasks. Pogue set up a camera at the wardroom window to photograph a cloud of barium vapor released from a rocket as part of an experiment to study the Earth's magnetic field. He and Gibson also took turns monitoring each other's reactions as subjects in a vestibular function experiment.

In addition to these scientific duties, the crew also made a nine-minute television tape for the news media to demonstrate in-orbit exercise routines. Carr explained using the ergometer and the "Thornton treadmill" while exercising on them. The Thornton treadmill, a simple yet ingenious device conceived by Scientist-Astronaut William Thornton, consisted of a sheet of slippery Teflon fixed to the floor. The astronaut, wearing only socks, would walk on this surface. At the same time, a bungee cord harness pressed him d, simulating gravity and providing essential exercise for the leg muscles—muscles not adequately stressed by the bicycle. The device proved highly effective, so much so that no astronaut could use it for more than a few minutes. It became a valuable addition to the crew's exercise regimen.

Throughout the day, Carr and Pogue intermittently took photographs through the wardroom window, capturing images of various sites listed by Mission Control. This activity was part of a program to systematize what had previously been informal observations of cloud patterns, ocean currents, and geological features. These photographic efforts were later supplemented by detailed visual observations and descriptions, further enriching the collected scientific data.

Meanwhile, back on Earth, flight controllers and Control Moment Gyro (CMG) experts diligently worked to understand the limits of Skylab's maneuvering capabilities with only two functional control moment gyros. The challenge was immense, but their efforts were crucial in ensuring the mission could proceed despite the setbacks, and their work would ultimately play a key role in the crew's ability to continue their observations and experiments in the weeks ahead.

The photograph, taken with a hand-held 70-mm Hasselblad camera, revealed a striking symmetrical bow-wave cloud pattern downwind of Gough Island, a small mountainous island in the South Atlantic. The island itself, visible in the lower right of the image, remained clear. This image, SL4-137-3632, was taken during intense activity aboard Skylab as the crew grappled with maintaining the station's orientation.

Positioning the Skylab workshop for astronaut William Pogue's photography of a barium cloud experiment posed significant difficulties. The effort saturated the Control Moment Gyroscopes (CMGs), crucial for maintaining the station's orientation without using thrusters.

As a result, considerable fuel had to be expended to return the station to its solar inertial attitude. This unexpected situation led to the cancellation of the next day's maneuvers so that engineers could analyze the problem more thoroughly.

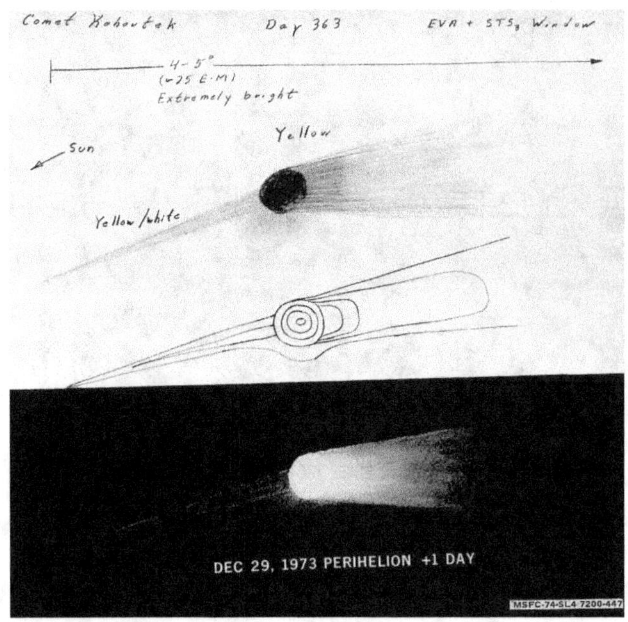

This pencil sketch of the Comet Kohoutek made by Skylab 4 astronaut Edward Gibson illustrates the crew's collective impressions of the comet's appearance on December 29, 1973. An early discovery of a large comet in an orbit to reach close to the Sun at the end of 1973 prompted NASA to initiate Operation Kohoutek, a program to coordinate widespread observations of the comet from ground observatories, aircraft, balloons, rockets, unmanned satellites, and Skylab. NASA

At the evening press briefing, reporters asked Flight Director Donald Puddy for an estimate of how much Earth resources data might be lost due to the maneuvering issues.

However, Puddy was optimistic, refusing to concede that any data would be lost. He expressed confidence that the complexities of maneuvering with only two functional gyros would soon be mastered, ensuring that all mandatory observation sites would be covered before the mission's conclusion.

That evening, after a long and challenging day, Commander Gerald Carr sat down at 9 p.m. to deliver the evening status report. He detailed the crew's sleep patterns, exercise routines, food and water intake changes, and the clothing used throughout the day. A series of questions and answers were exchanged between the ground control and the spacecraft, focusing on flight plans and the status of various systems. After briefly summarizing the day's news headlines, the CapCom signed off just after 10 p.m.

William R. Pogue at the controls of the Apollo Telescope Mount

By November 30, the guidance and control experts felt more confident in understanding the new constraints imposed by the saturated gyros. They executed a complex Earth-resources pass that day, which the astronauts completed flawlessly. The amount of attitude-control fuel used closely matched predictions, indicating that the new procedures were effective. However, two days later, an attempt to conduct two sequential passes again saturated the gyros and consumed more thruster gas than anticipated. Engineers returned to their computers and simulators, spending an additional two days devising new procedures to manage the station's orientation more efficiently.

During the week of November 26, as flight planners began to accelerate the pace of the workday, each astronaut was asked to respond to a questionnaire about the habitability of the Skylab workshop. One question, which invited comments on unanticipated problems, prompted Carr to reflect on the frantic first two weeks of the mission. He noted that much of the unanticipated trouble stemmed from the crew's lack of adequate training for zero-gravity maneuvering. "When you get up here, it's a whole new world," Carr remarked. "Everything we did took two or three times as much time as we thought it would take. We fooled ourselves." He candidly admitted that, despite their intentions to take activation slowly and methodically, the crew allowed themselves to be rushed, driven to exhaustion by the demands of the mission. Carr acknowledged that while they voiced concerns about being rushed, they failed to take control and prioritize their tasks as they had planned.

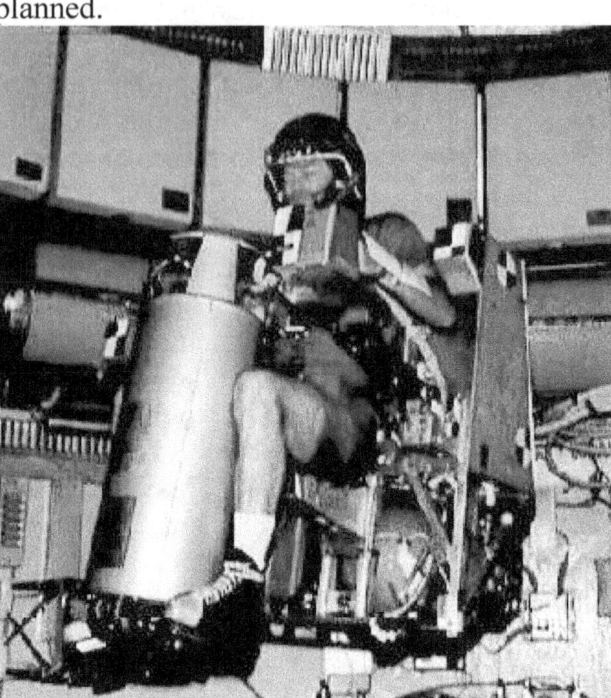

Gerald P. Carr flying the Astronaut Maneuvering Unit

These reflections, however, went unnoticed by the flight planners, who were focused on pushing the third crew to match the pace set by the previous one. They continued to shorten task durations and decrease the time between planned activities, assuming that the crew's proficiency steadily increased. Despite this, flight directors observed that crew performance was still not meeting their expectations. On December 5, both the flight director and the crew physician noted signs that the astronauts were no longer as rushed as they had been. However, the very next day, Carr expressed frustration with the demanding schedule, which required the crew to work up to 16-hour days. "We wouldn't be expected to work a 16-hour day for 85 days on the ground," Carr pointed out, "so I really don't see why we should even try to do it up here." That night, the flight director informed reporters that the crew was now being scheduled for 27 man-hours of experiment work per day—a slight increase from the nominal schedule but still less than what the previous crew had accomplished under Commander Alan Bean.

This narrative vividly described the challenges faced by the Skylab crew as they navigated the complexities of long-duration spaceflight. It highlighted the tensions between the demands of mission control and the realities of working in the unforgiving space environment, offering a glimpse into the human aspects of this historic mission.

On the 28th day of the mission, December 13, 1973, program officials gathered to review the progress and performance of the Skylab mission. This assessment was crucial as they evaluated both the spacecraft systems and the crew's ability to sustain the planned 84-day duration. That afternoon, during a press conference, Bill Schneider, the Skylab Program Director, outlined the mission's accomplishments up to that point. The crew had logged 84 hours of solar observations, completed 12 Earth-resource passes, and conducted 80 photographic and visual Earth observations. They had successfully carried out all scheduled medical experiments, along with numerous corollary studies, student experiments, and science demonstrations. In addition, the astronauts had completed three major repair tasks, showcasing their versatility and resilience.

Despite these successes, concerns lingered. The solar X-ray telescope was experiencing issues with a jammed filter wheel, and there were intermittent signs of distress in one of the remaining Control Moment Gyros (CMGs), a crucial component for maintaining the station's orientation. These potential problems were being closely monitored, but Schneider remained optimistic. "We're GO for our 60-day mission, open-ended to 84," he declared unless something unforeseen occurred.

However, the press had its concerns, particularly regarding the crew's performance. Reporters questioned why the crew seemed slower than previous teams, why mistakes were being made, and how their performance compared to the first two Skylab crews. With Kenneth Kleinknecht, the Skylab Program Manager, Schneider firmly denied any higher incidence of errors on the third mission. They refused to draw direct comparisons between the crews, instead pointing to the hundreds of changes to the flight plan that had made the third crew's tasks significantly more challenging. Kleinknecht placed some of the blame on the ground personnel who had approved so many alterations. Still, he also praised Commander Gerald Carr, Pilot William Pogue, and Science Pilot Edward Gibson, asserting they were doing "an outstanding job" under the circumstances.

The press conference took a more confrontational turn when a reporter mentioned the "vomiting incident" earlier in the mission. The incident, in which one of the astronauts became ill, had led to an unguarded discussion among the crew, which some perceived as an attempt to cover up the event. The reporter questioned Schneider on whether he suspected that other matters were being withheld from flight controllers or physicians. Schneider dismissed these concerns, explaining that the channel B tapes, which recorded internal communications, were filled with admissions of error, and the medical conferences had been frank and open. He emphasized that Carr and Gibson had decided to report the incident to management and had preserved the physical evidence, indicating their transparency.

Both Schneider and Kleinknecht vigorously defended the crew, and the reporters eventually moved on from the subject. Nevertheless, despite the officials' assurances, tensions within the crew were evident. That same week, each astronaut expressed frustration, signaling underlying problems that could not be easily dismissed.

On December 12, Pogue voiced his frustration over the tight scheduling of experiments, complaining bitterly to channel B. He had just missed capturing a couple of photographs because he had been forced to set up a camera quickly. Addressing the principal investigator, Pogue remarked, "This was going to happen again [and again] until the word gets through to the Flight Activities Officers that they will have to give us time to get from one point in the spacecraft to another. I don't know how we're going to get this across to [them] unless you [principal investigators] put your foot down and stomp it hard." Carr echoed Pogue's sentiments two days later, expressing similar concerns to channel B. Their complaints highlighted the growing strain on the crew as they struggled to keep up with the demanding schedule imposed by mission control.

These exchanges painted a picture of a crew under pressure, trying to manage the high expectations of a complex mission while grappling with the realities of life in space. The tension between the crew's capabilities and the mission's demands was becoming increasingly apparent, signaling challenges to be addressed as the mission progressed.

On December 27, 1973, Skylab's astronauts captured a striking image of Comet Kohoutek as it passed behind the Sun, utilizing the white-light coronagraph, designated as experiment S052. This sophisticated instrument, designed to create an artificial eclipse, allowed the crew to observe dim celestial objects in close proximity to the Sun. The photograph, taken by the High Altitude Observatory, depicted the comet during this rare event, where its brilliant tail contrasted against the solar backdrop.

The Apollo Telescope Mount (ATM), a key feature of Skylab, was primarily designed to keep the observatory's instruments centered on the Sun. However, tracking Comet Kohoutek required additional efforts, as the ATM had to be carefully oriented a few degrees away from the Sun. This task was particularly challenging since the astronauts relied solely on the coronagraph display for visual guidance. Initially, locating the comet proved difficult, but with persistence, the crew confidently navigated the telescope.

In anticipation of the comet's appearance, an extravehicular activity (EVA) was scheduled for Christmas Day, followed by another four days later. These spacewalks were crucial not only for reloading the ATM film but also for capturing images of Comet Kohoutek. The astronauts were assigned additional repair tasks during these EVAs, including pinning open a malfunctioning aperture door on the ultraviolet spectroheliograph and freeing a jammed filter wheel in the X-ray telescope.

On Christmas morning, after exchanging holiday greetings with Mission Control, astronauts Jerry Carr and William Pogue commenced the lengthy preparations for their spacewalk. Once outside, they began by taking various exposures of the comet with the coronagraphic camera. Carr then reloaded the ATM cameras and secured the experiment door that had been causing issues.

Encouraged by fellow astronaut Edward Gibson, Carr took a moment to appreciate the breathtaking view from the Sun-facing end of the telescope mount.

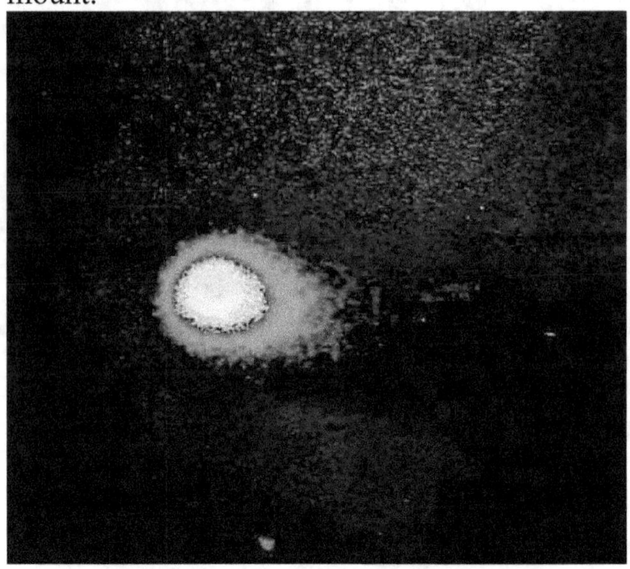

Comet Kohoutek's hydrogen halo was clearly evident in the far-ultraviolet camera photograph taken from Skylab on Christmas Day, 1973.

Next, Carr and Pogue attached the electronographic camera to capture additional images of the comet. Although neither could see the comet directly, they aimed the camera toward the region where it was expected to appear and followed the prescribed sequence of exposures. As the EVA progressed, Carr moved to the central workstation on the telescope mount to address the jammed filter wheel.

This task was challenging, as there had been no opportunity to practice this repair on Earth. Using a flashlight and a large mirror similar to those used by dentists, Carr carefully located the filter holder and confirmed it was stuck between two positions. With a steady hand, he used a screwdriver to nudge the wheel into an open position, ensuring no filter blocked the view.

During this delicate operation, Carr slipped momentarily, causing the shutter to close and bending one of the screwdriver's thin blades. Concerned that he might have damaged the instrument, Carr paused his efforts until radio contact with Mission Control was reestablished. After discussing the situation with the ground team, it was decided to bend the shutter blades out of the way, leaving the aperture fully open. With this issue resolved, Carr successfully moved the filter wheel to the correct position, concluding their planned work. By the time they returned inside the airlock, 6 hours and 54 minutes had passed.

The following day, December 28, Czech astronomer Lubos Kohoutek, the comet's discoverer, graced NASA's Mission Control in Houston with a brief yet highly publicized 11-minute conversation with the crew of Skylab. Although the exchange didn't yield any groundbreaking insights, it served as a symbolic gesture emphasizing the immense significance of the comet's observation. Despite not being a comet expert, Kohoutek handled the attention with remarkable poise and good humor, becoming a central figure in the American media's event coverage.

On December 29, during a third EVA dedicated explicitly to observing the comet, astronauts Gibson and Carr finally got an unobstructed view of Kohoutek. Gibson provided a brief description to Mission Control as the comet passed into the airglow just after orbital sunset. The two astronauts then retrieved samples of materials from outside the spacecraft, set up cameras for additional comet photography, and captured images after completing another orbit around the Earth. Gibson later detailed the comet's size, orientation, and the prominent spike extending toward the Sun.

Over the next few days, the crew devoted significant time to observing Comet Kohoutek, fully using the ATM instruments while the comet remained near the Sun. As the comet began its rapid departure from the Sun, observation shifted to other instruments, with the final sightings occurring on January 5, 1974. It was noted that the comet, after its journey through the inner solar system, would not return for another 75,000 years.

Despite the additional workload brought on by the comet's observation, the Skylab crew managed to stay on top of their assignments during the busy month of December. While Jerry Carr occasionally voiced concerns about the demanding schedule, the crew remained focused and committed, successfully balancing their comet-watching duties with their regular mission tasks. As the holiday season approached aboard Skylab,

the crew found time to bring a bit of festive cheer to their orbiting home. Using packing material from food storage cans, they fashioned a crude Christmas tree, adorning it with makeshift ornaments. Despite this brief respite, tensions simmered beneath the surface, stemming from a growing disconnect between the crew and the flight planners on the ground.

The flight planners, skilled in maximizing every minute of the astronauts' day, took pride in the sheer volume of scientific data their meticulous schedules could yield. To them, the efficiency of these plans was a testament to their expertise. However, the astronauts held a different view. They believed their primary responsibility was to produce high-quality scientific results, not merely to churn out large quantities of data according to a rigid schedule. The inflexible nature of the daily routines, where even the smallest housekeeping tasks were tightly scheduled, left the crew feeling constrained and frustrated. To them, it seemed as though the flight plans were dictating their every move, leaving little room for the autonomy they believed was essential for meaningful work.

By Christmas, the crew—Commander Jerry Carr, Scientist-Astronaut Edward Gibson, and Pilot William Pogue—decided that something needed to change. They agreed that they had to establish a better understanding with the flight planners regarding how the mission should be conducted. On the evening of December 28, after the daily status report was transmitted, Carr hinted at the brewing discontent in a conversation with CapCom Richard Truly. He mentioned that he was preparing a special message for Mission Control, which he intended to transmit on a private channel before retiring for the night.

Carr's message, recorded on the onboard system, was a six-minute plea for an honest discussion about the mission's progress at its halfway point. In his message, he expressed concerns that the crew's approach, which was deliberately slower and more measured than the previous Skylab crew, might be falling short of expectations. He questioned whether the flight controllers were worried about the amount of free time the crew was requesting, particularly for exercise, and whether this was seen as unreasonable. Carr emphasized that he was open to a private discussion but was prepared to address these concerns openly. The central question he posed was, "Where do we stand? What can we do if we're running behind and need to get caught up? We want some straight words on just what the situation was right now."

In hindsight, Carr regretted waiting so long to address these issues. "We swallowed a lot of problems for a lot of days because we were reluctant to admit publicly that we were not getting things done right," he later reflected. This reluctance, he acknowledged, was an ordinary human failing. His crewmates, Gibson and Pogue, wholeheartedly agreed with this assessment.

The need for a candid exchange of views was not lost on the ground personnel either. Robert Parker, one of the mission's key figures, recalled that the open communications channel often inhibited the ground team from being entirely forthright with the crew. No one who spoke directly to the astronauts ever suggested they were underperforming. Parker noted, "We just very seldom found ourselves capable of calling a spade a spade." This hesitation grew more pronounced as media reports began to suggest that the third Skylab crew was slower and more prone to errors than their predecessors.

The tension between the crew's desire for autonomy and the ground team's focus on efficiency highlighted the challenges of maintaining a productive and harmonious relationship during extended missions in space. Resolving these issues would be crucial not only for Skylab's success but also for the future of long-duration human spaceflight.

The tension between the Skylab 4 crew and Mission Control had been building for some time, with Houston becoming increasingly defensive about the astronauts, feeling that the crew was being unfairly criticized. Commander Gerald Carr, determined to address the growing concerns, called for "straight words" from Mission Control, a request that resonated deeply with the team on the ground. On the evening of December 29, 1973, Mission Control acknowledged Carr's message, recognizing the need for a candid and constructive dialogue. The following day, flight planners sent a detailed teleprinter message to the crew, outlining their perspectives and proposing an air-to-ground discussion for the evening of December 30th. While intended to resolve immediate issues, this conversation was more significant as it represented an important step in improving communication and mutual understanding between the crew and the ground team.

During the discussion, one of the first points raised by Richard Truly, the CapCom (Capsule Communicator), was that Mission Control was not fully aware of Commander Carr's intention to adopt a more deliberate pace of work than the previous Skylab crew. The flight planners had initially attempted to push the third crew to match the pace of the second, but when it became clear that this was unrealistic, they had scaled back the workload. To their surprise, when comparing the productivity of the two crews between the 15th and 30th mission days, they found no significant difference in output. This realization underscored the importance of allowing the crew to work at their pace, particularly given the extended duration of their mission.

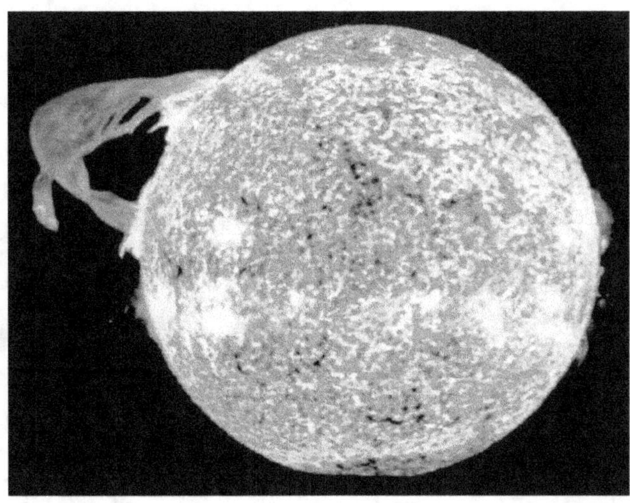

This photograph of the Sun, taken on December 19, 1973, during the third and final manned Skylab mission, shows one of the most spectacular solar prominences ever recorded, spanning more than 588,000 kilometers (365,000 miles) across the solar surface. The loop prominence gives the distinct impression of a twisted sheet of gas in unwinding itself. In this photograph the solar poles were distinguished by a relative absence of supergranulation network, and a much darker tone than the central portions of the disk. Several active regions were seen on the eastern side of the disk. The photograph was taken in the light of ionized helium by the extreme ultraviolet spectroheliograph instrument of the U.S. Naval Research Laboratory.

The conversation then turned to specific scheduling issues, notably the challenge of fitting physical exercise into the daily routine. Carr insisted on the need for sufficient time to cool down and clean up after his workouts on the ergometer, as he found it unbearable to rush from one task to another while still feeling grimy and overheated. The 90 minutes initially allocated for exercise each day had caused significant scheduling difficulties, prompting Mission Control to propose splitting it into two 45-minute sessions. Carr, however, expressed his dissatisfaction with this solution, emphasizing the physical and psychological strain of having to endure two rushed exercise sessions each day.

Another sensitive topic was the crew's need for uninterrupted personal time. The astronauts valued having some free time after waking up in the morning and again at the end of the day to relax and unwind, particularly important given their 12-week mission in space. While Mission Control was willing to allocate an uninterrupted hour before bedtime, they reserved the right to interrupt this time if a valuable scientific opportunity arose. Carr acknowledged this compromise but voiced the crew's growing frustration with the frequency of such interruptions. Truly, in response, recognized the importance of balancing scientific goals with the crew's well-being and reassured Carr that they would strive to be more considerate in future planning.

Dr. Lubos Kohoutek, discoverer of the Comet Kohoutek, was seen in the Mission Operations Control Room in the Mission Control Center during a visit to JSC. He was talking over a radiotelephone with the Skylab 4 crewmen in the Skylab space station in Earth orbit. Professor Kohoutek, a well-known Czechoslovakian astronomer who works at the Hamburg Observatory in West Germany, discussed the comet with Astronauts Gerald P. Carr, Edward G. Gibson, and William R. Pogue. Dr. Zdenek Sekania, who accompanied Dr. Kohoutek on the visit to JSC, was on the telephone in the left background. Dr. Sekania was with the Smithsonian Observatory in Cambridge, Massachusetts.

As the conversation drew to a close, Truly offered words of encouragement, acknowledging that the crew had successfully handled the demanding workload of the past few weeks. He hinted that Houston still aimed to maximize scientific output, urging the crew to communicate any gaps in their schedule that might allow for additional work. He emphasized the value of discussing flight plan issues during live air-to-ground communications rather than relying on voice dumps so that the crew could directly engage with the planning team. Carr agreed to this approach, marking a new phase in the relationship between the Skylab 4 crew and Mission Control, one based on open communication and mutual respect.

This dialogue between Carr and Mission Control was not just about resolving the immediate challenges of scheduling and workload. It was a pivotal moment in the Skylab 4 mission, highlighting the complex dynamics of long-duration spaceflight, where the physical and psychological well-being of the crew had to be carefully balanced against the mission's scientific objectives. The discussion ultimately led to a better understanding between the crew and the ground team, setting the stage for a more collaborative and successful continuation of the mission.

During the 20-minute communications gap that followed the initial discussion with Mission Control, Commander Gerald Carr took the opportunity to consult with his fellow crew members, William Pogue and Edward Gibson. Together, they crafted a response that balanced the crew's needs with the demands of the mission. Carr remained firm on the importance of having some quiet time at the end of each day, a crucial element for maintaining their mental well-being during the grueling 12-week mission. However, he also showed a willingness to compromise by agreeing to consider breaking up their exercise periods if it would help alleviate the scheduling difficulties. Carr further suggested that non-time-critical activities, such as some of the corollary experiments and most housekeeping tasks, could be performed at the crew's discretion. This approach would allow the astronauts to exercise judgment in their daily routines, offering a reprieve from the rigid, automaton-like existence they had endured for the past six weeks.

The 55-minute discussion concluded with Richard Truly expressing Mission Control's satisfaction with the outcome. He relayed that Dr. Christopher Kraft, Director of the Johnson Space Center, and Deke Slayton, Chief of Flight Crew Operations, had listened in on the conversation and were pleased with how the crew was handling the mission. Truly remarked that they had made what felt like "a million dollars" that night, an indication of the value placed on the open and honest exchange that had taken place. While the exact benefits of the conversation were not immediately clear, there was a palpable sense of relief among all involved. The successful dialogue demonstrated that candid discussions could be held without causing significant issues, and it reassured the crew that their concerns were being taken seriously. This realization significantly boosted crew morale, as they now felt confident that any difficulties could be quickly addressed and that mission planners were responsive to their needs and preferences.

Reflecting on the situation, it became evident that the delay in reaching this level of candor resulted from ground personnel not fully understanding that the third crew required a different approach than the previous two. Unlike some other astronauts, Carr was not one to readily express dissatisfaction, which may have contributed to the initial communication challenges.

Although he had vowed before the launch that he would speak up if Mission Control pushed the crew too hard, Carr's mishandling of Pogue's illness on the first day of the mission had put him on the defensive. He felt compelled to make up for that early mistake by pushing the crew to produce results. By the end of the mission, Carr accepted some responsibility for the communication breakdown but also criticized the flight planners for not allowing the crew sufficient time to adjust. "Obviously, they were not thinking," he later remarked. "They were just coloring squares and filling in checklists. That was no way to operate a mission."

Daily housekeeping involved the stowage of biologically active trash, such as food cans and urine waste, in the waste tank. Here, Carr opens the trash airlock through which bagged trash would be passed into the waste tank.

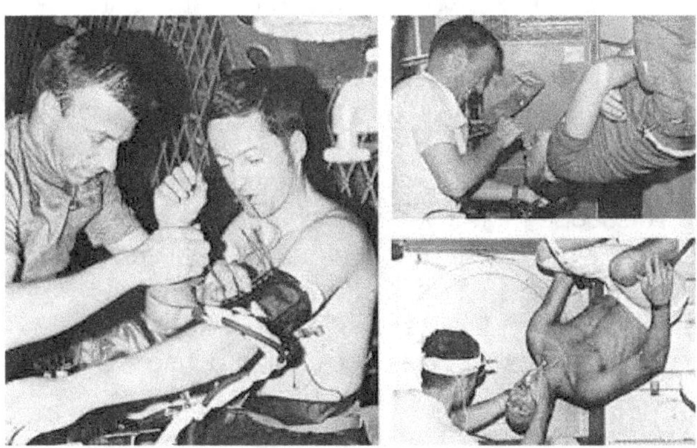

Throughout the three manned periods, the medically trained astronauts conducted frequent and thorough examinations of each other, to assure their continued good health. All flight crewmen remained in excellent health throughout.

Astronaut Gerald P. Carr, Commander for the Skylab 4 mission, jokingly demonstrates weight training in zero-gravity as he balances astronaut William R. Pogue, pilot, upside down on his finger.

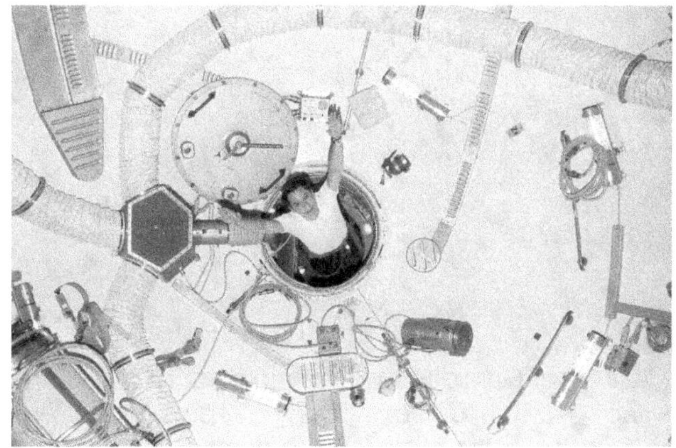

Scientist-Astronaut Edward G. Gibson, science pilot for the Skylab 4 mission, demonstrates the effects of zero-gravity as he sails through airlock module hatch.

Scientist-astronaut Edward G. Gibson, Skylab 4 science pilot, stands at the Apollo Telescope Mount (ATM) console in the Multiple Docking Adapter (MDA) of the Skylab space station cluster in Earth orbit. This picture was taken with a hand-held 35mm Nikon camera. The ATM console was one of the busiest areas of the space station during the 84-day third manned Skylab mission, as Comet Kohoutek and solar activity were closely followed by the ATM and monitored by the crewmen from the ATM console. As Gibson demonstrated during a television transmission on Dec. 5, 1973, the ATM console controls several instruments on the solar telescope. Joining Gibson for the record-setting Skylab 4 mission were astronauts Gerald P. Carr, commander, and William R. Pogue, pilot. Photo credit: NASA

Astronaut Gerald P. Carr, commander for the Skylab 4 mission, demonstrates the effects of zero-gravity as he floats in the forward dome area of the Orbital Workshop of the Skylab space station while in Earth orbit.

Carr and Gibson look through the length of the station from the trash airlock.

Skylab -- February 1974 Scientist-astronaut Edward G. Gibson has just exited the Skylab extravehicular activity hatchway. Astronaut Gerald P. Carr, Skylab 4 commander, took this picture during the final Skylab spacewalk that took place on Feb. 3, 1974. Carr was above on the Apollo Telescope Mount when he shot this frame of Gibson. Note Carr's umbilical/tether line extending from inside the space station up toward the camera. Astronaut William R. Pogue, Skylab 4 pilot, remained inside the space station during the spacewalk by Carr and Gibson.

In the aftermath, members of the Mission Control team dplayed the significance of this discussion and the issues that had led up to it, focusing instead on the overall success of the Skylab 4 mission. However, at the time, everyone was relieved that the air had been cleared. The impact of the conversation was almost immediately apparent. Just two days later, Flight Director Neil Hutchinson noted a marked improvement in the crew's performance. The astronauts appeared more alert, were actively looking ahead in the daily flight plan, and were organizing their activities to optimize their work schedule. Remarkably, they managed to stay ahead of the flight plan for the entire day, demonstrating the effectiveness of the new mission management approach.

The Pogue Seiko, a 'Seiko Automatic-Chronograph' Cal. 6139, the first automatic chronograph in space, used by Bill Pogue

84 Days Around The World

In early January, the crew of Skylab 4—Commander Gerald Carr, Pilot William Pogue, and Science Pilot Edward Gibson—found themselves approaching a significant milestone in spaceflight history. By January 4, 1974, they had surpassed Pete Conrad's cumulative spaceflight duration record, which had taken Conrad four separate missions to achieve. However, it was not until January 25 that the first all-rookie crew in eight years would officially set a new world record for time spent in space, eclipsing the second Skylab crew's record. Despite these impending accolades, Carr, Pogue, and Gibson were not focused on setting endurance records. For them, these achievements were secondary to their primary goal: completing the mission as planned.

With their earlier tensions with Mission Control now resolved, the crew approached their work with renewed enthusiasm and dedication.

January 10 offered the astronauts a rare day off, a welcome respite in their demanding schedule. Although "day off" in space still meant some structured activities, only about a third of their time was formally scheduled, allowing them the freedom to pursue their interests. Gibson, deeply fascinated by solar observations, spent almost the entire day studying the sun. Pogue and Carr, on the other hand, found themselves drawn to the wardroom window, where they spent much of the day making observations, taking photographs, and simply marveling at the ever-changing views of Earth below. Like the crews before them, they were captivated by the beauty and dynamism of the planet, which offered a constantly shifting panorama from their vantage point in space.

While the astronauts enjoyed their brief respite, mission managers on Earth were engaged in a crucial review of the mission's progress at the 56-day mark. Their primary concern was whether both the crew and the spacecraft were fit to continue for the full 84-day mission originally planned. On January 11, Bill Schneider, Skylab Program Director, announced the decision: the mission was "GO" for 84 days. Although the formal approval was technically granted weekly, there was little doubt that the crew would complete the full 12-week duration. The only significant concern was the status of the control moment gyroscope, an essential piece of equipment showing signs of distress. Even if the gyroscope failed, there was no immediate danger to the crew. They would have ample time to retrieve the Apollo Telescope Mount (ATM) film, pack up their command module, and depart from the workshop in an orderly manner.

The gyroscope issue, however, was becoming increasingly worrisome. Engineers on the ground suspected that the problem stemmed from inadequate lubrication of the wheel bearings. To mitigate the risk of a complete failure, they began to carefully manage the spacecraft's maneuvers carefully, aiming to reduce the stress on the gyroscope's bearings. Starting in late December, engineers also took over manual control of the bearing heaters, maintaining temperatures within the allowed limits' upper range. The hope was that by keeping the oil warmer, it would become less viscous and flow more easily into the bearings, thus improving lubrication. Despite these efforts, little more could be done to address the issue.

As a result of the gyroscope's instability, experiments requiring precise maneuvering of the spacecraft had to be meticulously planned. Earth-resources photography, in particular, became a challenge. The passes over Earth had to be perfectly aligned before being added to the flight plan, and weather conditions in late December and early January did not always cooperate, leading to some setbacks in this area. Nevertheless, as the mission reached its 56-day milestone, the crew had completed roughly two-thirds of the experiment program. They were well on their way to fulfilling the mission's scientific objectives despite the challenges posed by the ailing gyroscope and the complexities of long-duration spaceflight.

As Skylab's third and final mission progressed, the crew faced the inevitable task of managing the station's waste. The trash airlock, which had once been the liquid-oxygen tank on the S-IVB stage of the Saturn V rocket, now served as the repository for the station's refuse. As the mission drew to a close, the need to pack away the remaining bags of trash became a priority. Astronaut William Pogue, ready to assist, positioned himself with his feet firmly against the ceiling, preparing to use his body weight to seal the hatch. This effort would allow fellow astronaut Gerald Carr to securely stow the last of the bags securely, ensuring the airlock could be properly closed.

Inside Skylab, the scene was a mix of routine and urgency. A rare view from near the hatch of the airlock module provided a glimpse of the workshop's interior. Through a passageway, one could see a crew member busily stashing trash bags, a task that had become an unglamorous yet essential part of daily life aboard the station. Two spacesuits hung nearby while another crew member attended to tasks on the upper deck, a silent testament to the continuous efforts that kept Skylab operational.

Beyond the daily chores, the crew remained deeply engaged in the scientific objectives of their mission. The solar observations, one of the key experiments, were progressing well, with a significant amount of observing time and photographs accumulated. However, the sun had been disappointingly quiet, with low solar activity. The sun's corona had sh some activity, but it mainly occurred while the crew was asleep, missing opportunities to capture these events in real-time. Solar scientists back on Earth, including Ed Gibson, were particularly eager for a flare—a sudden and intense burst of solar radiation—to occur. Capturing a flare from its inception to its peak was a rare opportunity that had eluded scientists thus far. With only four weeks left in the mission, Gibson knew his chances were dwindling.

In early January, solar scientists predicted that some active regions of the sun, previously out of view, would rotate back into focus. Gibson, ever vigilant, expressed his desire to dedicate significant time to a "flare wait" mode, staying ready to capture pre-flare activity as soon as it began. On January 10th, Robert MacQueen, the principal investigator for the coronagraph experiment, discussed the strategy with Gibson for the remaining weeks. Both scientists shared the hope that the sun would finally present them with the solar activity they had been waiting for. This was their last opportunity after more than a decade of preparation, and MacQueen granted Gibson the flexibility to adjust the preplanned programs at his discretion. Following this conversation, the entire Skylab crew participated in a general science conference with experimenters 'representatives in Houston. These conferences, scheduled several times throughout the mission, were typically held on the crew's days off. They provided a platform for experimenters to brief the astronauts on the various science programs, discuss strategies for the upcoming days, and incorporate the astronauts 'insights into experiment planning.

Daily instructions were sent up to Skylab via teleprinter, but these conferences offered a more interactive way to bridge the gap between the astronauts and the mission planners on Earth. The discussions helped ensure the crew understood the scientific objectives and mitigated any feelings of isolation that might arise during their extended time in space.

As January progressed, the Skylab crew faced new challenges. By mid-month, the Earth's position in its orbit, combined with the high inclination of Skylab's orbital plane, meant that the spacecraft remained in continuous sunlight for 46 revolutions—a situation known as Skylab's "midsummer day." The crew took special measures to reduce the load on the cooling systems, a critical concern given the sustained exposure to sunlight. Mission Control recommended that the crew avoid showering during this period to prevent increased humidity, though they did not enforce this suggestion strictly. Nevertheless, the temperatures within the workshop began to rise, reaching a peak of 28°C by January 18th.

The heat posed a particular challenge for Ed Gibson, whose sleeping compartment was not fully protected by the improvised solar shields that had been installed earlier in the mission. As a result, he relocated his sleep restraint to the cooler airlock, where the temperature was more tolerable. This move introduced a new operational constraint, as the teleprinter in the airlock was noisy, and Mission Control sought to minimize its use while Gibson slept.

On January 20th, CapCom informed Gibson that observers had detected two small solar flares in an active sun region within six hours. Despite the initial excitement, subsequent reports from Houston dampened hopes that anything significant might occur. However, Gibson remained optimistic, noting that the region still appeared promising, and he continued to monitor it closely. From this point onward, Gibson would take the lead at the console, with both Carr and the mission's scientists determined to ensure that, should any interesting solar activity arise, they would be ready to capture it.

During the second mission, Mission Control had relaxed a long-standing rule, allowing individuals other than CapCom to communicate directly with the astronauts. This change facilitated more direct and immediate collaboration between the crew and the scientists, a crucial adjustment as they worked together to make the most of their remaining time aboard Skylab.

As the Skylab mission neared its conclusion, astronaut Ed Gibson became increasingly focused on maximizing the opportunities for solar observations, particularly the elusive solar flares he had been so eager to capture. The flight plans were occasionally adjusted, and duties were exchanged among the crew, allowing Gibson to spend more time at the control and display panel, anticipating solar activity. Despite the sun's apparent inactivity as reported by ground-based observers, Gibson remained optimistic, confident that a flare was imminent.

During one of his afternoon watches, Gibson's determination grew. He was so convinced that a flare was about to occur that he offered a playful bribe to his fellow astronaut Gerald Carr, hoping to secure another orbit at the panel. "I've already promised the commander some butter cookies when we get back if I could have the orbit," Gibson joked with Houston, adding that he would be willing to pay the price of a bottle of Scotch if it meant he could continue his observations. The light-hearted banter masked the serious intent behind Gibson's request—he was determined not to miss a moment of potential solar activity.

As the hours passed, Bill Pogue was scheduled to take over the ATM (Apollo Telescope Mount) on the next orbit. However, when Houston sent up some instructions from the solar scientists, Pogue humorously pointed out a slight complication: "Ed has the MDA hatch barricaded up there." Undeterred, Gibson remained at the panel, and his perseverance finally paid off. Just before communications with Houston were temporarily lost, Gibson announced, "I think this time we finally got one on the rise." Elated, he immediately went to the channel B recorder, dictating a detailed 23-minute event description. Once communications were restored, he repeated his report over the air-to-ground channel, ensuring Houston had all the details. That night, Gibson went to bed with a sense of fulfillment, knowing that his efforts had been rewarded.

Throughout the remainder of the mission, the crew faced concerns about one of Skylab's gyroscopes, which had been malfunctioning intermittently. The gyro's instability prompted Program Director William Schneider to order the prime recovery ship to prepare for an early retrieval of the crew. Fortunately, the gyro settled down and continued to function adequately despite operating at a reduced speed. The possibility of gyro failure briefly brought Skylab back into the spotlight, but public interest in crewed spaceflight had waned by this stage. On January 23, the major television networks made a significant decision: for the first time since live coverage began with Gemini 6 in 1965, they would not broadcast the return of a crew from space live. This marked a turning point as the excitement and novelty of space exploration began to fade from public consciousness.

Despite lacking media attention, the Skylab crew remained committed to their mission. On January 31, they held their second televised press conference of the mission, during which they expressed their strong belief in the value of Skylab and the scientific data they had collected. The crew argued that the program had demonstrated the indispensability of human presence in orbital science, which allowed a level of productivity and flexibility that automated systems alone could not achieve. Gibson was confident enough to predict that, while space stations and crewed planetary expeditions were still far in the future, they were clearly possible "when the American people choose to make the effort."

Carr, reflecting on the mission's lessons, emphasized the importance of habitability in the design of future space stations. He pointed out that it was not just a matter of having functional quarters and work areas but also of creating a space where astronauts could find some personal comfort and solitude—a place they could call home, even if only temporarily. When asked about the public's apparent lack of interest in the mission, Carr remarked, "Well, I think people just get used to things and take them for granted. As long as things stay rather routine in the space program, public interest will stay pretty low." The press conference was brief, leaving little time for all the questions. However, CapCom Dick Truly later relayed four questions submitted by a sixth-grade science class in upstate New York during the next orbit. These questions, surprisingly penetrating, touched on topics that even professional journalists had overlooked. One question, directed at Bill Pogue, asked whether he "felt more of a man now, as compared with before you left?" Pogue, caught off guard by the philosophical nature of the question, hesitated to delve into its implications but did admit that he felt he had become a more efficient astronaut after 77 days in space.

Another question from the students asked whether the astronauts missed female companionship, a topic that had not been addressed before. Gibson, surprised by the directness of the question, quipped, "What grade did you say that was, Dick?" before admitting, "Obviously, yes."

As February began, the mission entered its final phase. February 1 marked the last full day dedicated to experiment work, including an earth-resources pass, various medical experiments, and a final attempt to observe the comet Kohoutek. On February 2, Ed Gibson completed his last observations from the ATM console. The following morning, Carr and Gibson ventured outside the station to retrieve the ATM film carriers and collect samples from some particle collection experiments. Gibson took several photographs, carefully documenting the condition of the twin-pole sail after its long exposure to the harsh space environment.

With the mission drawing to a close, the crew turned their attention to the significant task of shutting down the workshop and preparing Skylab for its eventual reentry into Earth's atmosphere. On the evening of January 31, Houston sent up a lengthy list of changes to the deactivation and reentry checklists. The following day, Carr was greeted with an overwhelming 15 meters of teleprinter paper detailing these modifications. Updating the checklists by hand became an ongoing task that filled the crew's idle moments over the next two days, providing ample material for lighthearted banter. That evening, Carr greeted Bruce McCandless, who was coming on shift, with a joke: "I understand you're going to teleprinter up the Old Testament tonight."

Even as the deactivation process continued, the crew focused on completing the remaining medical experiments. On February 4, Carr conducted zero-g flammability tests, which had been postponed until the end of the mission due to concerns about contamination from the residues being exhausted into space. While seemingly routine, these final tasks were vital in ensuring that the scientific objectives of the Skylab mission were fully realized, even as the crew prepared to return to Earth.

The effect of gravity on flame behavior in space provided a unique opportunity for scientific study aboard Skylab. Under the influence of gravity on Earth, the hot gaseous products of combustion rise due to convection, which allows cooler air with additional oxygen to enter the flame and mix with the fuel, sustaining combustion. However, convection was absent in the microgravity space environment, causing a flame's corona to form a spherical shape. This spherical flame quickly consumes the available oxygen around it, diminishing the flame until more oxygen diffuses into the area. This phenomenon was meticulously studied in Experiment M479, which provided insights into fire behavior in space—a crucial consideration for future space missions.

During their time on Skylab, the astronauts made significant observations, some of which were tied to the spacecraft's passage through the South Atlantic Anomaly. Earth's magnetic field was weaker in this region, allowing more cosmic radiation to penetrate the atmosphere. Astronaut William Pogue took the opportunity to observe the light flashes he experienced while Skylab passed through this anomaly. These light flashes, previously observed during Apollo missions, were hypothesized to result from cosmic rays interacting with the retina. However, Pogue's experiment suggested a strong correlation with the South Atlantic Anomaly, challenging the prevailing cosmic-ray hypothesis. Strapped into his sleep restraint, Pogue meticulously noted the time, direction, and shape of each flash, contributing valuable data to the understanding of space radiation effects on human vision.

As the mission neared its conclusion, the crew faced the logistical challenge of packing up and preparing for their return to Earth. Like travelers struggling to fit souvenirs into an overstuffed suitcase, the astronauts ran out of space in their command module. Commander Gerald Carr humorously reported that, despite his best efforts to rearrange the contents, he had to force five Earth resource tapes into a command module locker to

make them fit. Meanwhile, Edward Gibson encountered similar difficulties with the trays containing the mission's urine and blood samples, highlighting the practical challenges of life in space.

NASA's plans for Skylab after the final mission were a topic of speculation. Neil Hutchinson, a NASA flight director, explained that while it was technically possible to revisit Skylab, it was highly unlikely due to the lack of atmosphere, power, and food aboard the station. Additionally, the systems onboard Skylab were expected to deteriorate, making the prospect of reactivating and reusing the workshop both risky and impractical. Despite this, before departing, the crew used the Apollo thrusters to boost Skylab's orbit, extending its operational life by another five to eight years. This decision was made in anticipation of the Space Shuttle program, which could potentially retrieve some of Skylab's components for further study.

As the third crew (Skylab-4) departed the space station after 84 days in the orbiting laboratory. A smiling Skylab seemed to wink good-bye for the job well done.

Left: Splashdown of Skylab 4, ending the longest crewed mission to that time. Right: The

Chapter 7 - Skylab Unmanned

On February 8, 1974, the Skylab 4 crew—Carr, Gibson, and Pogue—prepared for their return to Earth. As they transitioned into the command module, they faced an unexpected challenge during reentry.

Film retrieval was the final act before departure. Here, film was removed from the solar observatory cameras. A retractable boom made handling in space easy.

At 9:36 a.m. Houston time, Carr fired the service module's propulsion engine to initiate their descent. However, nine minutes later, when Carr attempted to maneuver the spacecraft using the hand controller, he was shocked to find no response to yaw and pitch commands.

This sudden loss of control, which had not been indicated during pre-flight checks, caught the crew off guard. After a brief moment of confusion, Carr quickly switched to a backup system, regaining control of the spacecraft.

It was later discovered that four circuit breakers had been inadvertently opened, disabling the yaw and pitch thrusters—a reminder of the importance of maintaining proficiency through repeated simulations during long missions.

Following a successful reentry, the crew splashed down in the ocean, waiting for about half an hour before being retrieved by recovery teams. Despite the challenging conditions of their mission, none of the astronauts experienced seasickness, a testament to their resilience and preparation. As they looked back at the workshop they had called home, they knew that Skylab would continue orbiting Earth for several more years, a silent testament to the achievements of human space exploration.

The return to Earth's gravity after months in space was a jarring experience for the Skylab astronauts. Edward Gibson, in particular, was acutely aware of the sudden burden gravity placed on his body. The weight of his head felt immense, and even the simple act of moving his arms required a significant effort, as though he were still in the early stages of re-entry. William Pogue experienced a similar shock when he retrieved a camera from its locker while they were still descending under parachutes. The camera, which had felt nearly weightless in space, now weighed thirty-five or forty pounds. After taking just one photograph of the parachutes, Pogue struggled to hold the camera, fearing he might drop it due to its unexpected heaviness. Unable to return it to the locker, he held onto the cumbersome device until the command module splashed down in the ocean.

Now unoccupied, the space station circled the Earth at its orbital attitude. While unmanned, it operated at reduced power and with many of its systems either inoperative or operating at reduced capacity. But Skylab was now a fully operational space station, its scientific value well established.

As the astronauts began their first round of postflight medical tests, NASA officials in Houston held the traditional post-mission press briefing. Administrator James C. Fletcher highlighted the importance of Skylab's accomplishments, emphasizing that the program had transitioned crewed spaceflight from the spectacular realm into a new phase that could be described as almost businesslike, if not entirely routine. William Schneider, the Skylab Program Director, provided an overview of the mission's achievements, noting that every experiment had exceeded pre-mission expectations, with some surpassing them by more than 200%. However, he stressed that this was the beginning: "Our portion of Skylab has been completed. The science phase has just begun." Skylab had demonstrated that the only limits to space research were human resolve and technical knowledge, not the ability of astronauts to work in space.

Following the crew's departure from Skylab, engineers conducted various tests on the main power system's batteries to assess the extent of deterioration during the mission. They also tested the Apollo Telescope Mount (ATM) computer's memory by unloading and reloading it, finding that the system still functioned perfectly. Despite their efforts, they could not revive the failed control moment gyro. Still, they measured bearing friction on the remaining gyros as they wound d, attributing the failures to inadequate lubrication. On the afternoon of February 9, flight controllers maneuvered Skylab into an attitude stabilized by the gravity gradient, with the docking adapter pointed away from Earth, and then shut off the station's power. With the mission over, Mission Control fell silent, the remnants of activity swept away like cigar ashes.

Results

As Schneider had indicated, the missions were only the beginning of Skylab's extensive science program. Principal investigators quickly began processing the vast amounts of data and material the crews had collected. The five solar telescopes onboard captured nearly 103,000 photographs and spectra, supplemented by 68,000 images from the H-alpha cameras. Earth-resource instruments had generated piles of photographs and kilometers of magnetic tape, rich with detailed information. Medical researchers had gathered 18,000 blood pressure measurements, 200 hours of electrocardiograms, and many food, urine, and fecal samples for biochemical analysis.

While only a fraction of this data had been accessible during the missions, most related to the astronauts' health, Houston's medical team was crucial in monitoring and maintaining crew health throughout the missions. They continuously advised program managers on the astronauts' physical condition using telemetered data, medical conferences, and regular reports. Any signs of unfavorable trends or sudden changes could have led to the curtailment of a mission. For the other experimenters, the bulk of their data—film, tape, and samples—had to wait until the crews returned to Earth. After each of the first two missions, quick-look assessments were conducted to suggest changes or additions to experiment plans for the subsequent flights.

The detailed and laborious evaluations of the data collected during Skylab's missions continued for years. Even during the later flights, preliminary results were presented at scientific meetings, and by the end of 1974, several major symposia had been held to summarize Skylab's findings. In late August of that year, medical investigators convened in Houston for three days to discuss the data from all the missions. These investigations were of paramount importance to the future of crewed spaceflight, as they provided crucial insights into humanity's ability to adapt to space conditions and remain healthy over long durations. The knowledge gained from Skylab would lay the groundwork for future explorations, pushing the boundaries of what was possible in space.

Science Accomplishments

During Skylab's missions, remarkable strides were made in various scientific fields, particularly in solar physics, life sciences, engineering, technology, astrophysics, and Earth observation. The planned and actual accomplishments of these experiments reflected the rigorous efforts of the astronauts and scientists involved, with certain deviations occurring due to the unique challenges of spaceflight.

One of the most significant areas of investigation was in life sciences, where experiments were designed to understand the effects of prolonged weightlessness on the human body. A symposium on life sciences revealed that, overall, the Skylab missions had provided critical insights into human physiology in space despite some ongoing challenges, such as motion sickness.

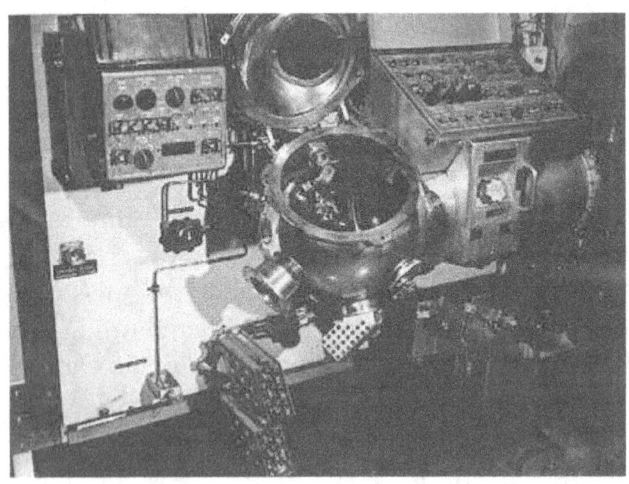

The materials processing facility aboard Skylab was a compact unit which permitted a number of significant experiments to be carried out during the three manned periods.

This issue, prevalent in early stages of the missions, affected five out of the nine Skylab astronauts. Interestingly, only the first crew and astronaut Ed Gibson, who was on the final mission, were spared from these symptoms. Notably, astronaut Joe Kerwin experienced seasickness while awaiting recovery in the command module, highlighting the unpredictability of space-related ailments.

To address motion sickness, Skylab carried out an experiment that involved a specially designed chair to induce motion sickness through rapid, multidirectional head movements. Although astronauts on Earth were susceptible to this test, none reached the same level of discomfort while in orbit. This experiment underlined the individualized nature of space sickness, which, despite using medication, remained an unpredictable problem. However, all crew members eventually adapted within the first week of their missions, and the illness did not recur, although Skylab's findings were insufficient to understand this complex condition fully.

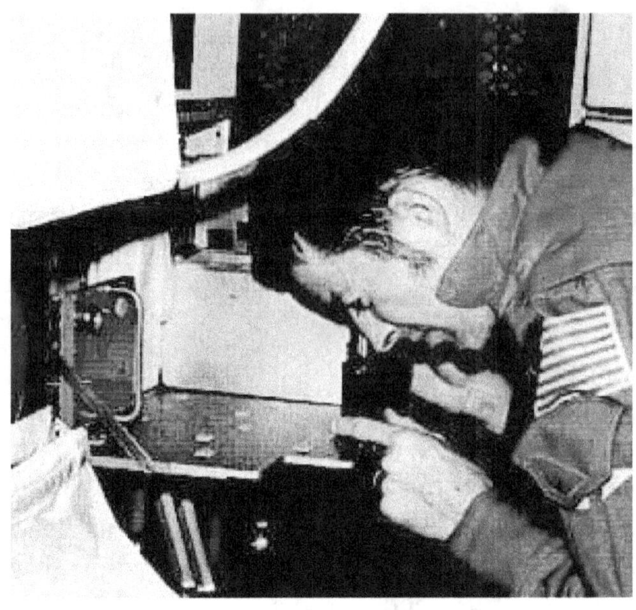

Experiments proposed by high school students in a nationwide competition were carried out during the three manned periods. Here Kerwin studies the growth in space of bacteria and spores.

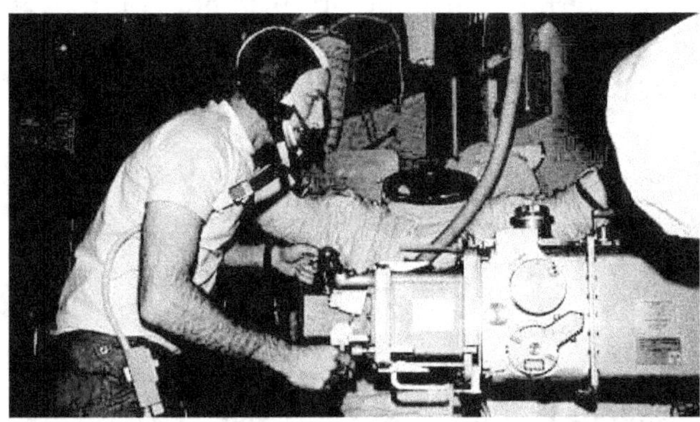

Skilled operators were required for much of the experimental equipment. Astronaut Bean was shown obtaining stellar ultraviolet spectra.

Skylab provided clearer results in other domains, such as mineral balance studies. The study showed that all crew members excreted more calcium in their urine during spaceflight, along with elevated levels of hydroxyproline, an amino acid linked to bone metabolism. This suggested a loss of structural material in bones typically subjected to compressive loads on Earth.

X-rays before and after the missions confirmed these findings, showing decreased bone density, particularly in weight-bearing bones. Even though the third crew increased their exercise routines, losing calcium and nitrogen—indicative of muscle mass loss—continued unabated throughout their mission.

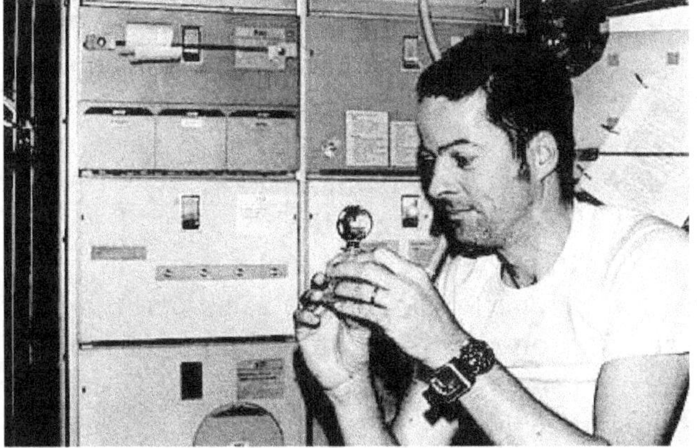

Joe Kerwin, the first scientist pilot, established the feasibility of performing a science demonstration using free-floating globules to investigate the damping of fluid oscillations, surface tension, coalescence of raindrops, and the general properties of fluids in the absence of Earth's gravity.

The flight crews' attention was required to operate the Earth resources cameras. Their ability to select targets for observation was invaluable.

While the amount of bone mineral lost after 84 days in space was not immediately concerning, the ongoing depletion suggested potential risks for more extended missions. Comparisons with bedridden patients on Earth, who experience similar conditions due to lack of mobility, indicated the possibility of irreversible damage to leg bones during missions lasting a year or more. Additionally, the high calcium concentration in the urine raised concerns about the potential formation of kidney stones, a risk that needed careful management on future long-duration flights.

Cardiovascular experiments aboard Skylab yielded complex but largely encouraging results. Using a bicycle ergometer and metabolic analyzer, scientists observed that the astronauts' tolerance for exercise did not diminish during the missions. However, postflight tests indicated that the astronauts had adapted to weightlessness, as they could no longer perform at their preflight levels of physical efficiency. The first crew took nearly three weeks to regain their preflight exercise capacity, while subsequent crews required less than a week, demonstrating a quicker readaptation process with each mission.

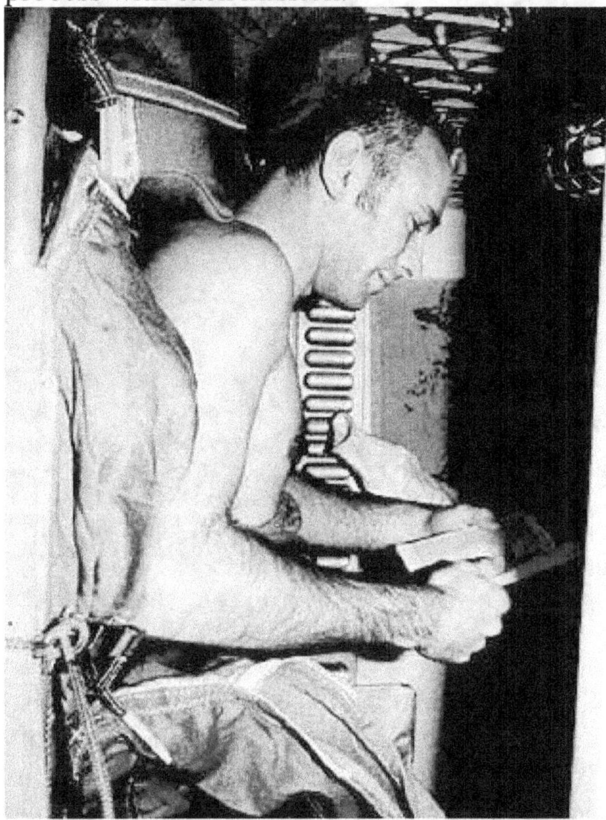

Some pastimes were considerably different from those enjoyed on Earth, but reading was a popular diversion. Here, Bean relaxes in his sleep restraints as he reads.

The lower-body negative-pressure experiment, designed to assess changes in cardiovascular function during extended periods of weightlessness, proved more challenging in space than anticipated. During the first mission, astronaut Joe Kerwin had to terminate two tests prematurely due to the experiment's stressful nature. After 28 days, adaptation appeared minimal, raising concerns among cardiovascular experts. Despite these initial challenges, the decision to continue the experiment on subsequent missions was justified, as the longer flights demonstrated that astronauts gradually built up tolerance to the testing, particularly after the first 30 to 50 days. While the first crew required almost three weeks to return to their preflight cardiovascular responses, the adaptation period was shorter for the later crews.

Kerwin prepares to sleep. In the zero-gravity atmosphere, beds were unnecessary. Crewmen zipped themselves into sleeping bags stretched against the wall.

Skylab's medical investigations also contributed significantly to understanding how weightlessness affects the human body. Various studies measured leg volume, body volume changes using stereophotographs, and alterations in hormonal and hematological profiles. Despite inconsistencies in the data, these investigations painted a detailed picture of the physiological changes that occur in space. Before Skylab, aerospace medical researchers had developed a working hypothesis to explain the effects of weightlessness. Upon entering microgravity, they believed body fluids shifted toward the upper body, leading to distended veins, puffy eyelids, and nasal congestion—symptoms reported by all astronauts. The body's sensors interpreted this as increased blood volume, prompting hormonal adjustments to reduce fluid levels.

This triggered a complex chain of physiological reactions, resulting in a new equilibrium characterized by a decrease in red blood cells, plasma, and potassium concentration. However, Skylab's medical data did not fully align with this hypothesis, revealing lower-than-expected hormone levels and unusual electrolyte balances. This suggested that the effects of weightlessness on the human body were more intricate than previously understood.

Skylab's legacy in life sciences was a groundbreaking discovery, albeit with some unresolved questions. The data collected during these missions laid the foundation for future research, highlighting the need for continued investigation into the challenges of long-duration spaceflight, particularly as humanity looks toward even longer journeys in space.

Additional experimental work was deemed necessary before a comprehensive understanding of adaptation to weightlessness could be achieved. Although no physiological changes had been observed to prevent weightless flights lasting up to nine months, the possibility of extending such missions indefinitely remained uncertain. Critical issues, particularly those concerning motion sickness and bone deterioration, needed to be thoroughly investigated before considering crewed missions of up to a year or longer.

During a panel discussion after the three-day medical symposium, several outside experts speculated on the implications of Skylab's findings. Most agreed that Skylab had resolved many key questions about human survival in orbit and the process of readapting to Earth upon return. However, the results also opened up new avenues for research, with participants offering innovative ideas for future investigations. Some experts envisioned a new generation of space laboratories where only a few occupants would need to be astronauts in the traditional sense. One suggestion was to send "professional subjects" into space— normal individuals without responsibility for managing the spacecraft. These individuals could allow their physiological systems to deteriorate deliberately, enabling scientists to test compensatory measures, both preventive and therapeutic.

In discussing the challenges posed by bone loss in space, one expert suggested that the physical qualifications for astronauts might need to be reconsidered. Recognizing the dual need for astronauts to function in zero gravity and during reentry, this expert postulated that "sedentary, skinny, small individuals" might have physical advantages in such an environment. He even proposed the idea of selecting legless amputees as astronauts since many of the medical issues associated with spaceflight, such as bone loss and cardiovascular strain, were linked to the legs. Despite the diverse perspectives, there was unanimous agreement that Skylab's medical investigations had raised as many questions as they had answered—a hallmark of impactful research. The consensus was clear: further exploration in space was necessary for more definitive answers.

Among the numerous experiments conducted, the mineral balance studies were the only ones that could be effectively simulated on Earth. Prolonged bed rest was found to adequately mimic the conditions of spaceflight, making it a valuable tool for understanding the effects of weightlessness on bone mineralization.

While the medical investigations had their share of complexities, astronomers were equally, if not more, overwhelmed with the data they had gathered. The sheer volume of collected photographs and spectra required months of cataloging, classifying, and calibrating, with interpretation expected to take longer. Despite the enormous task, astronomers began publishing preliminary results even before the second mission was complete. A mere month after the first crew returned, researchers from American Science and Engineering submitted a brief description of their X-ray data to a professional journal, with other investigators quickly following suit.

Though the astronomers did not hold an all-encompassing seminar like their medical counterparts, assessments of the solar physics programs were presented at several professional meetings. Notably, on December 3, 1973, just three weeks into the third crew's mission, Leo Goldberg, a prominent figure in the field, discussed the significance of some early data from the Apollo Telescope Mount (ATM) at the 141st meeting of the American Astronomical Society. Goldberg, then the director of the Kitt Peak National Observatory in Arizona, delivered the prestigious Henry Norris Russell lecture titled "Research with Solar Satellites." Having initially been the principal investigator for the Harvard solar instrument, Goldberg experienced early conflicts with NASA officials over the ATM management and was skeptical about the effectiveness of using humans as observers in space. His lecture underscored the evolving understanding and the significant contributions of Skylab's solar research to the broader field of astronomy.

Groundbreaking discoveries, persistent questions, and a profound impact on future space exploration marked Skylab's scientific legacy. The mission addressed immediate challenges and set the stage for continued research, pushing the boundaries of human knowledge both in space and on Earth.

Goldberg's initial skepticism regarding the value of crewed spaceflight in scientific research began to dissolve as he reviewed the early results from Skylab. The delay in launching Skylab, initially seen as a setback, ultimately became a fortuitous event that allowed crucial improvements. These advancements transformed Skylab from what might have been just another exercise in crewed spaceflight into one of the most significant milestones in the history of solar physics.

One of the most remarkable achievements was the stability of Skylab's orbital cluster, which maintained a precision of 2.5 seconds of arc. This stability was lauded as a groundbreaking engineering triumph. The spatial resolution achieved through the observatory's instruments was poised to revolutionize existing solar theories. Once a skeptic of the role of astronauts in space-based astronomy, Goldberg found himself convinced of their value. He had previously doubted whether humans could offer anything beyond minor adjustments and repairs to equipment in space, but Skylab's success changed his mind. The presence of astronauts onboard had proven to be indispensable.

Goldberg's enthusiasm for Skylab's contributions to solar research was widely shared among the scientific community. On August 22, 1974, during the American Astronautical Society's annual meeting in Los Angeles, E. M. Reeves of Harvard College Observatory summarized the profound accomplishments of the Apollo Telescope Mount (ATM) project. Reeves noted that every instrument aboard Skylab had either met or exceeded expectations. For instance, the quality and quantity of photographs from the coronagraph were unparalleled in the history of solar observation. Reeves was particularly impressed by the operational flexibility and responsiveness of the experiment management system, which allowed scientists to adapt quickly to new opportunities. One such opportunity was the study of Mercury during its transit across the Sun on November 10, 1973. The rapid transmission of data from remote stations in the communications network, combined with the remote-control capabilities of the Harvard instruments, enabled the collection of data that could be used to estimate the density of Mercury's atmosphere—an extraordinary achievement.

Among the most satisfied researchers were the scientists at the High Altitude Observatory in Colorado. Their white-light coronagraph revealed that the solar corona was far more dynamic than previously thought. The observatory recorded dramatic changes in the form and structure of the corona, sometimes over very short periods. During 227 days of observation, the coronagraph— designed to operate even when Skylab was uncrewed—recorded approximately 100 events known as "coronal transients." These transients, which could occur within minutes, sometimes involved the ejection of vast amounts of matter and energy into the corona. Nearly half of these transients were associated with solar flares or eruptive prominences, providing new insights into solar activity.

The intensity and variety of solar phenomena observed during Skylab's missions were particularly striking, considering that the observations were made during relatively low solar activity. Despite delays that forced the cancellation of plans to observe the Sun during its peak activity in 1969-1970, eight significant solar flares were recorded during Skylab's three missions. The last of these dubbed the "Gibson flare," was observed in detail after astronauts Garriott and Gibson identified a pattern of solar x-ray activity that preceded major eruptions. The simultaneous use of all ATM instruments allowed a comprehensive documentation of these flares and their impact on the solar corona.

By the end of 1974, solar astronomers were confident they had obtained the best observations of the Sun ever recorded from space. The correlation of data from x-ray, ultraviolet, and coronagraph observations would take years to fully analyze and interpret. Richard Tousey, a principal investigator from the Naval Research Laboratory, reflected on the immense effort required to develop Skylab and the challenges in interpreting the data. He believed the effort had been well worth it, a sentiment most of his colleagues shared. The solar observations retrieved by Skylab were extraordinary in both quantity and quality, and Tousey estimated that it would take no less than five years of work by large, competent teams to fully analyze the data— possibly even ten years.

Tousey, who had been involved in space research since the 1940s with experiments on V-2 rockets, was convinced that an uncrewed spacecraft could never have achieved the results of Skylab's crewed missions. Skylab had unequivocally vindicated using humans in space for scientific experimentation despite ongoing debates to the contrary. As the scientific community looked forward to the next solar maximum, Tousey advocated for flying the backup solar observatory, establishing nearly ready to launch. This observatory, he argued, represented a valuable resource that should not be overlooked, as there was still much to be learned about our Sun.

Skylab's extreme-ultraviolet spectroheliograph, S082A, captured a dramatic solar eruption, showcasing helium being ejected over 800,000 kilometers from the Sun's surface. To put this into perspective, the Earth, represented by a tiny black dot near the Sun's rim and beneath the helium arch, was dwarfed by the scale of this solar activity. The instrument responsible for this remarkable observation was constructed by the U.S. Naval Research Laboratory and Ball Brothers Research Corporation, demonstrating the sophisticated capabilities of Skylab's solar observational tools.

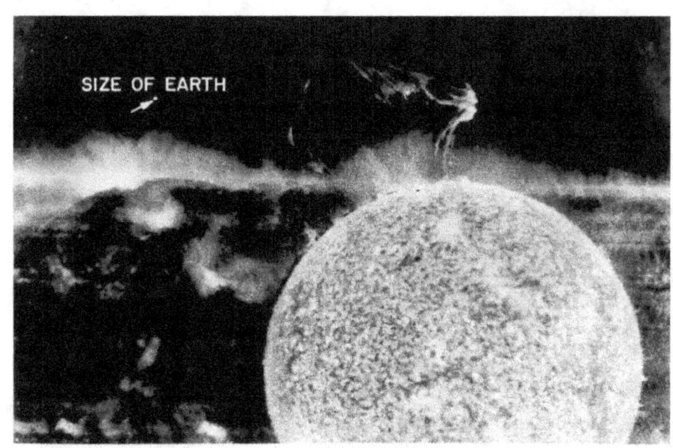

(July 1973) --- A huge solar eruption can be seen in this Spectroheliogram obtained during the Skylab 3 mission by the Extreme Ultraviolet Spectrograph/Spectroheliograph SO82A Experiment aboard the Skylab space station in Earth orbit.

SO82 was one of the Apollo Telescope Mount experiments. The SO82 A instrument covers the wavelength region from 150-650 angstroms (EUV regions). The magnitude of the eruption can be visualized by comparing it with the small white dot that represents the size of Earth. This photograph reveals for the first time that helium erupting from the sun can stay together to altitudes of up to 500,000 miles. After being ejected from the sun, the gas clouds seem to have come to a standstill, as though blocked by an unseen wall. Some materials appear to have been directed back toward the sun as a rain, distinguished by fine threads. At present it was a challenge to explain this mystery--what forces expelled these huge clouds, then blocked its further progress, yet allowed the cloud to maintain its threads. Both magnetic fields and gravity must play a part, but these curious forms seem to defy explanation based on magnetic and gravitational fields alone. The EUV spectroheliograph was designed and constructed by the U.S. Naval Research Laboratory and the Ball Brothers Research Corporation under the direction of Dr. R. Tousey, the principal investigator for this NASA experiment. On the left may be seen the sun's image in emission from iron atoms which have lost 14 electrons by collision in the sun's million-degree coronal plasma gas.

Despite the scientific success, there was little hope for a follow-up mission, as the possibility of launching a second Skylab had been dismissed long before. The earth-resource experiments conducted aboard Skylab stood out from the other research programs in several ways. Unlike the medical and solar experiments, which had more focused objectives, the earth-sensing experiments encompassed a broader range of instruments, a larger number of investigators, and a diversity of goals. This complexity made it challenging to quickly assess the overall value of the earth-sensing data.

At the Skylab Results Symposium in Los Angeles in August 1974, researchers began to present early findings. Four teams of investigators showcased the scope and potential of the earth-resources program. A group from the University of Kansas highlighted the promise of microwave instruments for measuring soil moisture from orbit, a crucial parameter for understanding agricultural and hydrological processes. Meanwhile, geologists from the University of Wyoming evaluated earth-terrain and multispectral photographs for mapping geological and agricultural features. They concluded that Skylab's instruments, particularly the high-resolution photographs, were superior to those on the Landsat satellite for certain applications. However, they noted that both satellite and aerial photography were necessary to achieve comprehensive results.

Among the various instruments aboard Skylab, the multispectral scanner drew significant attention. This device covered 13 wavelength bands in the visible and infrared regions of the spectrum, allowing for detailed analysis of land use. Researchers at Purdue University utilized data from this scanner and the S190A multispectral photographs in a computerized land-use classification program. By isolating the characteristic spectra of different land uses, they were able to classify areas with high accuracy into categories ranging from residential and commercial to grassland, farmland, and woodland. The data from Skylab's scanner were comparable in quality to those from Landsat's multispectral scanner, which operated in only four wavelength bands. Similar successes were reported by the U.S.

Geological Survey, which used the data to study Florida's swamplands, and by General Electric, which applied the data to analyze geological features in New Mexico.

Later that year, a conference in Huntsville brought further reports on the success of Skylab's sensors. The multispectral scanner continued to be a focal point of interest, but geophysicists also shared promising results from the radar altimeter. This instrument was capable of measuring the shape of the Earth's surface with remarkable accuracy, particularly the surface of the oceans. One of the most impressive findings was detecting local sea-level variations, including a 20-meter depression near Puerto Rico, likely caused by a local gravity anomaly. The radar altimeter also revealed correlations between its data and subsurface geological features, clearly mapping the profile of the continental shelf off the coast of Georgia and Florida.

While preliminary results indicated that Skylab's earth-sensing instruments had performed as expected and had provided valuable data, the broader scientific community was slow to adopt this information. Many users preferred to rely on data from Landsat, which had been launched in July 1972. Landsat's appeal lay in its consistent, repetitive coverage of the same ground track every 18 days at the same local time, a feature not available with Skylab. This familiarity and consistency made Landsat the preferred choice for many researchers despite the groundbreaking achievements of Skylab's earth-resources program.

In mid-1975, a NASA-sponsored symposium on Earth resources gathered scientists and researchers to discuss the results of various experiments conducted in space. Among the 166 reports presented, only 29 focused on the findings from the Skylab missions. However, the experiments aimed at Earth resources did little to demonstrate the value of having humans in space for such tasks. These experiments were added to the Skylab program late, so the instruments used were not optimized for human interaction. The astronauts 'involvement in data collection was largely mechanical; they tracked specific sites using the viewfinder on the infrared spectrometer,

pressed buttons, and recorded operational sequences on channel B. There was little room for human judgment in choosing alternative sites or methods of operation.

Despite this, the astronauts proved their value in space by performing essential maintenance tasks vital to the mission's success. Although the instruments had not been designed with maintenance in mind, the crews adapted and managed to perform critical repairs. For instance, astronauts Pogue and Gibson performed a major repair on the microwave antenna. Additionally, the crews regularly cleaned tape recorder heads, replaced a faulty tape recorder, and installed an improved detector on one of the infrared instruments during their time in orbit.

The value of human observation from orbit was further validated by a special program developed for Skylab's third crew. A team of 19 scientists devised a plan for visual and photographic observations of Earth's surface features, highlighting the importance of an intelligent observer in space. This program was intentionally flexible; scientists provided the crew with general guidelines on key areas of interest, such as ocean currents, geological formations, and regions affected by drought in Africa. A book summarizing what the astronauts should look for and what they might expect to see served as their guide.

During the mission, the crew's observations were only loosely scheduled, allowing them to exercise judgment in identifying and documenting features of interest. Weekly conferences with scientists on the ground enabled adjustments to the observation plan, ensuring the program remained dynamic and responsive to new discoveries. Gazing out the window became a cherished activity for the astronauts, particularly as it gained scientific significance. Equipped with two cameras, various lenses, film, and 10-power binoculars, they spent countless hours at the wardroom window, observing and photographing Earth.

While not always quantifiable, the results of these observations underscored what proponents of crewed space missions had long believed: humans possess an unparalleled ability to discriminate between important and unimportant features, select critical elements from a vast landscape, and respond effectively to unexpected events. The crew's ability to track and describe ocean currents over distances of up to 3,500 kilometers, recognize upwelling eddies of cold water within warm currents, and capture such phenomena in photographs were achievements that could not have been programmed into automated sensors.

NASA's habitability experiments aboard Skylab further demonstrated the feasibility of living and working in space for extended periods. Caldwell Johnson, who led the habitability experiment, had ample reason to be satisfied with the results. In-flight evaluations by the astronauts, along with movies, videotapes, and post-flight debriefings, revealed no serious errors in designing the space station's living and working environment. While there were still aspects of habitability to be optimized and minor irritations to be addressed, Skylab clearly showed that humans could adapt to the space environment without becoming disoriented or facing significant challenges from the lack of gravity.

Life aboard Skylab became simply another work environment, one to which each crew member adjusted relatively easily. They all found the experience enjoyable. Some tasks were even easier in the microgravity of space. For example, moving massive objects required minimal effort, provided there were adequate handholds for control. However, small objects, such as hand tools, screws, and other small parts, presented more of a challenge, as they tended to float away unless adequately secured.

In essence, Skylab's missions provided invaluable insights into humans' capabilities and limitations in space, laying the groundwork for future long-duration spaceflights and the continued exploration of Earth's resources from orbit. The astronauts' contributions went beyond mere button-pushing; their ability to observe, adapt, and maintain the space station proved crucial to the success of the mission and the advancement of space science.

The Skylab crews quickly discovered an unexpected convenience in the orbital environment. Small objects that might have been lost in the spacious workshop inevitably found their way to the screen covering the intake of the ventilation system, carried there by the gentle air currents circulating throughout the station. This made it relatively easy to retrieve misplaced items, ensuring that nothing of importance would be permanently lost in the microgravity setting.

While Skylab's wardroom and experiment area had been meticulously designed with a uniform architectural layout featuring a clearly defined "floor" and "ceiling," the astronauts found that such orientation was unnecessary once they were in orbit. The preflight advantage of this design for assembly and testing quickly became irrelevant in space, where a consistent up-and orientation held no real value. What mattered more was having a reference axis at each workstation, with all related instruments aligned in a single direction. This made tasks more intuitive and manageable in a weightless environment.

In contrast, the multiple docking adapter—a section of Skylab where equipment had been arranged somewhat randomly due to space constraints—proved equally functional. The astronauts adapted easily to working in this area, with the shift from one workstation to another requiring only a simple adjustment in orientation. Ed Gibson, one of the astronauts, praised the docking adapter for its efficient use of space, noting that, unlike the workshop, it made full use of the available walls and ceiling, which were often underutilized in the more structured areas of the station.

Interestingly, both Jerry Carr and Ed Gibson experienced an odd sensation in the docking adapter. When Carr entered the compartment from the command module feet first, he felt he was positioned at a great height, with a curious need to be cautious as if he might fall all the way "d" to the workshop below. Gibson reported a similar sensation when using a specific foot restraint that positioned him above the airlock hatch. This was the only place within the entire Skylab cluster where he felt any sense of height, a feeling that was otherwise absent in the microgravity environment.

One area where Skylab revealed a clear need for further development was in mobility and restraint in zero gravity. Preflight simulations had been limited, as only a few experiments could be conducted in the brief periods of weightlessness afforded by the zero-g aircraft. While mobility in space was generally excellent, allowing the astronauts to move quickly, certain challenges arose in narrow spaces, such as the hatches leading to the airlock or docking adapter. Here, the crew's feet would often bump into the sides of the passageways, occasionally triggering poorly placed or insufficiently protected switches.

Restraint posed a more significant challenge. The triangular metal gridwork used as flooring throughout the workshop and the triangular cleats on the crewmen's shoes provided good security when locked into place. However, maintaining a stable position became complicated in the waste management compartment, where smooth surfaces were implemented to facilitate cleaning. Straps on the floor, intended to secure the feet, proved ineffective, making tasks in this area more cumbersome.

Despite these challenges, Skylab's design and operation were largely successful, though not without minor deficiencies. One common issue reported by all crews was the lack of a dedicated workbench for conducting maintenance and small repairs. Improvisation became necessary, with the ventilation screen in the forward dome often serving as a makeshift workspace. The experience underscored the importance of adaptability and ingenuity in space, as the astronauts continually found ways to overcome the station's limitations and make the most of their time in orbit.

One of the most fascinating scientific endeavors during the Skylab missions was crystal growth in the microgravity space environment. The results of these experiments provided valuable insights into materials science, which were unachievable under Earth's gravitational pull.

Among the crystals gr, one stands out—a 20-mm crystal of germanium selenide, the largest ever produced on Earth or in space as of 1974. This achievement was a milestone in crystal growth and a testament to the unique conditions aboard Skylab. Each crystal carried its story, with the leftmost crystal grown during the third crew's tenure and the middle crystal during the second mission. Interestingly, all crystals from the second mission exhibited a ring-shaped groove. A spacecraft maneuver likely caused this anomaly during the crucial cool period, indicating how

sensitive and intricate these space-based experiments were.

NASA's commitment to advancing scientific knowledge was evident when, in a demonstration, NASA Administrator James C. Fletcher explained the crystal-growing process to President Gerald Ford. This moment, captured alongside Howard Johnson, chairman of MIT, illustrated the intersection of science, policy, and public interest in the nation's space program.

However, the Skylab missions were not solely about scientific discoveries like crystal growth. They also served as a proving ground for the habitability of spacecraft, shedding light on the challenges astronauts faced while living and working in space. Although Skylab was well-designed for its purpose, the astronauts encountered several practical issues. Bill Pogue, one of the astronauts, expressed frustration with the locker numbering system and the poor latches on lockers and film vaults. The need for a more efficient stowage system and a proper workspace became evident as the missions progressed.

Despite these frustrations, Skylab provided a functional living environment, and the astronauts quickly adapted, learning to work around the station's limitations. This adaptability was crucial, as the missions also tested new technologies and methods for working in space, including a significant experiment focused on astronaut maneuvering outside the spacecraft.

As the Space Shuttle era approached, NASA engineers were keen to experiment with different concepts for astronaut maneuvering units. Skylab's upper dome, a spacious area 6.5 meters in diameter and height, became the perfect testing ground. Three concepts were tested: a large backpack, a small hand-held gas pistol, and a foot-controlled unit designed to leave the hands free for work.

The backpack unit, though bulky, was the most sophisticated, equipped with fourteen cold-gas thrusters that provided control over motion along three axes and allowed rotation in all directions. The unit also featured gyroscope stabilization, which helped maintain the astronaut's orientation. Five crew members tested this backpack unit extensively during the second and third Skylab missions, accumulating nearly 14 hours of flight time. One of the astronauts, Owen Garriott, found that learning to operate the unit was straightforward, mastering the controls in less than an hour despite having no prior experience.

The backpack unit's experiments demonstrated its potential utility for various tasks, such as station-keeping and inspecting spacecraft. In one notable test, an astronaut used the unit to approach a spinning object in Skylab's upper dome, synchronize with its rotation, grasp it, and then reduce its spin to zero—an operation that could be crucial for recovering tumbling objects in space.

In contrast, the other two maneuvering units tested— the hand-held gas pistol and the foot-controlled unit—were less effective. The hand-held unit was difficult to control accurately, often causing unwanted rotation alongside translational movement. It was deemed useful only for short, simple maneuvers. The foot-controlled unit, with thrusters beside the astronaut's feet, faced similar issues, as it could not produce simple linear motion without causing rotation.

Beyond maneuvering tests, Skylab was also host to a wide array of other experiments, including the highly publicized observations of comet Kohoutek and various student projects. Although Skylab's instruments were not designed explicitly for comet observation, NASA's comprehensive program to observe Kohoutek included ground-based observatories, airborne telescopes, and satellites, all contributing valuable data to the mission.

These diverse experiments, ranging from crystal growth to astronaut maneuvering, were critical in pushing the boundaries of space exploration. They provided scientific insights and practical lessons to inform the design and operation of future space missions, such as those conducted aboard the Space Shuttle. Despite its challenges, Skylab proved to be a vital stepping stone in humanity's journey into space, showcasing both the potential and the complexities of living and working beyond Earth's atmosphere.

While the instruments aboard Skylab contributed valuable observations, their impact was somewhat overshadowed by the more extensive data gathered from other sources during the same period. Nevertheless, Skylab's experiments yielded some notable successes, particularly in far-ultraviolet astronomy. One of the most significant achievements was made by the far-ultraviolet electronographic camera, which successfully detected a cloud of hydrogen enveloping the comet Kohoutek. This discovery offered a deeper understanding of the comet's composition and behavior. Additionally, the photometric camera on Skylab provided critical insights by capturing periodic exposures that revealed a significant dimming of Kohoutek after it passed perihelion. Combined with sketches and visual data collected by the crew, these observations added to the body of knowledge about this celestial event, even if they did not dominate the overall scientific effort.

The student experiments aboard Skylab were a fascinating aspect of the mission, though they produced mixed results. Given their late inclusion in the program, it was anticipated that not all would be successful. Several experiments faced setbacks due to equipment failure or operational constraints, while others managed to yield usable data. For instance, a planned observation of Jupiter using the X-ray telescopes had to be canceled due to power limitations that prevented the necessary maneuvering of Skylab. An alternative proposal to observe an x-ray source in the Veil Nebula was also unsuccessful, as Skylab's instruments lacked the sensitivity and pointing accuracy required for such a task. Similarly, efforts to detect ultraviolet radiation from pulsars and to study x-rays from stars of various spectral types were hindered by similar challenges.

Despite these difficulties, some experiments with living organisms aboard Skylab produced intriguing results. Students discovered differences in bacterial colonies grown in space compared to those on Earth, indicating that microgravity had a noticeable effect on their development. Rice seedlings, another subject of experimentation, exhibited unusual anomalies during their growth, further underscoring the unique biological responses to the space environment. Perhaps the most memorable student experiment involved the common cross spider, Araneus diadematus, establishing taken aboard Skylab to study its web-spinning abilities in weightlessness. Initially, the spiders struggled to create webs, but after some adjustment, they eventually produced nearly normal webs. Unfortunately, the experiment could not be extended further as the spiders died shortly after the initial observations, likely due to starvation or dehydration.

The student experiments, though they did not lead to groundbreaking scientific discoveries, were significant in other ways. These projects, marked by originality and often sophisticated design, had a profound impact on the students and their teachers. NASA's interest in their ideas provided an enlightening experience, exposing them to real-world scientific investigations and the challenges that accompany them. For many students, the experience was invaluable, teaching them not only about the possibilities of scientific research but also about the realities of failure. The opportunity to see their experiments fl in space was a unique educational experience that left a lasting impression on all who participated.

For NASA, these student experiments demonstrated that simple, low-cost experiments could be effective and provide meaningful data, even within the constraints of a busy mission schedule. The mixed results also highlighted the importance of adequate training for both the crew and operations personnel, particularly when dealing with time-sensitive and delicate experiments. Despite the challenges, the student experiments on Skylab were a testament to the value of educational engagement in space exploration, inspiring a new generation of scientists and engineers.

Arabella, a common cross spider, spins an earthly web aboard the second Skylab mission in 1973 after initial disoriented attempts. The experiment, Web Formation in Zero Gravity, part of the Skylab Student Project, was submitted by Judith Miles, a junior at Lexington High School in Lexington, Massachusetts. The Marshall Space Flight Center had program management responsibility to develop Skylab hardware and experiments, including the Skylab Student Project.

Assessing Skylab Science:

Skylab, despite the uncertainties that clouded its early development, steadfastly fulfilled its primary purpose: placing humans into orbit to conduct groundbreaking scientific work. In this, it was undeniably successful. Some scientists even argued that the creation of a second Skylab would have been justified, if only to continue the significant work initiated by the first. However, in an era of shrinking space budgets, NASA faced tough decisions and could not afford to revisit that path. The three Skylab missions, however, cleared the way for the agency's future endeavors, particularly the Space Shuttle program. The backup Skylab hardware—a fully functional replica of the orbiting station—was removed from storage in 1976 and became one of the most striking exhibits at the National Air and Space Museum.

The medical results from Skylab were particularly significant, breaking down many of the barriers to extended crewed spaceflight by demonstrating that humans could adapt remarkably well to a zero-gravity environment.

The astronauts maintained their ability to function effectively for weeks on end, provided they received proper attention to environmental factors. They could also readapt to Earth's conditions with surprising speed upon return. While long-term spaceflight problems remained unresolved, Skylab laid the groundwork for future research, proving that spacefarers did not need to be exceptionally conditioned athletes; healthy individuals could safely embark on orbital missions without undue risk.

One of the more debated aspects of Skylab was the value of having humans in space as scientific observers. Some questioned whether the funds spent on life support systems might have been better allocated to more advanced uncrewed equipment. However, the scientists who participated in Skylab missions strongly advocated for the presence of humans. Astronomers who had previously worked solely with uncrewed satellites were impressed by the Skylab crews 'performance and the support personnel on the ground. The astronauts 'ability to react swiftly to unexpected solar phenomena was instrumental in the success of the Apollo Telescope Mount (ATM) experiments. The same held true for the Earth-observation program, where having a human in orbit, trained to identify objects of interest and remain alert for unfamiliar features, proved invaluable to Earth scientists across various disciplines.

In hindsight, Skylab's ambitious experiment program might have been slightly too heterogeneous, leading to operational difficulties, crowded training schedules, and occasional crew errors. While these challenges were ultimately overcome, and valuable experience was gained, individual experiments might have benefited from a more focused approach. However, the political environment in which Skylab was developed gave managers little choice. Skylab was tasked with more experiments than might have been optimal as the first crewed program in years, the first multipurpose space station, and the testing ground for human utility in space. The Earth-resources package and the student experiments exemplify this—both were timely responses to public demand and played a crucial role in broadening

support for crewed spaceflight, even if they added to the complexity.

While the specific results of many of Skylab's experiments would take years to integrate into the broader scientific landscape, the program established beyond doubt that humans have a place in space science. Had Skylab failed or left critical questions unanswered, the future of crewed spaceflight would have been bleak. However, its success ensured that humans would remain at the forefront of America's space exploration efforts.

As Skylab's final crew prepared for their departure, Commander Gerald Carr took a critical step to ensure the space station's longevity. Before undocking, Carr activated the Apollo spacecraft's attitude-control thrusters for three minutes, gently propelling Skylab into a higher orbit, between 433 and 455 kilometers above Earth. This maneuver was intended to extend Skylab's operational life, albeit temporarily.

After the crew safely returned to Earth and post-mission engineering tests were completed, flight controllers began shutting down Skylab. The station's atmosphere was vented and oriented in a gravity-gradient-stabilized attitude, with its docking adapter pointed away from Earth. Most of Skylab's systems were powered down, though it retained the ability to respond to telemetry signals whenever its solar panels were illuminated. Theoretically, the station remained accessible, with a suited astronaut able to enter if necessary, but no such missions were planned or seriously considered. Skylab, which had cost $2.5 billion, was essentially relegated to space debris, destined for a fiery reentry into Earth's atmosphere.

In the mission's final days, calculations based on solar activity and atmospheric density suggested that Skylab would remain in orbit for just over nine more years. The station was expected to descend slowly at first, losing about 30 kilometers by 1980, but then accelerating its descent, losing another 100 kilometers by the end of 1982. By March 1983, the massive 75,000-kilogram structure would reenter Earth's atmosphere and disintegrate, with some fragments potentially surviving to impact the surface.

NASA, however, held onto the hope that the newly developed Space Shuttle could intervene. If the Shuttle program proceeded as planned, an early mission might attach a propulsion module to Skylab, boosting it into a higher, safer orbit. Yet, this plan was fraught with uncertainty; any delays in Shuttle development could lead to an uncontrolled reentry of Skylab, presenting a significant public relations challenge for NASA.

As the years passed, Skylab's fate became increasingly precarious. From 1974 onwards, NASA faced significant budget cuts, which slowed the progress of the Shuttle program. Meanwhile, the Soviet space program continued to advance, with cosmonauts breaking Skylab's endurance records and Soviet officials discussing plans for permanent space stations in orbit. By early 1977, the Shuttle Orbiter Enterprise was being readied for landing tests, and NASA began considering potential missions, including one to save Skylab.

In early 1977, NASA's Headquarters directed the Johnson Space Center (JSC) and the Marshall Space Flight Center (MSFC) to draft schedules and budget estimates for a Shuttle mission to boost Skylab into a higher orbit. However, JSC's optimism was tempered by the numerous technical challenges. Rendezvousing and docking with an inert Skylab had not been fully explored, and studies suggested that such a mission could not be undertaken before the fifth test flight of the Shuttle, anticipated for late 1979. Compounding the problem was the approaching solar maximum, predicted for 1980-1981, proving to be much more intense than expected. Increased solar activity heated the Earth's upper atmosphere, raising its density at Skylab's orbital altitude and accelerating the station's descent.

In March 1977, Marshall's experts urged NASA Headquarters to initiate a study contract by midyear to define the booster stage required for the Skylab mission. In response, NASA set the fifth Shuttle test flight as the target mission and established September 1, 1977, as the deadline for a final decision. This timeline would allow just over two years to develop the necessary hardware. As the centers continued gathering data, the urgency of the situation grew.

By September 1977, the decision to proceed was made, and by November, Marshall awarded a $1.75 million contract to Martin Marietta Corporation. The contract focused on analyzing and designing a teleoperator retrieval system, which would be carried in the Shuttle's cargo bay and used to attach a yet-to-be-designed propulsion module to Skylab's docking port. Given the tight timeline, the project relied heavily on already developed and qualified hardware, adhering to the pragmatic approach that had characterized Skylab's construction and operation.

Within a month of setting the ambitious schedule to re-boost Skylab, it became evident that the timeline might be insufficient. In December, at a meeting of the American Geophysical Union, Howard Sargent, the chief forecaster for the National Oceanic and Atmospheric Administration (NOAA), delivered unsettling news. Sargent reported that the ongoing sunspot cycle was the second most intense in a century. Based on a model different from the one NASA had used, this forecast sparked criticism of the space agency's approach. Sargent and other experts argued that NASA's model was inaccurate. When questioned by journalists about the likelihood of success for the Skylab re-boost mission, Sargent ominously suggested that NASA was "in a pile of trouble" if it was relying on Skylab remaining in orbit long enough for the Shuttle to reach it under the current schedule.

The discrepancy between NASA's predictions and those of NOAA provided ammunition for critics of crewed spaceflight. However, the truth was more nuanced—no single method for predicting sunspot activity had universal acceptance among solar scientists. Ironically, Skylab's scientific observations, unavailable in 1974, would later contribute to refining these predictive models. NASA's scientists, working with more extensive observational data, had predicted less sunspot activity than their counterparts at NOAA. Sargent and his colleagues, however, questioned the reliability of some of the early observations from the 17th century that NASA had included in their analysis, arguing that these data points reduced the accuracy of NASA's predictions. The space agency's decision to disregard NOAA's 1976 forecasts led some cynics to speculate that Marshall Space Flight Center, which still harbored hopes of repurposing Skylab, might have been motivated to display the station's precarious future. However, this speculation was never substantiated, and it was equally plausible that more immediate concerns overshadowed Skylab.

In early 1978, Skylab was abruptly thrust back into the public eye, not by its developments but by events within the Soviet space program. The uncrewed Soviet satellite Cosmos 954 suffered a systems failure and reentered Earth's atmosphere over northern Canada, scattering debris from its nuclear-powered electrical module across a vast area. The module, which contained 45 kilograms of uranium highly enriched with the fissionable isotope uranium-235, triggered an urgent and intensive search for radioactive fragments. This alarming event, coming so soon after the debate over Skylab's fate, intensified public concern about what might happen when Skylab itself reentered the atmosphere.

NASA's public affairs office reassured the global audience that Skylab contained no radioactive material and would not descend below 278 kilometers before October 1979. However, this reassurance did little to calm anxieties, as it also revealed that Skylab's expected orbital lifespan had been shortened by nearly four years from earlier estimates.

The situation took a more serious turn when the U.S. State Department, mindful of the global reaction to Cosmos 954, sought clarification from NASA on their plans for Skylab. Given the worldwide interest and potential diplomatic repercussions, mainly if Skylab debris were to fall on populated areas, the State Department was understandably concerned. Skylab's orbital path took it over 90% of the world's population, meaning the risk of human injury, while small, was not negligible—a significant consideration for a public agency attuned to the sensitivities of the late 1970s.

In response to these growing concerns, NASA immediately began assessing the condition of Skylab's systems, recognizing that the stakes were higher than ever as they worked against the clock to address the station's impending reentry.

As Skylab orbited the Earth, its fate hung in the balance. The space station, now a derelict, required immediate intervention if NASA hoped to either re-boost it for future use or safely deorbit it at a controlled location. However, the ability to control Skylab from the ground was severely limited. The best-case scenario allowed ground controllers to adjust the station's attitude, thereby modulating atmospheric drag, but they could not increase its altitude. In the most favorable conditions, this control could potentially extend Skylab's orbital life by up to five months—a period that might, just might, give enough time for Shuttle engineers to prepare a mission to re-boost the station.

In late February, an eight-man team comprising four engineers from NASA's Marshall Space Flight Center and four from the Johnson Space Center (JSC) deployed to Kindley Naval Air Station in Bermuda. This site was critical, as it was the only tracking station still capable of transmitting the UHF signals required to operate Skylab's aging telemetry equipment. During this period, NASA Administrator Robert Frosch faced intense scrutiny during budget hearings before the Senate Space Committee. Frosch candidly explained the challenges NASA faced, emphasizing the difficulties of predicting Skylab's orbital decay, which were compounded by the unpredictability of solar activity. He hoped that the teleoperator retrieval system, a proposed solution to save Skylab, could be developed and launched by October 1979. Still, he conceded that the chances of Skylab remaining in orbit until then were no better than fifty-fifty.

William C. Schneider, another NASA official, detailed the ambitious reboost mission to the Senate. The plan involved launching a 4,540-kilogram teleoperator unit aboard a Shuttle orbiter, primarily composed of fuel tanks and engines. An astronaut would guide this unit to dock with Skylab's multiple docking adapter, where its thirty-two 100-newton thrusters would then push the station into a higher orbit. The schedule was tight—design studies were ongoing, with fabrication and assembly slated to begin in six months. The completed module was expected to arrive at Cape Canaveral by early September 1979, ready for an October launch on the third Shuttle test flight. However, this timeline was incredibly ambitious, especially given that the first Shuttle orbiter had yet to be launched.

In March 1978, the engineers stationed in Bermuda successfully made their first contact with Skylab. Collaborating with the North American Aerospace Defense Command (NORAD), they used radar to locate the station and aimed a radio signal at it, receiving a brief response. Skylab transmitted data about its condition for two minutes before falling silent again. It was determined that the station was rotating at approximately ten revolutions per hour, and as its solar panels rotated out of sunlight, radio transmissions ceased. The engineers 'first priority was to recharge Skylab's batteries. Since they could only transmit commands briefly during each orbital pass, this task required considerable time. Nevertheless, within a week, they had charged two batteries, determined the station's attitude, and confirmed that the onboard computer could still assist in controlling the spacecraft.

With the basics under control, the team focused on more critical systems, including the control moment gyros, the thruster attitude control system, and the attitude-sensing rate gyros. Gaining control of these systems would enable flight controllers to maintain Skylab in a minimum-drag attitude, thus conserving its altitude while the future of the Shuttle mission remained uncertain. Depending on developments, they could either continue minimizing drag to extend Skylab's life or increase it to influence the station's eventual reentry point.

Balancing these requirements was a complex task, requiring more than the efforts of a skeleton crew at a remote location. Consequently, in June, a control center was hastily assembled at the Johnson Space Center, staffed by two teams of flight controllers. Shortly after that, additional tracking stations in Madrid, Goldstone, California, and near Santiago, Chile, were integrated into the network. By early June, the JSC team had activated two functioning control moment gyros and used them to stabilize Skylab in a low-drag attitude that kept the batteries charged. This was no easy feat; one of the gyros, which had caused concern during Skylab's final crewed mission, again exhibited signs of stress—its wheel speed

decreased, and motor currents increased. Complicating matters further, the refrigeration systems that cooled the batteries in the airlock module were failing—one had nearly lost all its cooling fluid, and the other was only marginally reliable. Juggling these demands required meticulous planning, and the teams worked in ten-hour shifts throughout the summer.

In July, the mission nearly faced disaster when a spurious telemetry signal caused the onboard computer to shut down the control moment gyros and activate the gas thrusters, consuming a significant portion of the remaining propellant before the Houston team regained control.

Meanwhile, NASA Headquarters established the Skylab Contingency Working Group on July 25, 1978, to coordinate interagency planning for Skylab's eventual reentry. This group, led by William G. Bastedo, included representatives from NASA and the Departments of State, Justice, and Defense. The group was tasked with various responsibilities, from monitoring Skylab's condition to ensuring that foreign governments were kept informed of its status.

As the effort to save Skylab continued, the financial costs mounted. By June 1, 1978, NASA had already spent $750,000 on the dying station, and it anticipated spending at least $3 million more by the end of the year. Despite the mounting costs and the uncertain odds of success, NASA pressed on, driven by the hope of saving Skylab or ensuring its controlled reentry.

Chris Kraft, the director of the Johnson Space Center (JSC), did not shy away from expressing his skepticism regarding the ongoing efforts to save Skylab. Publicly, Kraft deemed the mission futile, doubting that the aging systems onboard the space station would function long enough to allow for a controlled reentry. This doubt implicitly dismissed any realistic hope for the re-boost mission. While acknowledging that his engineers were obligated to do everything possible, Kraft believed the resources spent on Skylab were misallocated. He argued that were it not for the heightened public concern over falling spacecraft debris—intensified by the recent Cosmos accident—NASA might not have invested so much effort into Skylab's final days. In Kraft's view, these funds would have been far better spent on the Shuttle program, establishing struggling with delays due to insufficient funding.

In contrast, Robert Frosch, NASA's Administrator in Washington, was under significant pressure from the White House and the State Department. Despite the low probability that Skylab's reentry would result in harm, Frosch reaffirmed NASA's commitment to the mission, determined to continue the effort even if the chances of Skylab hitting anyone were exceedingly small.

As summer transitioned into fall, the operation in Houston, now under the direction of Charles Harlan from JSC's Flight Control Division, intensified. With the addition of the tracking station in Chile, Skylab's orbit was fully monitored throughout each of its revolutions. By October 1978, Harlan had expanded his team, organizing five flight control teams to work around the clock in three shifts. While a few of these controllers had experience from the crewed Skylab missions, most were new to such high-stakes operations.

In September, the Skylab working group experienced a dry run when the uncrewed satellite Pegasus 1 reentered Earth's atmosphere. This exercise primarily tested impact prediction models and interagency coordination using orbital data provided by NORAD. The group's efforts culminated in the uneventful reentry of Pegasus over the southwestern coast of Africa on September 17, 1978. From this, the team set goals for managing Skylab's eventual reentry, informed by their experience with Pegasus.

Initially, the flight controllers approached their task with limited confidence, wary of the aging systems onboard Skylab. However, as they delved deeper, they found these systems more robust than expected. This discovery ignited a newfound enthusiasm among the controllers, who saw the mission as significant and challenging. Early in the summer, they had confirmed that Skylab's onboard computer could be used effectively, prompting control-system engineers at Marshall to devise new programs to manage the spacecraft's attitude without relying on the gas thrusters. The remaining fuel in these thrusters had to be conserved, as it would be critical if the reboost module successfully docked with the workshop.

Monitoring the batteries was a constant concern. As those in use began overheating, the controllers switched to backup batteries. Occasionally, all the batteries would warm up simultaneously, necessitating the brief activation of the cooling system to restore normal temperatures. The challenges increased as Skylab's orbital plane shifted relative to the Sun, altering the dynamics of power generation and thermal control, thereby requiring continuous adjustments.

The Pegasus 7 satellite, launched on February 16, 1965, was part of a test mission involving a boilerplate Apollo spacecraft. The Pegasus satellites were equipped to measure and report on micrometeoroids at orbital altitude, providing valuable data for future space missions. The success of these earlier missions provided a framework for the Skylab controllers as they navigated the complex and evolving challenges of their current task.

Through these efforts, the team in Houston demonstrated an unwavering commitment to managing Skylab's final days, even as the odds of success seemed slim. The operation became a testament to the resilience and ingenuity of NASA's engineers, who found themselves reinvigorated by the very challenges that had initially appeared insurmountable.

In the months leading up to Skylab's final days, countless hours were dedicated to devising and testing new strategies to maintain control over the space station. However, in one of the critical control moments, gyros, showing signs of wear, began to slow down even further in November. This posed a significant threat to the mission, as the failure of this gyro would severely complicate efforts to manage Skylab's attitude and extend its operational life.

Faced with this challenge, the engineers implemented a maneuver used during Skylab's third crewed mission, albeit with uncertain results. The plan involved turning the workshop around to expose the ailing gyro to sunlight, hoping the increased warmth would encourage the lubricant in its bearings to flow more freely, thereby improving its performance. Though this method had previously yielded ambiguous results, the gyro operated far beyond the conditions encountered during the crewed missions, making the maneuver a calculated risk worth taking.

To the engineers' relief, this maneuver, combined with their meticulous efforts, allowed both the troubled gyro and its companion to continue functioning until Skylab's reentry. Their perseverance and ingenuity in the face of deteriorating systems demonstrated the remarkable commitment and resourcefulness of the NASA teams working to maintain control over Skylab during its final orbiting days.

Skylab's Last Days

In December 1978, after a year-long struggle to keep Skylab in orbit, NASA faced the inevitable conclusion that the space station could not be saved. Despite the nearing completion of the teleoperator propulsion module, various setbacks, particularly with the Space Shuttle's main engines, had caused significant delays in crucial tests. These delays showed that the re-boost mission, originally scheduled for October 1979, would be unfeasible. On December 15, Dr. Robert Frosch, NASA Administrator, informed the President that Skylab's reentry was unavoidable but assured him that NASA would do everything possible to control the descent and minimize risks to populated areas.

By December 19, John Yardley, the Office of Space Transportation Systems Associate Administrator, communicated the decision to the press. It became clear that the Shuttle could not launch the reboost mission until March 1980. However, Skylab's orbit was decaying rapidly, and the increasing difficulty in maintaining its attitude control meant that any attempt to save the station would be futile.

In the following weeks, Houston's flight control teams worked tirelessly, operating around the clock in three shifts, to maintain Skylab in a low-drag attitude. This effort was crucial as policymakers deliberated on the best strategy for managing the station's reentry. Options were limited, and the stakes were high, given the potential dangers of uncontrolled descent. The decision was ultimately made to allow Skylab to reenter the Earth's atmosphere, primarily focusing on ensuring the safety of those on the ground.

Once the reentry plan was finalized, a detailed strategy was sent to key government agencies, including the Departments of State, Defense, Justice, and the Federal Preparedness Agency, for their input. On January 9, 1979, a critical meeting with the North American Aerospace Defense Command (NORAD) occurred, establishing the necessary radar tracking requirements and formalizing technical liaison procedures. NORAD's radar data would be transmitted to NASA field centers and to a newly established coordination center within Yardley's office, which would be tasked with directing the reentry.

The preparations for Skylab's reentry were meticulous, rivaling those for the return of an Apollo mission. One of the primary objectives of the Skylab team during this period was to ensure that NASA maintained a consistent message regarding the station's reentry. This was particularly important because NASA and NORAD used different models to predict reentry times, and discrepancies in public statements could lead to confusion and panic.

This precaution proved to be prudent. Just a few months later, in March 1979, the nation witnessed a nuclear reactor accident in Pennsylvania, which nearly necessitated the evacuation of thousands of people. The situation was exacerbated by conflicting statements from various experts regarding the level of danger, underscoring the importance of clear and consistent communication in times of crisis.

As the countdown to Skylab's reentry continued, NASA, alongside NORAD and other agencies, remained vigilant, closely monitoring the situation to ensure that the risks to human life and property were minimized when Skylab finally reentered the Earth's atmosphere. Though marked by challenges and the ultimate loss of the station, the final chapter of Skylab's mission was handled with the utmost care and coordination, reflecting NASA's commitment to safety and its responsibility to the public.

As the inevitability of Skylab's reentry became apparent, the media eagerly anticipated what was certain to be a spectacular and newsworthy event. Just as they had with Comet Kohoutek, reporters and headline writers seized the opportunity to sensationalize the upcoming reentry, turning it into a much-discussed topic nationwide and beyond. This media frenzy led to some bizarre and humorous side effects. In Washington, two enterprising computer specialists launched a business called "Chicken Little Associates," offering to provide real-time estimates of Skylab's risk to any individual—for a fee. Their service, implying that NASA's predictions were unreliable, garnered considerable publicity, particularly in international circles.

Adding to the unusual responses to Skylab's impending reentry, a group from the Brookline Psychoenergetics Institute in Massachusetts attempted to increase Skylab's altitude using telekinesis. In a coordinated meditation session that spanned several eastern states, they endeavored to push Skylab higher into orbit. However, their efforts, unsurprisingly, had no detectable effect, as confirmed by NORAD's radars.

Meanwhile, serious preparations for Skylab's reentry continued in Washington and Houston. On January 30, 1979, Bastedo's team completed NASA's official reentry plan and submitted it to the White House. March provided NASA with a chance to refine their procedures when the High Energy Astronomy Observatory (HEAO) 1, a NASA satellite, made its return to Earth. This event served as a trial run for the communication links between NORAD, Washington, and Huntsville, ensuring all systems were ready for Skylab's reentry.

Further preparations took place in late April when the Skylab group, alongside NORAD, the Marshall Space Flight Center (MSFC), and the Johnson Space Center (JSC), participated in the reentry tracking of a Soviet rocket body. This real-world exercise was used as a final rehearsal to assess the readiness of all teams involved in the Skylab operation. By June, a comprehensive paper simulation was conducted as the last preparatory step before Skylab's descent.

In early February, following a decision to return Skylab to a solar-inertial attitude, the intensity of operations at the control center in JSC began to ease. This attitude made power management simpler, allowing round-the-clock monitoring to be temporarily suspended. Many flight control personnel were reassigned to their regular duties. Although controlling Skylab's attitude in this mode was relatively straightforward despite the increased drag, it was anticipated that maintaining control would become more challenging as the station continued to lose altitude. From February through May, the control center maintained a watchful eye on Skylab while detailed plans were made for its final orbits.

Toward the end of April, NASA Headquarters issued its first official forecast for Skylab's reentry, using data derived from NORAD's tracking model. On April 25, with Skylab's altitude having decreased to approximately 320 kilometers, NORAD predicted a 50% probability that the station would reenter Earth's atmosphere by June 19, with a 90% chance of reentry occurring between June 13 and July 1. This forecasting format was consistently used throughout the remaining waiting period, as providing a more precise estimate was impossible until reentry was imminent. While Marshall's engineers employed a slightly different model, predicting reentry between June 15 and June 22, their estimates were never released to the public. Instead, NASA relied on NORAD's forecasts for all official statements.

As the reentry date approached, the media and public focused on Skylab, turning what was a meticulously planned and scientifically monitored event into a spectacle filled with serious concern and lighter moments of absurdity.

As Skylab's reentry loomed closer, the collaboration between NASA and NORAD became increasingly crucial, though the flight controllers were not particularly concerned with NORAD's predictions. The two organizations shared data, uncovering differences in how their computer models processed the information. NORAD's approach was relatively straightforward, extrapolating future positions based on recent observations. At the same time, NASA's method was more dynamic, continuously adjusting for changes in atmospheric density and the spacecraft's drag profile as Skylab descended. Despite these differences, NASA's lead engineer, James Harlan, correctly anticipated that the predictions from both models would converge as reentry approached, which ultimately proved true.

By the end of May 1979, NASA engineers and managers had agreed upon a strategy for controlling Skylab's reentry. The plan involved placing Skylab in a high-drag "torque equilibrium" attitude, where aerodynamic forces would be balanced by the control moment gyros as long as they remained functional. This configuration would apply a consistent retarding force on the station, allowing flight controllers to make more accurate predictions about the impact point. If necessary, they could reduce drag to adjust the reentry location. Once Skylab descended to an altitude of 140 kilometers, the plan was to induce an end-over-end tumbling motion, reducing drag to a known level and enabling a reasonably precise prediction of the impact site. Although theoretically possible to shift the impact by up to five orbits by varying the tumbling altitude, the systems were likely only capable of shifting the impact by one to three orbits—still a significant, yet challenging, maneuver.

However, returning Skylab to this torque-equilibrium attitude and complicated power management had prompted the Houston control center to resume 24-hour monitoring and control. During this period, NASA assigned a "hazard index" to each ground track Skylab covered. This index, ranging from 0 to 100, measured the potential danger based on the population density below each track. These indices provided a statistical basis for Harlan and his team to potentially shift Skylab's impact point to a less populated orbit in the final hours before re-entry. Yet, as Harlan later reflected, this approach, while logical, offered no guarantee of safety. The spacecraft could easily reenter over a densely populated area without causing any harm, or conversely, it could reenter over a sparsely populated area and still cause significant damage if it struck a vulnerable target.

In June 1979, NASA Administrator Robert Frosch testified before a House subcommittee, highlighting the agency's challenging position. He acknowledged the small risk of human injury—estimated at 1 in 152—and emphasized that any debris would be widely scattered across a large area. While Frosch tried to reassure the public that there was little to fear, he could not offer absolute certainty that no one would be harmed. His attempts to display the risks were met with skepticism, especially given that some of the fragments could weigh several hundred kilograms by the time they reached the Earth's surface.

This situation was further complicated by a growing shift in public attitude toward technology, particularly space technology. When the decision to launch Skylab was made in 1970, there was considerable optimism and confidence in such high-tech endeavors. However, by 1979, public sentiment had shifted, with many citizens increasingly unwilling to accept any risks associated with space programs—especially risks they had no control over and could not protect themselves from. The once-celebrated promise of space exploration had become a source of anxiety, reflecting broader societal concerns about the impacts of advanced technology on everyday life. This change in public perception placed additional pressure on NASA, as they faced the challenge of managing Skylab's reentry while addressing the growing unease among the public and lawmakers alike.

As Skylab's reentry loomed, concern spread across the United States, with congressmen and editors demanding answers. They questioned why Skylab had been launched without a means to control its eventual descent. NASA Administrator Robert Frosch could only explain that such measures had seemed prohibitively expensive at the time. By June 20, 1979, Skylab had descended to an altitude of 261 kilometers, having dropped 60 kilometers over the previous four weeks. The situation grew more precarious as the National Oceanic and Atmospheric Administration's North American Aerospace Defense Command (NORAD) periodically predicted the spacecraft's reentry. Initially, the median date predicted for Skylab's return to Earth was July 16, but by mid-June, this estimate had shifted to July 12. As reentry approached, the predicted window narrowed further, eventually settling between July 10 and July 18.

During this tense period, NORAD's predictions occasionally differed from NASA's, causing minor complications. Television networks, eager to cover the reentry event, frequently called NASA's Johnson Space Center (JSC) in Houston for updates. Charles Harlan, head of JSC's Public Affairs Office, found himself in the delicate position of needing to provide the media with a date in which NASA had confidence. To avoid any potential embarrassment, NASA advised the media to prepare a day or two before the official predictions.

As early July arrived, Skylab's descent accelerated rapidly. Now plunging into denser layers of Earth's atmosphere, the spacecraft became increasingly difficult to control, and its power supplies were under significant strain. On July 9, NASA activated the Skylab Coordination Center at its headquarters. This center was equipped with direct communication lines to NORAD, NASA field centers, the State and Defense Departments, and the Federal Aviation Administration (FAA), allowing for the rapid relay of information and orders. A closed-circuit television display from Houston depicted Skylab's ground track for several orbits and its current position. While news reporters and other nonessential personnel were kept out of the operations room, they were kept informed through periodic briefings and the television feed.

Chapter 8 - A Fiery Ending

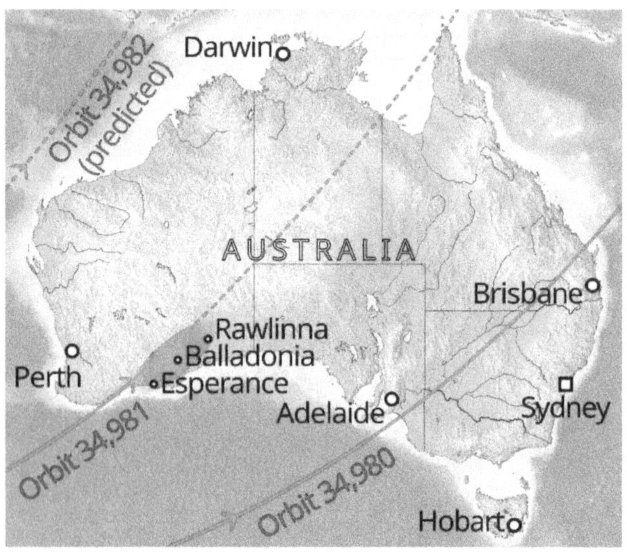

Equirectangular projection relief map of the Skylab re-entry site and final orbits, as predicted by NASA

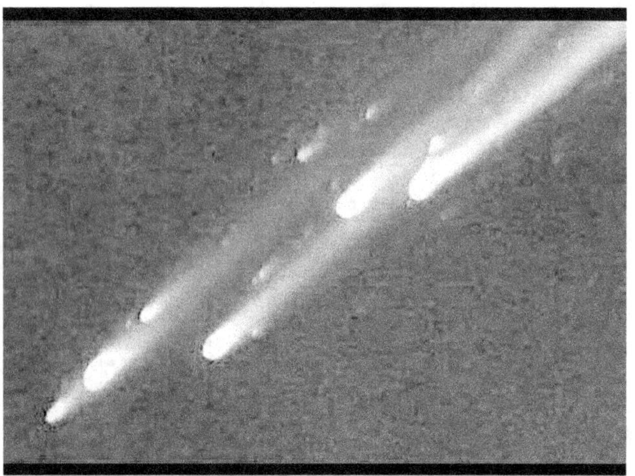

On July 9, the center predicted that Skylab would reenter Earth's atmosphere on July 11 between 2:10 a.m. and 10:10 p.m. Eastern Daylight Time (EDT), most likely during its 34,981st orbit. At that time, Skylab was orbiting at an altitude of 190 kilometers. By the next day, it had dropped another 17 kilometers, narrowing the reentry window further, now expected between 7:02 a.m. and 5:02 p.m. EDT on July 11.

In Houston, Harlan and his team prepared for the final decision. For several hours before reentry, computer models consistently predicted that Skylab would reenter on July 11. The only remaining question was the timing of the final tumbling maneuver, establishing crucial in determining where the debris would fall. Late on July 10, calculations indicated that Skylab would reenter on the best possible orbit for minimizing the risk to populated areas. This orbit would carry the spacecraft across southern Canada and the eastern coast of the United States before passing over a vast stretch of the Pacific Ocean, eventually making landfall in Australia.

However, early calculations of the debris pattern revealed a potential problem: if Skylab were to tumble at 140 kilometers as planned, the western edge of the 7,400 by 185-kilometer impact zone would slightly overlap North America. To mitigate this risk, NASA officials initiated the tumbling maneuver earlier, at an altitude of 148 kilometers. This adjustment moved the predicted impact area to a location about 1,300 kilometers south-southeast of Cape T, South Africa, a region halfway between North America and Australia and well south of major shipping lanes. The command was executed at 3:45 a.m. EDT, sending Skylab into an end-over-end spin.

But Skylab had one final surprise in store. Flight controllers expected the spacecraft to break apart before it passed over the United States' eastern coast, but radar operators at Bermuda reported a single, intact image. Over Ascension Island, NORAD's imaging radar confirmed that even Skylab's fragile solar arrays were still attached. However, the telemetry soon faltered, and communication ceased entirely as Skylab passed south of Africa. This unexpected resilience shifted the impact zone significantly to the east, raising concerns that some of the larger debris fragments might hit Australia.

At 12:37 p.m. EDT on July 11, NORAD confirmed Skylab's impact. Reports soon came in that the area southeast of Perth, Australia, had been showered with debris. Spectacular visual effects were reported as fiery fragments streaked across the early morning sky, accompanied by sonic booms and whirring noises. Despite the dramatic reentry, no injuries or significant property damage were reported. NASA had successfully managed Skylab's reentry without causing harm.

In an ironic twist, one Australian teenager turned the event into a profitable venture. A San Francisco newspaper had offered $10,000 for the first authenticated piece of Skylab brought to its office within 48 hours of reentry. Seventeen-year-old Stan Thornton, a beer-truck driver from the small coastal t of Esperance, found charred fragments in his backyard. He quickly bagged them, boarded the first flight to California, and claimed his prize.

After reentry, Harlan and his team reviewed the data. They concluded that they had miscalculated the drag during the tumbling maneuver by a small margin—only 4%—but this had been enough to shift the impact zone hundreds of kilometers farther east than anticipated. Fortunately, Skylab's reentry path had passed over the sparsely populated ranch country of Western Australia, sparing more densely inhabited areas.

With Skylab safely down, NASA began dismantling the makeshift control centers at its headquarters, Johnson Space Center and Marshall Space Flight Center. Harlan and his colleagues returned to their regular duties, focusing on the challenges of the Space Shuttle program. Five Marshall Space Flight Center engineers traveled to Australia to examine the recovered fragments, search for additional pieces, and determine the debris pattern. Despite some initial indignation from the Australian press, the NASA team was warmly received and given all possible assistance in their mission. Although a few Skylab pieces were displayed in nearby towns, further searches yielded no additional debris. Many fragments likely remained scattered across the vast, dusty outback, waiting to be discovered by a curious rancher or herder.

Meanwhile, three days after Skylab's reentry, two Soviet cosmonauts aboard Salyut 6 set a new endurance record in Earth orbit. The record they broke was not Skylab's. Still, one set only the previous year by another Soviet crew, underscoring the ongoing space race and the relentless pursuit of new milestones in human spaceflight.

In review, Skylab, NASA's pioneering space station, was a landmark achievement in space exploration that significantly advanced our understanding of extended human habitation in space. As a bold endeavor of the early 1970s, Skylab provided a wealth of knowledge that both validated and challenged early predictions about space research. The experiments conducted aboard the station spanned nearly every scientific discipline, offering crucial insights into the challenges and benefits of prolonged space missions.

The Skylab program commenced with its first mission, SL-1, on May 14, 1973, at 1:30 p.m. EDT, when a Saturn V rocket launched the space station into orbit. This marked the beginning of a groundbreaking era in space habitation. However, the initial mission faced significant technical difficulties, including damage to the station's micrometeoroid shield and solar panels during launch, which led to the station overheating and a reduction in power. Despite these challenges, the crew managed to repair Skylab, stabilize its temperature, and ensure its operability. The mission concluded on June 22, 1973, at 9:49 a.m. EDT, laying the groundwork for subsequent Skylab operations.

The second mission, SL-2, launched on May 25, 1973, at 9:00 a.m. EDT, using the smaller Saturn IB rocket. This mission was particularly notable for its extended duration and the successful establishment of operational protocols aboard Skylab. The crew spent 28 days aboard the station, conducting a variety of scientific experiments and proving the station's capabilities and endurance. They were recovered on June 22, 1973, at 6:19 p.m. EDT, marking the successful conclusion of their mission and providing valuable data for future endeavors.

On July 28, 1973, at 7:11 a.m. EDT, SL-3 began its mission, also utilizing the Saturn IB rocket. This mission, lasting 59 days, was a testament to Skylab's resilience and the crew's ability to adapt to the demands of prolonged space habitation. The mission was distinguished by extensive scientific research and the execution of numerous experiments, including solar astronomy and Earth observation. The crew's recovery on September 25, 1973, at 11:17 a.m. EDT, further demonstrated Skylab's capacity for supporting long-term missions.

The final mission, SL-4, launched on November 16, 1973, at 9:01 a.m. EST, continued to extend Skylab's operational period and allowed for additional scientific exploration. With a mission duration of 84 days, this final crew set new records in spaceflight duration and research output. The mission concluded with the crew's return on February 8, 1974, at 11:17 a.m. EDT, marking the end of Skylab's manned missions and the conclusion of a remarkable chapter in space exploration.

Throughout its operational period, Skylab traveled over 113.5 million kilometers (70.5 million miles) and completed 2,476 revolutions around the Earth. The space station maintained an orbital altitude between 431.5 and 433.7 kilometers (268.1 to 269.5 miles) with an inclination of 50 degrees and an orbital period of approximately 93 minutes.

Skylab's missions were crewed by a group of distinguished astronauts who played pivotal roles in the program's success. These included commanders Charles Conrad, Alan L. Bean, and Gerald P. Carr; pilots Paul J. Weitz, Jack R. Lousma, and William R. Pogue; and scientist-pilots Joseph Kerwin, Owen Garriott, and Edward Gibson. Each mission was a collaborative effort, with crews contributing significantly to the extensive scientific research and experimentation conducted aboard Skylab.

The allocation of crew time during the Skylab missions was meticulously planned, balancing essential daily routines with scientific exploration. During SL-1, for example, the crew's time was divided among sleep, rest, and off-duty periods, which accounted for 675.6 hours (34.7% of their time). Pre- and post-sleep routines, including meals, took 477.1 hours (24.5%), while housekeeping required 103.6 hours (5.3%). Physical training and personal hygiene consumed 56.2 hours (2.9%), with additional activities, including extravehicular activities (EVAs), occupying 232.5 hours (12.0%).

Scientific experiments were a major focus of SL-1. The crew dedicated 117.2 hours (29.9%) to solar astronomy, 71.4 hours (18.2%) to Earth observations, and 36.6 hours (9.4%) to material science. Other activities included life sciences and specific observations like those of the Kohoutek comet. The total time spent on these scientific endeavors was 145.3 hours, or 37.0% of the mission's total.

The subsequent missions followed a similar yet slightly adjusted distribution of time, with increasing focus on scientific research and experimentation. For instance, during SL-2, the crew dedicated 305.1 hours (28.2%) to solar astronomy and 223.5 hours (20.6%) to Earth observations. Extravehicular activities were also significant, with SL-2 featuring a standup EVA of 6 hours 29 minutes and additional EVAs on August 24 and September 22, totaling 1846.5 minutes.

SL-3, with its 59-day duration, saw further refinement in the crew's schedule, with 519.0 hours (33.2%) dedicated to solar astronomy and 274.5 hours (17.6%) to Earth observations. The final mission, SL-4, continued this trend, with even greater emphasis on maximizing scientific output. The mission's total time allocation and EVA activities were carefully documented, contributing to the overall success of the Skylab program.

Skylab's achievements were summarized in a comprehensive review that highlighted the wealth of data returned and the successful execution of its scientific and engineering objectives. The program provided substantial contributions to solar and Earth observations, with extensive imaging and data collection. The experiments conducted aboard Skylab generally met or exceeded their planned objectives, with notable increases in areas such as Earth observation passes, solar viewing time, and biomedical investigations.

One of Skylab's most significant contributions was its research on the effects of microgravity, which challenged earlier assumptions about the necessity of artificial gravity for long-term space missions. The unique environment of microgravity aboard Skylab allowed scientists to observe phenomena such as altered fire behavior, transformed cellular processes, and unique fluid dynamics that were not replicable on Earth. These insights led to advancements in technology and scientific processes, underscoring the importance of space research in enhancing our understanding of fundamental physics and practical applications.

As the Skylab program concluded, it became clear that Dr. Wernher von Braun's vision of space research revolutionizing our understanding of human capabilities and scientific processes had been both affirmed and redefined. Skylab's legacy continued to influence future space missions, providing a foundation for the ongoing exploration

of space and the continued pursuit of knowledge
beyond our planet.

Epilogue: Skylab's Enduring Legacy

The lessons learned from Skylab served as a pivotal foundation for NASA's next chapter in space exploration, leading directly to the development of Spacelab. This reusable laboratory became an integral part of the Space Shuttle program. From 1983 to 1998, under the management of the Marshall Space Flight Center, Spacelab facilitated 36 missions over 17 years, hosting more than 800 scientific investigations across various disciplines. Drawing on the experience gained from Skylab, Spacelab featured a modular design with external pallets and pressurized modules, providing a versatile platform for a wide range of experiments. This program advanced scientific knowledge and marked a significant shift toward international cooperation in space exploration.

In 1973, NASA and the European Space Agency (ESA) signed a Memorandum of Understanding to establish Spacelab, setting a new precedent for collaborative rather than competitive space endeavors. The European Space Agency played a crucial role in constructing and operating Spacelab, while Japanese and German space agencies sponsored dedicated research missions. This collaborative approach redefined the paradigm of space exploration, moving from a focus on nationalistic competition to a model of global partnership that brought together international researchers and engineers. The success of Spacelab laid the groundwork for an even more ambitious project: the International Space Station (ISS).

As NASA and its global partners refined the conduct of scientific research in orbit, the vision for a more extensive space station continued to evolve. This vision materialized in December 1998 with the launch of the first U.S.-built component of the ISS, the Unity node. The ISS represents the culmination of decades of planning and international collaboration, embodying a grandiose orbital laboratory that far surpasses the early designs envisioned by Dr. Wernher von Braun and his contemporaries. Supporting a crew of up to seven, the ISS provides an internal living and working space equivalent to a five-bedroom house and spans an external area roughly the size of a football field. Unlike von Braun's wheel-shaped station, the ISS was constructed as a modular structure collaboratively built by fifteen nations, including Russia. It stands as the largest and most ambitious peacetime multinational program in history.

The development and maintenance of the ISS drew heavily from the experiences of earlier projects like Skylab and Spacelab. The facilities at Marshall Space Flight Center, where Saturn V rockets were once assembled, now house teams dedicated to maintaining the ISS's environmental control and life support systems. These systems, designed to provide clean air and water for the station's crew, have also been adapted for potential use in future Mars habitats and have inspired innovations such as portable water purification systems for use in disaster relief on Earth. Thus, the ISS stands as a testament to the evolution of space research and international collaboration, embodying both the legacy of von Braun's early visions and the advancements made over decades of space exploration.

Marshall Space Flight Center's pivotal role in payload operations, which began during the Spacelab era, continues to be a cornerstone of space station science today. With the arrival of the Destiny Laboratory in 2001, Marshall established the Payload Operations Integration Center (POIC), which serves as the nerve center for managing science operations on the ISS. Staffed around the clock, every day of the year, the POIC coordinates closely with the mission control team at NASA's Johnson Space Center to ensure the seamless execution of scientific experiments and data collection.

Since its assembly was completed in the fall of 2011, the ISS has seen significant milestones, with Marshall playing an instrumental role in supporting the acceleration of crew science operations. Even during crew rest periods, the ground team at Marshall remains active, managing experiments and transmitting critical data to researchers worldwide. The ISS program has continued to evolve, with new investigations added during each six-month expedition, reflecting the station's dynamic and expanding research portfolio.

The International Space Station Program Science Office at the Johnson Space Center has documented numerous advancements derived from the research conducted aboard the ISS. These include breakthroughs such as using the Canadian robotic arm for intricate brain surgery, enhancing our understanding of combustion in space and rocket engines, developing new vaccines, and unraveling the mechanisms behind cellular cancerous transformations.

Today, the ISS orbits Earth as a symbol of unprecedented scientific and technological achievement, built through the collaborative efforts of multiple nations. Dr. Wernher von Braun's vision, articulated in *Collier's* over sixty years ago, envisioned a means to preserve peace and a step toward uniting humanity. While some of von Braun's technical details may have evolved, his foresight regarding the space station's role as a beacon of peaceful cooperation and a springboard for further exploration remains profoundly accurate. The ISS stands as a visible testament to humanity's commitment to peaceful, collaborative exploration of the cosmos, embodying the spirit of unity and innovation that von Braun so passionately envisioned.

Emeritus. His contributions have been spotlighted in documentaries on National Geographic, the History Channel, and other major networks. Barnes has authored several books, including "The Secret Genesis of Area 51" and "The CIA Area 51 Chronicles." He currently resides in Henderson, Nevada, continuing to influence aerospace, exploration, and literature, focusing on the formally highly classified of the CIA's era at Area 51.

About the Author

Thornton D. "TD" Barnes was a distinguished author, entrepreneur, and former military intelligence specialist. Born in Dalhart, Texas and raised on a ranch near Clayton, New Mexico and Dalhart, Texas, he cultivated a passion for exploration. After high school in Oklahoma, Barnes embarked on a ten-year military journey, initially serving in Korea as an intelligence specialist. While in the Army, he also specialized in missile and radar electronics, defending against Soviet threats and later attending the Artillery Officer Candidate School. An injury ended his military career, but Barnes soon transitioned to aerospace endeavors. He worked on pivotal projects at NASA's High Range in Nevada, including the X-15, the NASA NERVA nuclear rocket project, and atomic bomb testing at the Nevada Test Site. Furthermore, he participated in the CIA's Mach 3 A-12 Project OXCART and stealth projects at Area 51.

Barnes founded and led an oil and gas exploration company outside the aerospace sphere for over 40 years, delving into uranium and gold mining ventures. In retirement, he's dedicated to preserving Area 51's history, serving as president of Roadrunners Internationale and the Nevada Aerospace Hall of Fame Director

Bibliography

NASA -
https://images.nasa.gov/details/STS063-711-080
http://dayton.hq.nasa.gov/IMAGES/LARGE/GPN-2002-000078.jpg on the Wayback Machine at the Wayback Machine
http://grin.hq.nasa.gov/ABSTRACTS/GPN-2002-000078.html on the Wayback Machine at the Wayback Machine

https://www.nasa.gov/mission/skylab/

https://en.wikipedia.org/wiki/Skylab#

https://www.hq.nasa.gov/pao/History/SP-400/contents.htmhttps://ntrs.nasa.gov/api/citations/19840017669/downloads/19840017669.pdf

https://ntrs.nasa.gov/api/citations/19840017669/downloads/19840017669.pdf